t.40L

17

Energy Management

Paul W. O'Callaghan

Department of Applied Energy
Cranfield Institute of Technology

McGRAW-HILL BOOK COMPANY

London · New York · St Louis · San Francisco · Auckland
Bogotá · Caracas · Hamburg · Lisbon · Madrid · Mexico
Milan · Montreal · New Delhi · Panama · Paris · San Juan
São Paulo · Singapore · Sydney · Tokyo · Toronto

Published by
McGRAW-HILL Book Company Europe
Shoppenhangers Road·Maidenhead·Berkshire·SL6 2QL·England
Tel: 0628 23432; Fax 0628 770224

British Library Cataloguing in Publication Data
O'Callaghan, Paul W.
  Energy Management
  I. Title

  658.2

ISBN 0-07-707678-8

697 OCA

Library of Congress Cataloging-in-Publication Data
O'Callaghan, Paul W.
  Energy management / Paul W. O'Callaghan.
    p.   cm.
  ISBN 0-07-707678-8
    1. Energy conservation.   I. Title.
  TJ163.3.026   1992   92-15850
  658.2—dc20   CIP

12345 CL 96543

Typeset by Thomson Press (India) Ltd, New Delhi
Printed in England by Clays Ltd, St Ives plc

To
Menna, Catrin and Iwan

CONTENTS

PREFACE

The demand for energy, the consumption of fossil fuels and the concomitant release of pollutants continues relentlessly. In the industrialized world, the requirement for high rates of employment demands high rates of consumption of energy and raw materials. As a result, prices of fuels and electricity remain at artificially low levels, encouraging even greater demands.

This book is intended as a technical guide to assist in the joint objectives of saving energy and money whilst reducing environmental pollution. It is aimed at all consumers, but particularly those who are in a position to control the use and misuse of energy and materials.

World fossil fuel reserves, rates of consumption and projected 'lives' are first examined. The global 'greenhouse effect' and the possible effects on world climates are discussed.

A systematic and novel procedure for energy surveying, auditing and conservation is then developed and demonstrated. The fundamental concepts involved: thermodynamics; fluid flow; heat transfer; psychrometry; air conditioning; mass transfer; radiation and solar irradiation, are described comprehensively. The technologies of energy utilization: fuels and combustion, boilers, insulated pipework systems, building heat balance, comfort, climate, thermal insulation, waste heat recovery, thermal storage, heat pumps and refrigerators, are covered in depth.

Laws, rules and checklists for use when examining energy and materials flow systems are provided throughout.

Instrumentation, measurement, data analyses and computer-aided control systems are covered.

Techniques for economic evaluations, the construction of optimal investment schedules and effective monitoring and targeting procedures are presented. The uses of computational aids and software are also demonstrated, relevant equations and data are incorporated as well as two fully-documented energy management case studies.

Finally, current problems are reviewed and a sensible policy for the future is outlined.

ONE

ENERGY AND THE ENVIRONMENT

1.1 INTRODUCTION

In this chapter, world fossil fuel reserves, historical and current rates of consumption are reviewed and estimates of indigenous lives in geographical regions are made. The arguments for and against global warming are reviewed. The rates of production and accumulations of carbon dioxide in the atmosphere are calculated and correlations made with reported global mean temperatures and concomitant sea-level rises. The effects of other greenhouse gases are considered. It is concluded that, if present rates of global fossil fuel consumptions continue unabated, the world's fossil fuel store will be depleted by the year 2050. If global warming is occurring, this would be accompanied by a substantial rise in global mean temperature. The effects of various protocols for the reductions of emissions are examined. It is concluded that the careful course of action should cease the production and release into the atmosphere of the more damaging industrial and domestic greenhouse gases as soon as is practically possible and seek a sustained reduction in the rates of combustion of fossil fuels world-wide via energy management and conservation. This will result in better environmental protection and will extend the available time needed for the development of a future renewable energy economy.

1.2 WORLD FOSSIL FUEL RESERVES

According to the BP Statistical Review of World Energy,[1] proven world reserves of fossil fuels stood at an equivalent of 39.19×10^{15} MJ in 1987 (Tables 1.1 and 1.2).

It must be emphasized that the BP statistics are based upon proven and extractable reserves of fossil fuels, and do not include nuclear and renewable resources. Highest estimated undiscovered reserves and those not considered to be economically extractable are over four times this amount.[2]

Table 1.1 Proven world reserves of fossil fuels in 1987

Region or country	Oil 10^6 tonnes	Coal 10^6 tonnes	Gas $10^{12}\,m^3$
North America	5 200	270 421	8.1
Latin America	16 100	6 996	6.5
Western Europe	2 900	95 368	6.2
Middle East	76 500	0	30.7
Africa	7 400	65 907	7.0
Asia and Australasia	2 500	94 488	6.3
China	2 400	170 000	0.9
USSR	8 000	244 700	41.1
Others	200	78 267	0.8
Totals	121 200	1 026 147	107.6
United Kingdom	700	9 500	0.6

Table 1.2 Energy locked up in fossil fuel reserves ($10^{15}\,MJ$) (in this and subsequent tables small discrepancies in totals are due to rounding errors)

Region or country	Oil	Coal	Gas	Total
North America	0.236	7.788	0.311	8.335
Latin America	0.731	0.201	0.250	1.182
Western Europe	0.132	2.747	0.238	3.116
Middle East	3.473	0	1.179	4.652
Africa	0.336	1.898	0.269	2.503
Asia and Australasia	0.114	2.721	0.242	3.077
China	0.109	4.896	0.035	5.040
USSR	0.363	7.047	1.578	8.989
Others	0.009	2.254	0.031	2.294
Totals	5.502	29.553	4.132	39.187
United Kingdom	0.032	0.274	0.023 04	0.328

1 tonne oil $\equiv$ 45 400 MJ
1 tonne coal $\equiv$ 28 800 MJ
1 m^3 gas $=$ 38.4 MJ

1.3 WORLD ENERGY CONSUMPTION

Figure 1.1 shows depletion of the world fossil fuel reserves due to cumulative consumption since 1850. By 1977, 1.11×10^{16} MJ (21.3 per cent) of the fossil fuels, estimated in total to have been 5.2×10^{16} MJ at the outset (i.e. prior to industrialization), had been consumed. This fossil fuel was laid down principally via the anaerobic decomposition of vegetable matter mainly during the Carboniferous period occurring 345 million years ago and lasting some 100 million years.[3]

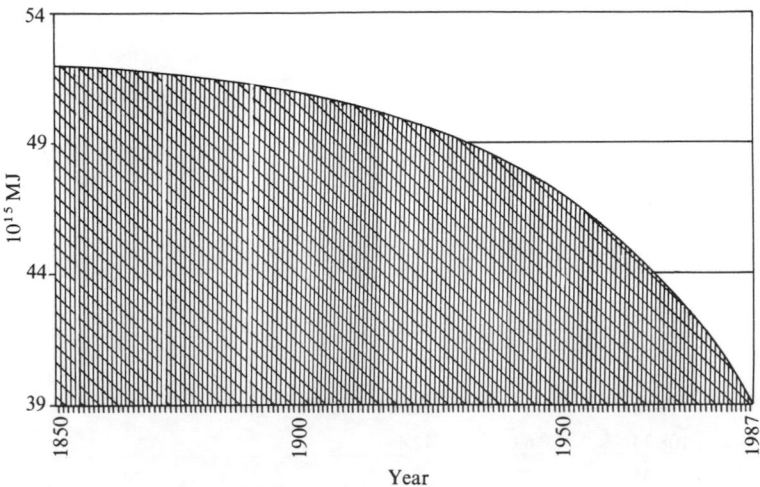

Figure 1.1 World fossil fuel depletion.

In 1987 alone, 3.12×10^{14} MJ (0.6 per cent or that which accumulated over 600 000 years) of the fossil fuel store was consumed (Tables 1.3 and 1.4). Figure 1.2 shows the breakdown of this consumption by region. It is noted that North America consumed 26 per cent of the world annual consumption during that year. Energy sources other than fossil fuels contributed 12 per cent during that year. Tables 1.5 and 1.6 show the energy consumed between 1977 and 1987: 3.2×10^{15} MJ (6.2 per cent of the store) was depleted during that decade.

Thus, by 1987, the proven fossil fuel reserves had been depleted by 1.43×10^{16} MJ (28 per cent).

Table 1.3 1987 fossil fuel consumption (10^6 toe)[†]

Region or country	Oil	Coal	Gas	Totals
North America	832.8	486.3	473.1	1 792.2
Latin America	220.6	22.7	73.4	316.7
Western Europe	585.2	259.0	206.7	1 050.9
Middle East	109.6	2.3	51.3	163.2
Africa	84.4	69.1	31.2	184.7
Australasia	32.8	41.8	18.1	92.7
Asia	186.0	170.0	33.5	389.5
China	103.9	553.4	12.8	670.1
Japan	208.1	68.5	36.4	313.0
USSR	449.2	378.9	520.2	1 348.3
Others	128.1	334.5	99.1	561.7
Totals	2 940.7	2 386.5	1 555.8	6 883.0
United Kingdom	75.2	67.3	50.0	192.5

[†] tonnes of oil equivalent

Table 1.4 1987 energy release from fossil fuels (10^{12} MJ)

Region or country	Oil	Coal	Gas	Totals
North America	37.80	22.07	21.47	81.36
Latin America	10.01	1.03	3.33	14.37
Western Europe	26.56	11.75	9.38	47.71
Middle East	4.97	0.10	2.32	7.40
Africa	3.83	3.13	1.41	8.38
Australasia	1.48	1.89	0.82	4.20
Asia	8.44	7.71	1.52	17.68
China	4.71	25.12	0.58	30.42
Japan	9.44	3.10	1.65	14.21
USSR	20.39	17.20	23.61	61.21
Others	5.81	15.18	4.49	25.50
Totals	133.50	108.34	70.63	312.48
United Kingdom	3.41	3.05	2.27	8.73

1 toe releases 45 400 MJ

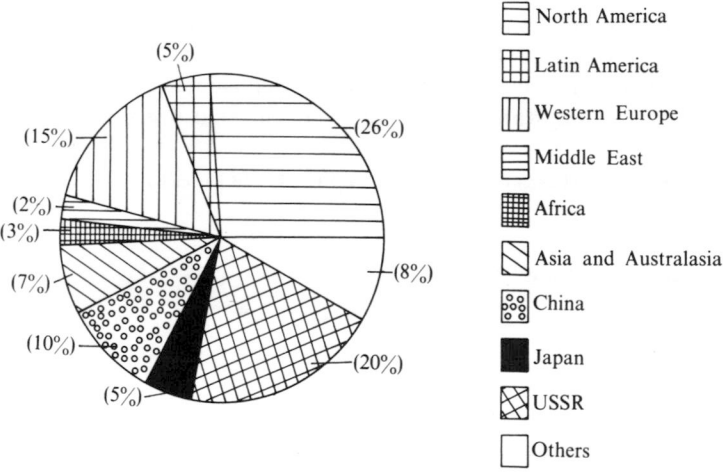

Figure 1.2 1987 fossil fuel consumption—breakdown by sector.

The rate of increase in the world use of fossil fuels has averaged 2.6 per cent per year since 1850[2]. Figure 1.3 shows, however, that the annual rate of growth from 1968 to 1973 (5.88 per cent) was cut to 4.14 per cent as a result of the first energy crisis in 1973, stemming from the Arab–Israeli war, and again to 2.79 per cent from 1983 onwards, resulting from the Iran–Iraq war, showing that world oil supply problems have an appreciable effect upon growth rates of fossil fuel consumption.

Table 1.5 World energy consumption between 1977 and 1987 (10⁶ toe)

Year	Oil	Coal	Gas	Total fossil fuels	Nuclear and Hydro	Grand total
1977	2 985.8	1 829.9	1 167.7	5 983.4	505.4	6 488.8
1978	3 082.8	1 860.4	1 212.5	6 155.7	557.2	6 712.9
1979	3 124.5	1 968.0	1 281.7	6 374.2	579.1	6 953.3
1980	3 011.7	2 001.5	1 297.2	6 310.4	605.9	6 916.3
1981	2 900.5	1 998.9	1 314.0	6 213.4	646.7	6 860.1
1982	2 825.4	2 044.8	1 330.1	6 200.3	689.7	6 890.0
1983	2 803.4	2 099.1	1 325.7	6 228.2	735.5	6 963.7
1984	2 836.2	2 178.3	1 425.2	6 439.7	795.4	7 235.1
1985	2 816.1	2 272.7	1 474.7	6 563.5	861.8	7 425.3
1986	2 899.3	2 318.3	1 487.2	6 704.8	893.4	7 598.2
1987	2 940.7	2 386.5	1 555.8	6 883.0	937.0	7 820.0
Total	32 226.4	22 958.4	14 871.8	70 056.6	7 807.1	77 863.7

Table 1.6 World energy consumption between 1977 and 1987 (10¹⁵ MJ)

Year	Oil	Coal	Gas	Total fossil fuels	Nuclear and Hydro	Grand total
1977	0.136	0.083	0.053	0.272	0.023	0.295
1978	0.140	0.084	0.055	0.279	0.025	0.305
1979	0.142	0.089	0.058	0.289	0.026	0.316
1980	0.137	0.091	0.059	0.286	0.028	0.314
1981	0.132	0.091	0.060	0.282	0.029	0.311
1982	0.128	0.093	0.060	0.281	0.031	0.313
1983	0.127	0.095	0.060	0.283	0.033	0.316
1984	0.129	0.099	0.065	0.292	0.036	0.328
1985	0.128	0.103	0.067	0.298	0.039	0.337
1986	0.132	0.105	0.068	0.304	0.041	0.345
1987	0.134	0.108	0.071	0.312	0.043	0.355
Total	1.463	1.042	0.675	3.181	0.354	3.535

Tables 1.7 to 1.14 show the estimated indigenous lives of the fossil fuels by region and type, assuming a constant rate of consumption at the 1987 level (Tables 1.7 and 1.8), at the 1987 increasing annual rate of consumption (2.79 per cent) (Tables 1.9 and 1.10) and for an increased annual rate of consumption of 5 per cent (Tables 1.11 and 1.12). For these three cases, the fossil fuels will be exhausted by the years 2112, 2041 and 2027, respectively.

Taking the United Kingdom alone, as a local example, if no imports or exports of fossil fuels occurred, this country would be self-sufficient until only 2024 (at 0 per cent rate of increase of consumption above the 1987 levels), 2013 (with 2.79 per cent growth) and 2008 (with 5 per cent growth).

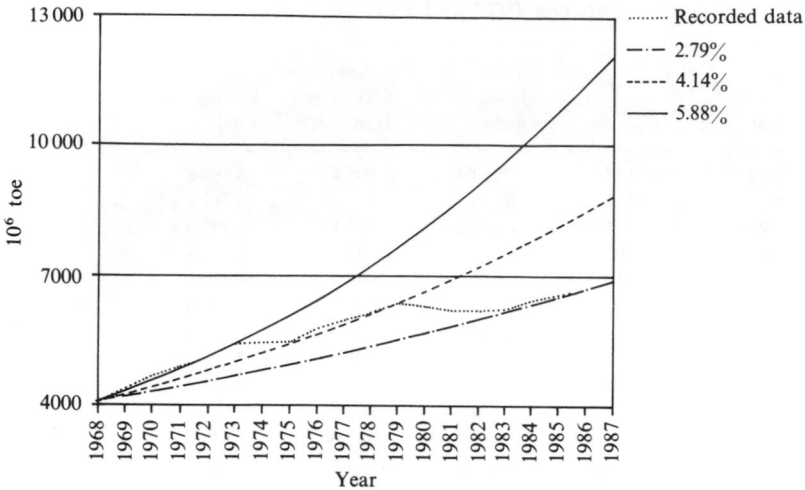

Figure 1.3 World fossil fuel consumption 1968–1987—trends of increasing use.

Table 1.7 Indigenous lives (years) at the 1987 consumption rate

Region or country	Oil	Coal	Gas	Total
North America	6.24	352.75	14.48	102.44
Latin America	72.98	195.50	74.90	82.20
Western Europe	4.95	233.58	25.37	65.31
Middle East	697.99	0	506.1	627.85
Africa	87.67	605.04	189.7	298.48
Asia and Australasia	11.42	283.00	103.2	140.53
China	23.09	194.87	59.47	165.65
Japan	0	0	0	0
USSR	17.80	409.68	66.82	146.84
Others	1.56	148.42	6.827	89.95
Totals	41.21	272.76	58.49	125.4
United Kingdom	9.30	89.545	10.149	37.57

Tables 1.13 and 1.14 illustrate the effects of energy conservation—if the consumption rate decreases by only 1 per cent each year, the life of the fossil fuel store can be extended infinitely, giving mankind time to develop a sustainable renewable energy economy.

The United Kingdom would, however, run out of indigenous reserves in 2031 if its fossil fuel consumption were reduced by only 1 per cent each year. This country would have to reduce fossil fuel consumption by 3 per cent each year to achieve an infinite indigenous life.

Table 1.8 Depletion year at constant 1987 rate of consumption

Region or country	Oil	Coal	Gas	Total
North America	1993	2339	2001	2089
Latin America	2059	2182	2061	2069
Western Europe	1991	2220	2012	2052
Middle East	2684	1987	2493	2614
Africa	2074	2592	2176	2285
Asia and Australasia	1998	2270	2090	2127
China	2010	2181	2046	2152
Japan	1987	1987	1987	1987
USSR	2004	2396	2053	2133
Others	1988	2135	1993	2076
Totals	2028	2259	2045	2112
United Kingdom	1996	2076	1997	2024

Table 1.9 Indigenous lives (years) for a consumption rate increasing annually by 2.79 per cent (1987 rate of increase)

Region or country	Oil	Coal	Gas	Total
North America	5.83	86.61	12.33	49.06
Latin America	40.35	67.76	40.99	43.31
Western Europe	4.70	73.30	19.45	37.70
Middle East	109.71	0	98.7	106.0
Africa	44.96	104.79	66.85	81.14
Asia and Australasia	10.05	79.42	49.28	57.90
China	18.07	67.66	35.54	62.74
Japan	0	0	0	0
USSR	14.65	91.58	38.24	59.18
Others	1.54	59.49	6.33	45.62
Totals	27.81	78.23	35.16	54.64
United Kingdom	8.39	45.50	9.06	26.05

Table 1.10 Depletion year for a consumption rate increasing at the 1987 level

Region or country	Oil	Coal	Gas	Total
North America	1992	2073	1999	2036
Latin America	2027	2054	2027	2030
Western Europe	1991	2060	2006	2024
Middle East	2096	1987	2085	2093
Africa	2031	2091	2053	2068
Asia and Australasia	1997	2066	2036	2044
China	2005	2054	2022	2049
Japan	1987	1987	1987	1987
USSR	2001	2078	2025	2046
Others	1988	2046	1993	2032
Totals	2014	2065	2022	2041
United Kingdom	1995	2032	1996	2013

Table 1.11 Indigenous lives (years) for a consumption rate increasing annually by 5 per cent

Region or country	Oil	Coal	Gas	Total
North America	5.56	59.95	11.16	37.13
Latin America	31.49	48.72	31.91	33.43
Western Europe	4.53	52.05	16.78	29.73
Middle East	73.39	0	67.01	71.28
Africa	34.50	70.54	48.17	56.72
Asia and Australasia	9.26	55.70	37.27	42.68
China	15.73	48.66	28.27	45.66
Japan	0	0	0	0
USSR	13.05	62.86	30.09	43.47
Others	1.54	43.67	6.01	34.93
Totals	22.92	55.00	28.02	40.65
United Kingdom	7.83	34.85	8.41	21.67

Table 1.12 Depletion year for a consumption rate increasing annually by 5 per cent

Region or country	Oil	Coal	Gas	Total
North America	1992	2046	1998	2024
Latin America	2018	2035	2018	2020
Western Europe	1991	2039	2003	2016
Middle East	2060	1987	2054	2058
Africa	2021	2057	2035	2043
Asia and Australasia	1996	2042	2024	2029
China	2002	2035	2015	2032
Japan	1987	1987	1987	1987
USSR	2000	2049	2017	2030
Others	1988	2030	1993	2021
Totals	2009	2042	2015	2027
United Kingdom	1994	2021	1995	2008

Table 1.13 Indigenous lives (years) for a consumption rate decreasing annually by 1 per cent

Region or country	Oil	Coal	Gas	Total
North America	6.37	infinite	15.32	212.29
Latin America	108.99	infinite	113.65	133.20
Western Europe	5.03	infinite	28.23	91.92
Middle East	infinite	0	infinite	infinite
Africa	150.13	infinite	infinite	infinite
Asia and Australasia	11.93	infinite	216.88	infinite
China	25.43	infinite	80.32	infinite
Japan	0	0	0	0
USSR	19.13	infinite	95.10	infinite
Others	1.56	infinite	6.99	157.91
Totals	49.77	infinite	78.49	infinite
United Kingdom	9.63	156.48	10.54	44.49

Table 1.14 Depletion year for a consumption rate decreasing annually by 1 per cent

Region or country	Oil	Coal	Gas	Total
North America	1993	infinite	2002	2199
Latin America	2095	infinite	2100	2120
Western Europe	1992	infinite	2015	2078
Middle East	infinite	1987	infinite	infinite
Africa	2137	infinite	infinite	infinite
Asia and Australasia	1998	infinite	2203	infinite
China	2012	infinite	2067	infinite
Japan	1987	1987	1987	1987
USSR	2006	infinite	2082	infinite
Others	1988	infinite	1993	2144
Totals	2036	infinite	2065	infinite
United Kingdom	1996	2143	1997	2031

1.4 HISTORICAL LIVES OF THE FOSSIL FUELS

Whilst the preceding analyses have been based upon the 1987 reserves and levels of consumption, the estimated lives of the fossil fuels at instantaneous annual levels of consumption change historically as new reserves are discovered and rates of consumption vary.

Tables 1.15 to 1.18 have been derived for coal, oil and gas over the period 1979–1989. Figure 1.4 shows that the 'current life' of coal *decreased* from 350 to 290 years over the 10-year period as the rate of annual consumption increased. Figure 1.5 shows that the 'current life' of oil *increased* from 28 to 44 years over the 10-year period as more reserves were discovered. Figure 1.6 shows that the 'current life' of gas *increased* from 48 to 56 years over the 10-year period as more reserves were discovered, although this trend starts to level out over the period.

Table 1.15 Historical lives of coal

	10^6 tonne	Reserves, 10^{16} MJ	10^6 toe	Consumption, 10^{13} MJ	Life at 'current' consumption, years
1979	1 083 460	3.120	1968	8.935	349.2
1980	1 077 929	3.104	2001	9.085	341.7
1981	1 072 399	3.089	1999	9.075	340.3
1982	1 066 869	3.073	2045	9.284	330.9
1983	1 061 339	3.057	2099	9.529	320.7
1984	1 055 808	3.041	2178	9.888	307.5
1985	1 050 278	3.025	2273	10.32	293.1
1986	1 044 748	3.009	2318	10.52	285.9
1987	1 039 217	2.993	2386	10.83	276.2
1988	1 033 687	2.977	2199	9.983	298.1
1989	1 028 157	2.961	2231	10.13	292.3

Table 1.16 Historical lives of oil

	10^9 barrels	Reserves, 10^{15} MJ	10^6 tonne	Consumption, 10^{14} MJ	Life at 'current' consumption, years
1979	650	4.013	3142	1.426	28.13
1980	655	4.044	3024	1.373	29.45
1981	690	4.260	2918	1.325	32.15
1982	690	4.260	2820	1.280	33.27
1983	690	4.260	2797	1.270	33.54
1984	703	4.340	2846	1.292	33.59
1985	703	4.340	2834	1.287	33.73
1986	702	4.334	2919	1.325	32.70
1987	900	5.557	2964	1.346	41.29
1988	910	5.618	3053	1.386	40.53
1989	1010	6.236	3098	1.406	44.33

Table 1.17 Historical lives of gas

	10^{12} m^3	Reserves, 10^{15} MJ	10^6 toe	Consumption, 10^{13} MJ	Life at 'current' consumption, years
1979	73	2.803	1271	5.770	48.57
1980	75	2.880	1286	5.838	49.32
1981	82	3.149	1302	5.911	53.26
1982	87	3.341	1299	5.897	56.64
1983	90	3.456	1310	5.947	58.10
1984	97	3.725	1412	6.410	58.10
1985	98	3.763	1459	6.624	56.81
1986	102	3.917	1480	6.719	58.29
1987	109	4.186	1562	7.091	59.02
1988	112	4.301	1641	7.450	57.72
1989	113	4.339	1707	7.750	55.99

Table 1.18 Historical lives of all fossil fuels

	Total reserves, 10^{16} MJ	Total consumption, 10^{14} MJ	Total life, years
1979	3.802	2.897	131.24
1980	3.797	2.865	132.51
1981	3.829	2.823	135.62
1982	3.833	2.798	136.95
1983	3.828	2.818	135.87
1984	3.847	2.922	131.66
1985	3.835	2.981	128.65
1986	3.834	3.050	125.72
1987	3.967	3.138	126.42
1988	3.969	3.129	126.82
1989	4.019	3.194	125.80

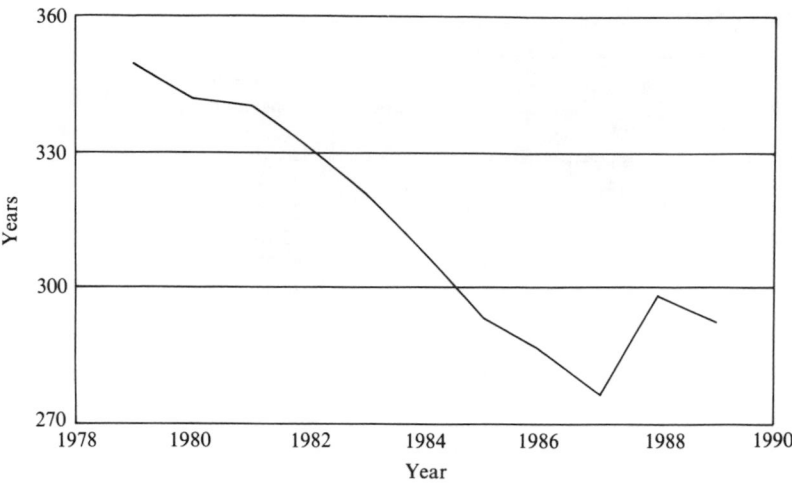

Figure 1.4 Historical 'lives' of coal at 'current' consumption rates.

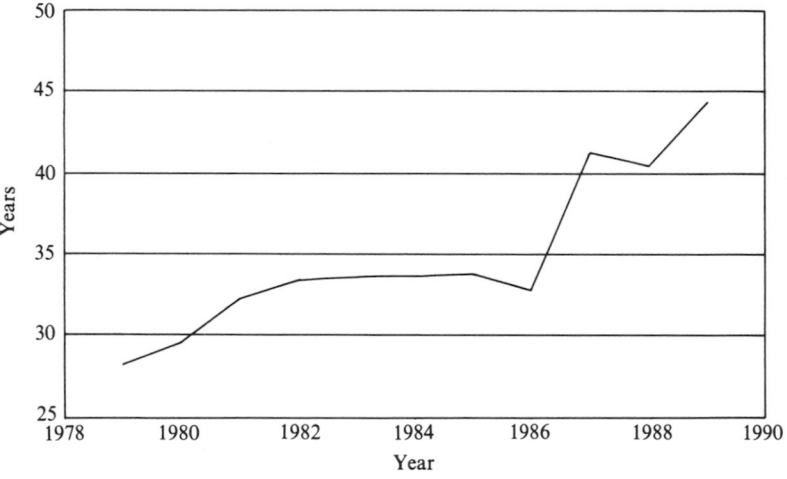

Figure 1.5 Historical 'lives' of oil at 'current' consumption rates.

The net result on the historical lives of the fossil fuels is seen in Figure 1.7, where, after peaking at 137 years in 1982, the total life of the fossil fuels declines to 126 years between 1986 and 1989.

1.5 THE GREENHOUSE EFFECT

Radiation

A body emits radiation when part of its internal energy is converted into electromagnetic waves. These waves travel in free space at the speed of light ($c = 2.997\,925 \times 10^8\,\mathrm{m\,s^{-1}}$) and transmit energy without the need for an intervening transport medium. When they encounter

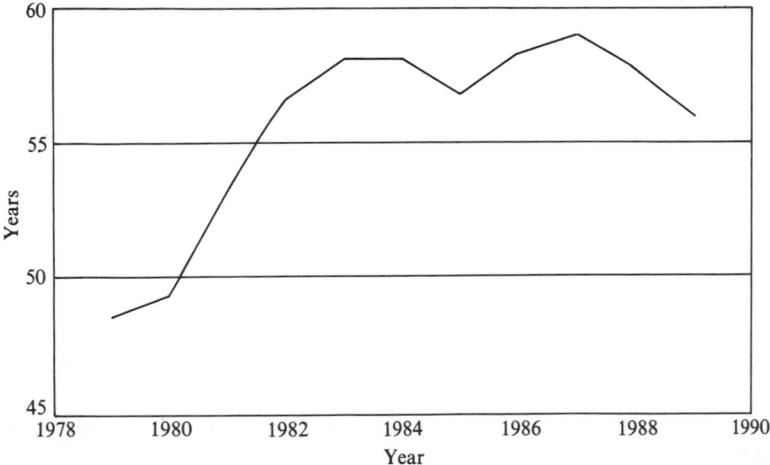

Figure 1.6 Historical 'lives' of natural gas at 'current' consumption rates.

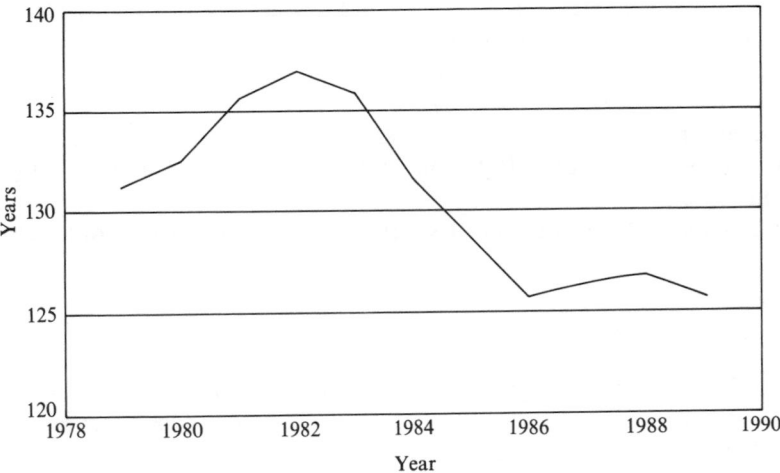

Figure 1.7 Historical 'lives' of all fossil fuels at 'current' consumption rates.

another body, part of the electromagnetic energy is absorbed and converted into internal energy in that body, raising its temperature. Electromagnetic waves are divided into classes according to their wavelengths, λ(m), and frequencies ω (cycles/s or Hertz). The wavelength of the radiation is defined as the ratio of the propagation velocity to the frequency via the relationship

$$\omega = \frac{c}{\lambda} \tag{1.1}$$

All bodies emit radiation, to which we are exposed to at all times. Humans can sense radiation in the band 0.1 to 100 μm, within which wavelengths in the range 0.38 to 0.76 μm, are detected as visible light.

The wavelength at which maximum emission occurs shifts with decreasing temperatures to longer wavelengths.

Absorption, reflection and transmission

When radiation encounters a surface, some of it is reflected, some absorbed and some transmitted, according to

$$\alpha + \rho + \tau = 1 \tag{1.2}$$

where α is the surface absorptivity, i.e. the fraction of incident radiation which is absorbed by the body, raising its specific heat; ρ is the surface reflectivity, i.e. the fraction of incident radiation which is reflected from the body; and τ is the surface transmissivity, i.e. the fraction of incident radiation which passes through the body.

The relative magnitudes of α, ρ and τ depend upon the material, its thickness and also the wavelength of the radiation.

Most solid materials absorb radiation in a very thin (less than 1 mm deep) surface layer. The exceptions are the translucent materials, i.e. glass, for which some transmission occurs at certain wavelengths. Many liquids and all gases are transparent over many radiation bands. Many common gases, such as oxygen, nitrogen, hydrogen and dry air, are practically transparent to thermal radiation over all wavelengths. Carbon dioxide, water vapour, sulphur dioxide, carbon monoxide, ammonia, hydrocarbon and alcohol vapours emit and absorb radiation only between narrow bands (see Table 1.19).

The *greenhouse gases*, carbon dioxide, CO_2, nitrous oxide, N_2O, the CFCs, methane, CH_4, tropospheric ozone, O_3 and water vapour, H_2O, all impede the transmission of long-wavelength radiation from the earth.

Stratospheric ozone inhibits the transmission of short-wavelength ultraviolet radiation coming from the sun.

Solar radiation

The sun is a fusion reactor in which hydrogen is being converted to helium, releasing nuclear energy. Internal sun temperatures are of the order of $10^7\,°C$, with an effective surface temperature, when viewed from the earth, $\approx 6000\,°C$. The solar energy flux at the sun's surface ($1.383 \times 10^6\,km$ diameter) is $70\,MW\,m^{-2}$. Most of this energy is radiated in the ultraviolet end of the spectrum.

Table 1.19 Emission and absorptance bands for common gases and vapours

Substance	Wavelength band $\lambda, \mu m$
Carbon dioxide	2.36–3.02
	4.01–4.80
	12.5–16.5
Water vapour	2.24–3.27
	4.80–8.50
	12.00–25.00

The earth (11 700 km mean diameter) intercepts solar energy at the rate of $1.362\,kW\,m^{-2}$ at the boundary of the upper atmosphere ($10^{21}\,MW$ in total). The current rate of fossil fuel consumption corresponds to about 0.001 per cent of the rate of solar energy received by the earth.[4]

The solar radiation incident at the perimeter of the earth's atmosphere contains 5 per cent ultraviolet, 53 per cent visible and 43 per cent infra-red radiation. When the sun is directly overhead in a cloudless sky, $1.025\,kW\,m^{-2}$ of specular radiation can reach the earth's surface. The remaining $0.337\,kW\,m^{-2}$ is reflected by gas molecules which filter out the shorter wavelengths and selectively transmit light in the blue end of the visible spectrum. The modified radiation contains 1 per cent ultraviolet, 39 per cent visible and 60 per cent infra-red radiation. Solar radiation which has been scattered, or absorbed and reradiated, by a combination of gas molecules, water vapour and dust particles, becomes white and diffuse. Layers of clouds reflect direct and diffuse 'sky' radiation away from the earth but also insulate against direct radiation losses from the earth's surface. Surfaces at terrestrial temperatures radiate energy in the infra-red end of the spectrum. At night, in the absence of cloud cover, up to $6000\,Wm^{-2}$ can be lost from the surface of the earth by direct radiation to space. In the steady-state, all solar radiation received by the earth is finally received by the vault of deep space.

The *greenhouse effect* is a thermal radiation rectifier which allows short-wave (or short-wavelength radiation) emanating from a high temperature source, such as the sun, to pass through a radiation barrier (the glass of a horticultural greenhouse or the blanket of greenhouse gases in the troposphere) but inhibits the passage of long-wave (or long-wavelength) radiation emitted by low temperature surfaces.

In the absence of the greenhouse effect of the atmosphere, the mean temperature of the earth would be considerably lower ($-18\,°C$) than the $15\,°C$ to which it has stabilized.

Pollution

Human activity releases pollutants into the troposphere. These give rise to the so-called greenhouse gases, which are threatening to disturb the thermal balance of the earth to a new equilibrium state.

Longwave radiation from the earth escapes through an infra-red window which occurs in the near infra-red at 6.5 to $14\,\mu m$. The greenhouse gases, mainly carbon dioxide at present, are filling this hole through the troposphere. Any change in the composition of the tropospheric atmosphere changes the thermal balance of the earth.

Simultaneously, the release of halocarbons are destroying the ozone layer, which absorbs ultraviolet radiation coming from the sun.

Carbon dioxide

Carbon dioxide in the atmosphere has built up from 270 ppm (parts per million) in pre-industrial times to 350 ppm today.[5] The distribution of carbon throughout the earth and its atmosphere is shown in Table 1.20.

The argument for global warming says that, even though the carbon released with fossil fuel combustion is small compared with the fluxes of the natural carbon cycle, the *extra* CO_2 emitted is enough to disturb the thermal equilibrium of the earth, stabilized at a global mean of $15\,°C$ in pre-industrial times.

Table 1.20 Distribution of carbon on earth

Location	Form	10^9 tonnes
Sediments	mainly in calcium carbonate	20 000 000
Oceans	dissolved bicarbonates and carbonates	40 000
Fossil fuels	hydrocarbons	12 000
Land	carbohydrates	1 760
Seas	carbohydrates	580
Atmosphere	CO_2	743

Concentrations of other greenhouse gases in the atmosphere: methane; nitrous oxide; tropospheric ozone and halocarbons (CFCs), are also building up.

The Montreal and Toronto protocols called for substantial cuts in emissions of CFCs and carbon dioxide to combat this build-up of greenhouse gases.

Global warming

Researchers at the University of East Anglia[6] have compiled a record of measured global temperatures from 1850 to the present day from weather reports, rainfall and temperature records and evidence of past climatic conditions locked in trees, ice cores and rocks.

Although natural variations in climate occur from changes in the sun's luminosity, volcanic eruptions, random factors and long timescale systematic drifts, the data suggests that an overall mean global temperature rise of 0.5 °C has occurred since 1850 (pre-industrial times). If this temperature rise has been caused by an increase in the carbon dioxide concentration in the atmosphere of 80 ppm, then a 1.5 °C to 4.5 °C extra temperature rise is predicted with future estimated increases in all the greenhouse gases. These predictions represent the highest temperatures for 100 000 years. The last ice age (18 000 years ago) was associated with temperatures only 4 °C colder than those prevailing today. The change in global temperature at the end of that ice age was accompanied by a doubling of CO_2 levels in the atmosphere, thought to be caused by a decrease of plankton in the oceans.

Climate models are based upon finite difference solutions of global meteorology. Atmospheric general circulation models are based upon the equations which determine the motion of the atmosphere (wind), its thermodynamics (temperature), and conservation of water (humidity and surface pressure). The values of these basic variables are kept at various locations over the globe and at various levels in the atmosphere. A nodal element is typically 5 degrees of latitude by 7 degrees of longitude.

The major uncertainties in the predictive models are the effect of the ozone hole in the stratosphere, speed of warming, the response of oceans and ice and the vegetable and animal responses.

Most scientists agree that global warming is inevitable and that we can only delay temperature rises to buy time.[6]

Effects of global warming

World warming would cause climatic changes, rises in sea levels, adverse effects on agriculture, altered rainfall patterns, increased algae growths, more frequent occurrences of extreme weather events (droughts, floods, hurricanes), food and water shortages, and an increase of pests.

1.6 SEEDS OF DOUBT

Almost everyone believes that global warming is occurring, although few have seen and analysed the raw data for global temperature rise and greenhouse gas build-up. This belief stems mainly from the observation and interpretation of isolated extreme weather events.

The belief in global warming is based upon six foundation stones:

1. Global carbon dioxide levels are increasing.
2. The global mean temperature is increasing.
3. Global sea-level is rising.
4. Increasing CO_2 causes global temperatures to rise.
5. Climate models predict future warming.
6. Atmospheric physics prove that the world must be warming.

Global carbon dioxide levels have increased from 270 ppm to 350 ppm over the past 150 years and are currently increasing annually by 1.5 ppm This evidence comes from ice core and deep-sea sediments from a few isolated expensive tests at discrete points of the globe, as well as direct measurements. There are few measurements available for the southern hemisphere and it is not fully understood how CO_2 moves from the northern to the southern hemispheres to equalize at 350 ppm globally.

The global mean temperature is increasing This is not accepted by some scientists who suggest that, because weather stations are largely situated in urban areas and heavily tarmacked airports, local micro-climates affect the temperature measurements. Heat island effects can result in temperatures 2–3 °C above that of the surrounding countryside.

Weatherstations are not evenly distributed. There are none in the 70 per cent of the earth's surface covered by water and very few in the southern hemisphere, which is mostly water. Infra-red measurements from space taken over the period 1980–1990 do not verify ground-measured data.

The long-term trend detected in temperature records depends upon the period selected for analysis. The global data shows that the world cooled by 0.3 °C from 1940 to 1980, during a period of intense post-war industrialization, whilst Britain and most of Europe also *cooled* during the 1980s.

There is much recorded evidence that northern Europe was hotter than today in medieval times, when grapes were grown and wine was produced in Britain.

Global sea-levels are rising Mean sea-levels are difficult to measure as waves can be tens of metres high and land masses also move up and down. Whilst a 0.1 m mean sea-level rise has been reported,[6] the reference points for sea-level are not discussed.

Sea-ice melt has also been reported, but this is based upon two single measurements taken ten years apart. Satellite observations show no change in the amount of sea-ice from 1978 to 1990. From Archimedes Principle, this would not in any case cause sea-levels to rise.

There is also no firm evidence of glacial melt.

Increasing CO_2 causes global temperatures to rise Data spanning 160 000 years extracted from bubbles in ice cores, fossilized tree rings, etc., indicate that global temperature and CO_2 levels move in step.[6] But the data correlations show that temperature falls before CO_2 levels fall and vice versa!

From 1950 to 1990, the data also show that warming occurs *before* CO_2 increases. This means that when the temperature falls, the sources for CO_2 decrease and the sinks for CO_2 increase. Absorption of carbon dioxide by the seas increases at lower temperatures. Thus, as the sea temperature increases, CO_2 is released into the atmosphere, *accelerating the greenhouse effect*. But this runaway and irreversible situation, which appears to be easily triggered, has not happened throughout the history of the earth.

If CO_2 increase causes temperature rise via the greenhouse effect, less heat will be radiated away and the earth will become hotter. This will result in more water being evaporated, more cloud cover, less solar gain and more temperate plants. These will absorb more CO_2 to help reverse the situation.

In the absence of fossil fuel combustion and neglecting the effects of the other greenhouse gases, as long as there are more plants than animals, the system is self-stabilizing. This suggests that CO_2 increase does not cause temperature rise via the greenhouse effect.

Climate models predict future warming These models usually try to predict world climates associated with a doubling of CO_2 levels in the atmosphere. Most predict that a 3–5 °C general increase will occur by the year 2100. The existing models are very coarse, make many simplifying assumptions and neglect feedback mechanisms. They differ from each other in their predictions and cannot accurately simulate past climates or even present climates from known data. They are reported to be in much need of funds to finance improvements.

Atmospheric physics prove that the world must be warming The underlying physics and chemistry of the atmosphere are not fully understood. In 1991, the UK Science and Engineering Research Council offered £ 150 m for research into atmospheric studies.

> Weighing the arguments for and against the occurrence of global warming results in significant uncertainty as to whether human activities are leading to climatic change.
> Nevertheless, needs for pure air, water, land, food and lifeforms, and for the conservation of energy and mineral resources, whilst avoiding pollution, poisons, unnatural products and unnecessary waste make energy and environmental conservation a laudable joint aim.

The following section examines present and future global releases of the greenhouse gases and the possible effects on global climates, assuming that a build-up of greenhouse gases in the atmosphere has occurred, that this has caused a rise in mean global temperature, and that the process will continue as more atmospheric pollutants are released.

1.7 SIMPLE MODEL FOR GLOBAL WARMING

1 tonne of oil (CH_2) releases 3142.8 kg CO_2.
1 tonne of coal (C_6H) releases 3616.4 kg CO_2.
1 m^3 of gas (0.8 kg)(CH_4) releases 2.2 kg CO_2.

1 toe coal is equivalent to 1.579 tonnes of coal (C_6H) and releases 5700.85 kg of CO_2.

1 toe gas is equivalent to 0.946 tonnes of gas or 1182.29 m^3 of gas which releases 2601.04 kg of CO_2.

Using these conversion factors, it is possible to translate the world fossil fuel consumption figures to the corresponding carbon dioxide released by region and fuel type (Tables 1.21 and 1.22).

Measurements taken between 1973 and 1985 at Alaska, Hawaii, Samoa, and the South Pole[6] indicate that atmospheric carbon dioxide concentrations rose on average from 327 ppmv (parts per million by volume) to 345 ppmv over that period (annual rate 1.5 ppmv). Gribbon extended this data to show that the trend has continued to 350 ppmv in 1989. Whilst the

Table 1.21 Total CO_2 locked up in fossil fuel reserves (10^{15} kg CO_2)

Region or country	Oil	Coal	Gas	Total
North America	0.016	0.978	0.018	1.012
Latin America	0.051	0.025	0.014	0.090
Western Europe	0.009	0.345	0.0143	0.368
Middle East	0.240	0	0.068	0.308
Africa	0.023	0.238	0.015	0.277
Asia and Australasia	0.008	0.342	0.014	0.363
China	0.008	0.615	0.002	0.624
USSR	0.025	0.885	0.090	1.000
Others	0.0006	0.283	0.002	0.285
Totals	0.381	3.711	0.237	4.329
United Kingdom	0.002	0.034	0.001	0.038

Table 1.22 1987 world CO_2 release (10^{12} kg) accompanying the combustion of fossil fuels

Region or country	Oil	Coal	Gas	Total
North America	2.617	2.772	1.230	6.620
Latin America	0.693	0.129	0.190	1.013
Western Europe	1.839	1.476	0.537	3.852
Middle East	0.344	0.013	0.133	0.490
Africa	0.265	0.393	0.081	0.740
Asia and Australasia	0.687	1.207	0.134	2.029
China	0.326	3.154	0.033	3.514
Japan	0.654	0.390	0.094	1.139
USSR	1.411	2.160	1.353	4.924
Others	0.402	1.906	0.257	2.567
Totals	9.242	13.605	4.046	26.893
United Kingdom	0.236	0.383	0.130	0.750

data suffer from seasonal cyclic variations, there is little random scatter and the increasing trend is clear. Analyses of air bubbles trapped in glacial ice[6-8] have indicated that carbon dioxide concentrations were stable at 270–280 ppmv over the period from 10 000 years ago to the 19th century.

Carbon dioxide enters the atmosphere via respiration, fermentation, decay, the burning of vegetable matter and the combustion of fossil fuels. It is 'fixed' to carbon via photosynthesis in land and sea plants and is dissolved in the seas. Until industrialization, the amount of carbon being fixed each year exactly balanced the amount being released, hence the constant 270 ppmv.[6] Since industrialization, carbon dioxide which was fixed in the fossil fuels over a period of 28 million years has been released into the atmosphere via the combustion of fossil fuels.

With the growth in the consumption of fossil fuels, some of the excess carbon dioxide has not been fixed or dissolved but has remained in the atmosphere.[6]

The radius of the earth is 5850 km. The thickness of the mixing layer in the troposphere is about 6.8 km: thus the volume of the atmospheric mixing layer is $3 \times 10^{18}\,m^3$.

The average temperature and pressure of the troposphere are 265.5 K and 72 000 Nm^{-2} respectively. The air mass within the mixing layer is therefore 2.8×10^{18} kg and 1 ppmm $CO_2 = 1.52$ ppmv at the mean temperature and pressure of the troposphere.

The atmospheric burden of CO_2 will therefore increase by 1 ppmv with the release of 4.256×10^{12} kg of CO_2.

Thus the annual rate of increase of carbon dioxide concentration of 1.5 ppmv corresponds to a gain of 6.37×10^{12} kg of CO_2, or 24 per cent of the 1987 amount of CO_2 released via fossil fuel combustion.

The excess concentration of carbon dioxide in 1989 over that amount pertaining in pre-industrial times (80 ppmv) corresponds to 3.4×10^{14} kg, or 28 per cent of the total amount released by the combustion of fossil fuels since 1850.

Global temperature rise

The university of East Anglia studies[6-8] suggest that global mean air temperatures rose by 0.5 °C over the period from 1850 to 1987 (annual rate 0.003 65 °C). Although there is random scatter (plus or minus 0.25 °C) and periodic variations (plus or minus 0.15 °C) in the data, an upward trend over the period is clear. The periodicity shows a recent increase of 0.2 °C between 1970 and 1987 (annual rate 0.0285 °C), and the periodicity plus scatter gives an increase of 0.5 °C between 1979 and 1987 (annual rate 0.0625 °C). If there were direct correlations between cumulative fossil fuel combustion, the accumulation of carbon dioxide in the atmosphere, and the mean global temperature, then the increase of the latter with time should follow a cumulative exponential increase. Furthermore, if the comparatively recent additions of CFCs and other industrial and domestic greenhouse gases are effecting global temperatures, this should show up as a change of slope in the data in recent years. This cannot be detected in the measured data. It is believed that the considerable damping effects of sensible heat absorption in the oceans, the evaporation of water and glacial melting cause the close-to-linear increase of temperatures observed.

Sea-level rise

As the world has become warmer, it has been suggested that the water in the seas has expanded and there has been some glacial melting,[7] which has resulted in a rise in mean sea-levels of 0.1 m over the period from 1850 to 1987.

Correlations

Global temperature rise is due to the combined impacts of (i) the greenhouse effect of carbon dioxide and other trace gases (methane, nitrous oxide and CFCs (industrial and domestic chlorofluorocarbons)) in the troposphere inhibiting longwave radiation from the earth's surface, and (ii) the increased reception of short-wave radiation from the sun, resulting from the destruction of the ozone layer in the stratosphere. The individual effects are difficult to quantify owing to the sparse and often conflicting data but it has been estimated[9] that carbon dioxide contributes about 50 per cent to global warming. Methane, nitrous oxide, tropospheric ozone, water vapour and the CFCs may thus double the effect of CO_2.

Thus, if the 0.5 °C global temperature rise were due solely to an additional 80 ppmv of carbon dioxide directly added to the atmosphere, the global temperature and concomitant sea level rise could be correlated to a first order with cumulative fossil fuel consumption. This approach neglects the other greenhouse gases (CFCs, methane, nitrous oxide, subtropospheric ozone and water vapour), but, as more complex analyses treat these in terms of their *global warming potential* with respect to carbon dioxide, should yield similar projections. The effects of ozone depletion in the stratosphere have not been included. In Fig. 1.8, the data has been extrapolated to indicate that, if world fossil fuel consumption continued to increase at an annual rate of 2.79 per cent (i.e. as in 1987), reserves would be depleted by 2050, the global temperature would rise to 17 °C and sea levels by over 0.4 m.[10]

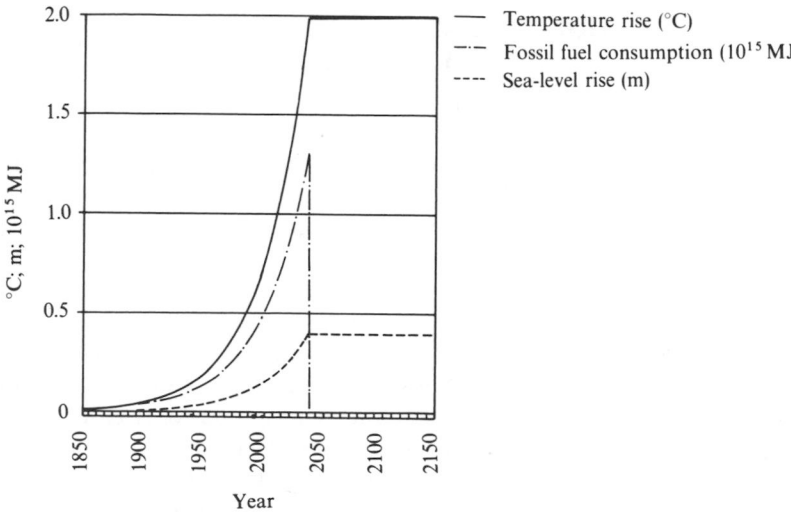

Figure 1.8 Greenhouse effect scenario assuming future 'business-as-usual'.

This method of projection also assumes that the proportion of carbon dioxide fixed by vegetation remains constant at its present level (24 per cent) over the period of the projection. At some unknown level of fossil fuel carbon dioxide emissions, however, the process of photosynthesis might become saturated, especially as large-scale deforestation continues. At the time of saturation, atmospheric carbon dioxide concentrations, global temperatures and sea levels would begin to rise at four times the rates used to produce Fig. 1.8, resulting in up to 8 °C global temperature rise.

Secondary effects, such as the reduced absorption of carbon dioxide by sea-water with increasing temperature, or the reduced reflectivity of the earth as ice melts, exacerbating the problem, or the reduced energy needs associated with increased environmental temperatures (e.g. reduced space heating requirements), helping to reduce the problem, have not been included. For example, the solubility of carbon dioxide in sea water reduces by 10 per cent for every 1 °C temperature rise accelerating the effect.

The situation could be much worse if the rates of production of the other greenhouse gases increase, or even remain at the current levels. CFCs are not fixed by vegetation. McElroy predicts that the greenhouse gases, other than carbon dioxide, will contribute an additional 180 per cent to global warming in the 2020s.

Figure 1.9 shows the effect of reducing release rates of carbon dioxide at an annual rate of 1 per cent, starting in 1990.

Levels of world fossil fuel consumption resulting from a 1 per cent annual reduction from 1990 are given in the following list. By the year 2000 the level would equate to that level which previously appertained in 1987, etc.

2000	1987
2025	1978
2050	1968
2075	1958
2100	1948
2125	1938
2150	1928

Figure 1.9 shows that, even with an annual reduction of 1 per cent, the global temperature would continue to rise to 16.7 °C by 2150. By 2050, 16.1 °C will have been reached.

In June 1988, the Canadian conference of world scientists produced the 'Toronto Protocol'. This called for a reduction in carbon dioxide emissions to 80 per cent of 1988 levels by 2005 as a 'global goal' to stabilize atmospheric levels.[10]

In 1987, the 'Montreal Protocol' called for a reduction in the world production of CFCs to the 1986 level in 1989, 80 per cent of this level in 1993 and 50 per cent of this level in 1998.[11]

In order to appreciate the effects of these reductions, it is necessary to examine in more detail the mix of the greenhouse gases and their individual increasing rates of release.

Knowing the CO_2 potential of the other greenhouse gases, it is possible to extrapolate forward, using the 1975–1985 trends given above to estimate future atmospheric levels. If the 0.5 °C temperature rise in global mean temperature from pre-industrial times is due to the cumulative addition of greenhouse gases to the troposphere, then, assuming that temperature increase correlates with this accumulation, Fig. 1.10 can be produced,[12] which reflects in

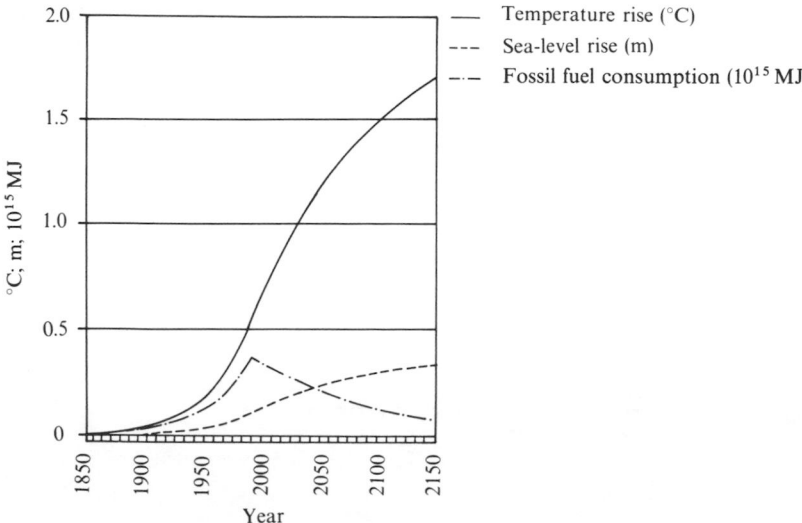

Figure 1.9 Greenhouse effect scenario assuming future reductions in fossil fuel consumptions.

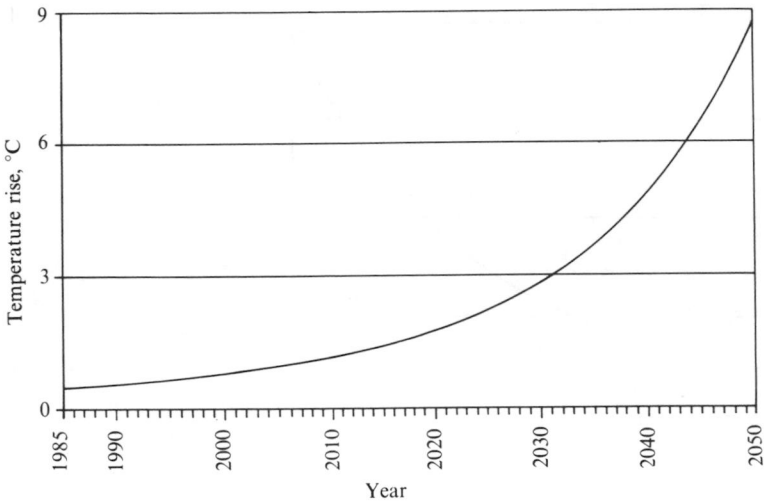

Figure 1.10 Greenhouse effect scenario assuming future rises in fossil fuel consumptions.

particular the rapidly increasing rates of release of CFCs. This suggests that rates of global release should not be allowed to continue on course as in 1987 if near-term global warming is to be avoided.

Figure 1.11 shows the effects of applying the Montreal Protocol alone—the situation is somewhat stabilized, but the global mean temperature continues to rise, albeit not to such dramatic levels in the near term. If the release of CFCs continues at the 1998 rate thereafter,

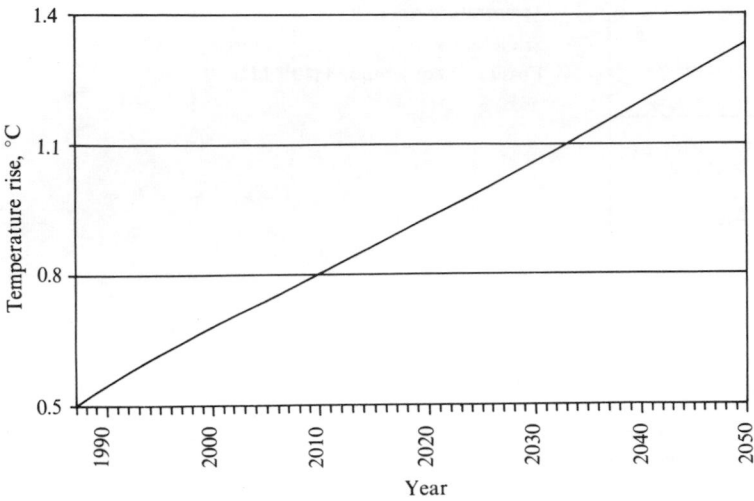

Figure 1.11 Greenhouse effect scenario assuming that the Montreal Protocol is obeyed.

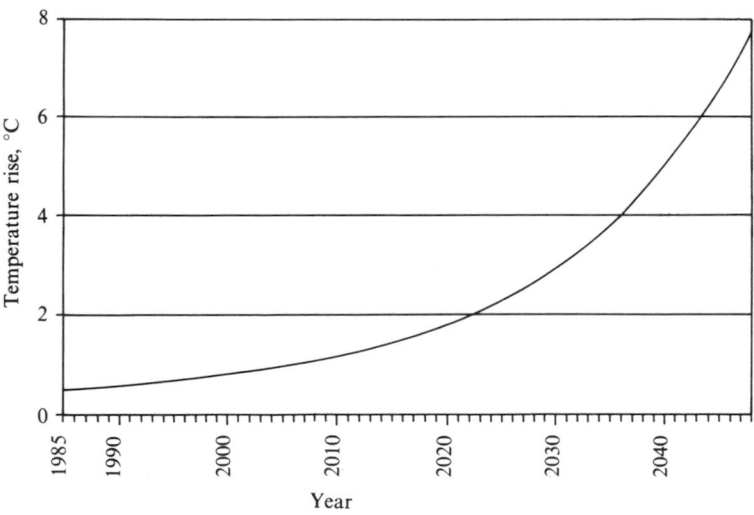

Figure 1.12 Greenhouse effect scenario assuming that the Toronto Protocol is obeyed.

mean global temperature continues to rise linearly with time. Satisfaction of the Toronto Protocol calling for a reduction of CO_2 emissions to 80 per cent of the 1988 level by the year 2005 does little in isolation if the rates of release of CFCs continue to increase from 1985 (Fig. 1.12). Figure 1.13 shows the effect of obeying both the Montreal and Toronto Protocols. In 1985, the aggregated breakdown of the sources of greenhouse gases was as given in Table 1.23.

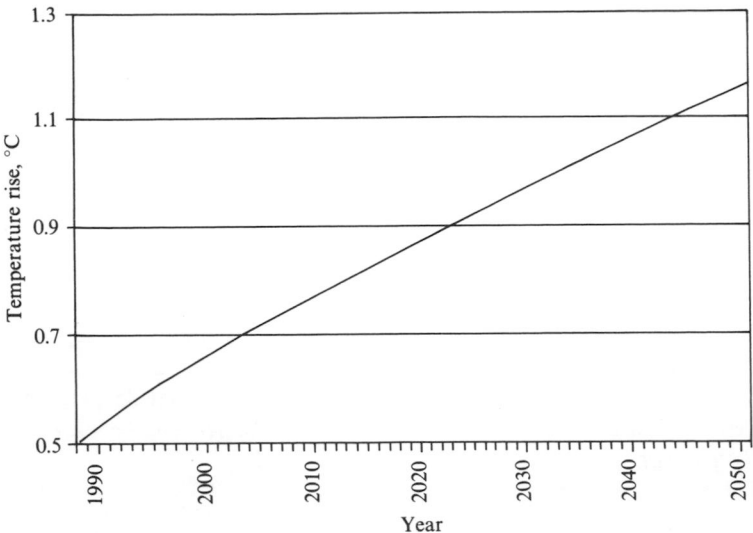

Figure 1.13 Greenhouse effect scenario assuming that both the Montreal and Toronto Protocols are obeyed.

Table 1.23 Breakdown of the sources of greenhouse gases, ppmm

	CO_2 equiv. by mass over pre-industrial times, ppm of CO_2	CFC Effect	Food production	Fossil fuel	Business as usual 2050
CFCs	40 (19%)	40			2290 (57%)
CO_2	122 (57%)			122	505 (13%)
CH_4	27 (13%)		27		69 (2%)
N_2O	6 (3%)		3	3	21 (0.5%)
O_3	20 (9%)	20			1149 (28%)
Total	215	60	30	125	4034

These figures for 1985 are broadly in line with those estimated by other workers[13,14]

1.8 GLOBAL ENERGY AND ENVIRONMENTAL MANAGEMENT

CFCs are used for aerosol sprays, blowing packaging materials, for industrial cleaning and in refrigeration systems. It appears that the safest course of action is to cease production of CFCs immediately, closely supervise the uses of those CFCs that have already been produced, and ban the release of CFCs into the atmosphere. This is already beginning to happen in the industrialized countries.

This should be accompanied by a planned steady and sustained reduction of CO_2 emissions from the fossil fuels, together with re-afforestation programmes. Carbon dioxide emissions may be reduced via flue gas clean-up and energy conservation. Flue gas clean-up is an expensive option. Energy conservation brings with it the added benefit of financial savings.

Table 1.23 categorizes the greenhouse emissions into:

- *The effects of industrial CFCs* These are direct greenhouse gases but also cause the destruction of stratospheric ozone and the accumulation of tropospheric ozone as a greenhouse gas.
- *The effects of food production* This is the methane produced from animals and the anaerobic decay of organic materials together with some nitrous oxide emanating from synthetic nitrogen-based fertilizers.
- *The effects of the combustion of fossil fuels* This is the carbon dioxide together with some of the nitrous oxide produced from fossil fuel exhausts.

Also shown in Table 1.23 is the projection (to the year 2050) which assumes release rates continuing to increase at the annual rates prevalent in 1985, and which results in the prediction of a mean global temperature of 9 °C, shown in Fig. 1.10. Table 1.23 shows that 85 per cent of this increase is due to the increased use and release of the CFCs. *Hence the primary importance of the Montreal Protocol.*

Industrial CFCs

If the addition of CFCs to the atmosphere is stopped, the problem becomes much more tractable.

Food production

The release of methane and the use of nitrogen-based fertilizers cannot be reduced substantially without taking the following actions:

- A widespread move to vegetarianism
- A return to organic farming methods
- A reduction in world population

Thus reductions in emissions of methane and nitrous oxide are unlikely in the near term.

Fossil fuel combustion

This leaves the third category resulting from fossil fuel combustion. An approximate average or typical breakdown of the use of fossil fuels in the industrialized countries is given in Table 1.24.

Table 1.3 shows that the bulk of fossil fuel is consumed by the industrialized regions of the world. It is the nations in these regions which must conserve energy, whilst developing countries might increase energy consumption to provide feedstock to growing economies.

Electricity production

Table 1.24 shows that the primary fuel (mostly coal) delivered to power stations accounts for 30 per cent of all fossil fuel use, although only 10 per cent of all energy is used in the form of electricity. Much of that is used for street lighting.

Table 1.24 Approximate breakdown of fossil fuel usage

Electricity	10%
Thermal losses*	20%
Industrial use	10%
Domestic heating	20%
Industrial heating	15%
Transport	25%

* accompanying electricity generation

It should be noted that every unit of electricity saved, for example, by attention to good lighting practices and the uses for industrial power, saves *three* units of primary fossil fuel.

Heating

Industrial and domestic heating accounts for a further 35 per cent of the fossil fuels. The need to heat buildings using fossil fuels directly can be eliminated technically by good solar-aided building design, the energy management and thermal upgrading of existing buildings, and by the use of combined heat and power and district heating schemes. The thermal losses from electricity generation would thus be cascaded to a secondary useful purpose. Good *energy management* of industrial processes might halve the high temperature industrial heating and other requirements.

Transportation

The third largest sector lies in the ever-increasing demands of transportation. Road tolls and carbon taxes might be applied to reduce emissions from vehicle exhausts. This, together with 'polluters-pay' policies may raise revenue for structural improvements to public transport systems, but where the economic and demographic infrastructures are highly developed, a substantial reduction in emissions from motor vehicle exhausts can be achieved only with sweeping sociological changes and population redistributions. Nevertheless, in the longer-term, a return to compact and relatively self-sufficient industrial village societies may be necessary.

Industry

All other industrial uses, that portion of the energy bill which results in the manufacture of products and commercial wealth, accounts for only 10 per cent of total usage.

1.9 CONCLUSIONS

The emission of CO_2 from the combustion of fossil fuels will eventually cease when the world's resources are depleted.

There are many opportunities for energy conservation in the industrialized countries. These, together with the increased use of nuclear power and alternative energy sources (including the use of biomass, the burning of organic waste) as well as flue gas clean-up operations and reafforestation schemes can substantially reduce CO_2 emissions.

As any financier knows, the earlier the investment the greater the net present value of the sum invested. Thus the more effective the action taken today, the less desperate the actions will be needed to be taken in the future.

Thus it makes long-term economic sense for a global programme of sustained energy and environmental management to be formulated and followed immediately.

This will allow more time for the development of a global renewable energy economy and, if significant global warming is occurring, for the world's climate to stabilize by natural processes.

REFERENCES

1. 'Statistical Review of World Energy', BP, June 1988.
2. O'Callaghan, P. W.: *Building for Energy Conservation*, Pergamon Press, Oxford, 1978.
3. Collin, P. H.: *Dictionary of Ecology and the Environment*, Peter Collin Publishing, Toddington, 1988.
4. Boyle, S. and Ardill, J.: *The Greenhouse Effect*, New English Library, Hodder and Stoughton, London, 1989.
5. HMSO: 'Greenhouse Effect', Report of House of Lords Select Committee on Science and Technology, Volume II—Evidence, HMSO 1989.
6. Gribbon, J.: *Hothouse Earth—The Greenhouse Effect and Gaia*, Bantam Press, London, 1990.
7. Embar, L. R., Layman, P. L., Lepkowski, W. and Zurer, P. S.: 'Tending Global Commons—The Changing Atmosphere', *Chemical and Engineering News*, **64**, 47, 16–35, November 1986.
8. Houghton, R. A. and Woodwell, G. M.: 'Global Climatic Change', *Scientific American*, **260** (4), 18–26, April 1989.
9. McElroy, M.: 'The Challenge of Global Change', *New Scientist*, 34–36, 28 July 1988.
10. Everett, R.: '35% Savings for No Extra Cost?', Paper presented at Mini-Power Stations Conference, Chatham House, London, 10 May 1990.
11. Cox, J. E.: 'The UNEP Agreement', ASHRAE Journal, **31**, November 1987.
12. O'Callaghan, P. W.: 'Energy Resources, CO_2 Production and Energy Conservation'. Paper presented at Cranfield Triple E Seminar, *Energy, Environment, Ecology*, 25 September 1990. To be published Applied Energy 1992.
13. Smith, P.: 'RIBA Conference Discusses Greenhouse Effect', Energy Management, 18–21, UK. DEn., April/May 1990.
14. Brookes, G.: *The Greenhouse Effect*, ESTA Energy Efficiency Year-book, UK, 1990.

FURTHER READING

Allaby, M.: *Green Facts: The Greenhouse Effect and Other Key Issues*, Hamblyn, London, 1989.
Anon: *Energy 2000: A Global Strategy for Sustainable Development* (World Commission on Environment and Development), Zed Books, London, 1987.
Anon: 'Chlorofluorocarbons–Professional and practical Guidance', Guidance Note, CIBSE, London, December 1989.
Bolin, B.: Greenhouse Effect Climatic Change and Ecosystems, SCOPE Report, Wiley, Chichester, UK, 1986.
Button, J.: *How to be Green*, Century Hutchinson, London, 1989.
Elkington, J. and Hayes, J.: *The Green Consumers Supermarket Guide*, Gollanz, London, 1989.
Elkington, J. and Hayes, J.: *The Green Consumer Guide*, Gollanz, London, 1989.
'Friends of the Earth': Memoranda to the House of Commons Select Committee on Energy on the Greenhouse Effect, 1989.

Hodgson, P.: (ed.), *Dictionary of Ecology and the Environment*, Peter Collin Publishing, Toddington, 1988.

Lovelock, J.: *The Ages of Gaia*, Oxford University Press, Oxford, 1988.

Mead, W. J.: 'Energy and the Environment: Conflict in Public Policy', American Enterprise Institute for Public Policy Research, 1987.

Neal, P.: *The Greenhouse Effect and Ozone Layer*, Dryad, London, 1989.

O'Hara, G. and Sweeney, J.: *Atmospheric Systems: Introduction to Meteorology and Climatology*, Oliver and Boyd, Edinburgh, Scotland, UK, 1986.

Porteous, A.: *Dictionary of Environmental Science and Technology*, Open University Press, Milton Keynes, 1991.

Ramanathan, V.: 'The Greenhouse Theory of Climate Change: A test by inadvertent global experiment', *Science Magazine*, **240**, 293–299, 15 April 1988.

Richardson, S.: *Basics of Climatology*, Edward Arnold, London, 1986.

Robinson, P. J., and Henderson-Sellers, A.: *Contemporary Climatology*, Longman, Harlow, UK, 1986.

Steinberg, M., Cheung, H. G. and Horn, F.: 'A Systems Study for the Removal, Recovery and Disposal of Carbon Dioxide from Fossil Fuel Power Plant in the USA', BNL Report DOE/CH/00016, 2 December 1984.

Thurlow, G. (ed.): 'Technological Responses to the Greenhouse Effect', preprint of the Watt Committee on Energy Twenty-six Consultative Conference, Royal Geographical Society, London, 24–25 April 1990.

Watson, R. T., Prather, M. J. and Kurylo, M. J.: 'Present State of Knowledge of the Upper Atmosphere 1988, An Assessment Report', NASA Reference Publication 1208, August 1988.

TWO

ENERGY MANAGEMENT AND CONSERVATION

2.1 ENERGY MANAGEMENT

The ability of any nation to survive economically depends upon its ability to produce and manage sufficient supplies of low-cost safe energy and raw materials.

It has been seen that the world consumption of limited fossil fuel resources currently increases annually by 3 per cent. Projections of this trend shows that all known reserves will be exhausted in the first half of the coming century. Any sustained attempt to reduce rates of energy consumption, even as little as 1 per cent per annum, ensures an effectively eternal future supply as the nations of the world move slowly towards renewable energy economies.[1] Over the past ten years, a further 6.2 per cent of the world fossil fuel store has been consumed, the 'heat limit'[2] has been reached and overtaken, and global climate may be becoming affected. Assuming that the production and release of CFCs can be reduced and eventually eliminated, the long-term solution to these problems is to *institute firm, systematic and effective energy and environmental management*. Governments, industrialists, commercial organizations, public sector departments and the general public have now become aware of the urgent requirements for the efficient management of resources and energy-consuming activities. Most organizations in the materials, manufacturing, retail sectors and in the service industries have created energy management departments, or have employed consultants, to monitor energy consumptions and to reduce wastage.

Energy management is a technical and management function the remit of which is to monitor, record, analyse, critically examine, alter and control energy flows through systems so that energy is utilized with maximum efficiency.[2] It embraces the disciplines of engineering, science, mathematics, economics, accountancy, design and operational research, computation and information technology.

The energy manager must also be responsible for the day-to-day management of fuels and deliveries, boiler houses, distribution systems, building services, plant, process equipment, polluting exhausts, effluents and waste.

Great strides in the now well-established profession of energy management have been made over the past ten years. This chapter summarizes the energy management procedures which have emerged:

2.2 ENERGY SURVEYING AND AUDITING

An energy audit is a fundamental part of any energy management programme of any organization which wishes to control its energy costs.[3] The construction of a complete and detailed energy audit is an intricate, tedious but necessary procedure so that major energy use activities can be identified.

The consumption of fossil fuel energy involves five basic processes:

- Energy release via combustion, in which chemical energy is converted to thermal energy
- Conversion of energy to alternative forms (i.e. thermal to mechanical and vice versa)
- Energy distribution to places of use
- Energy utilization for a specific purpose
- Energy rejection to the environment

The *energy audit* is a balance sheet of energy *inputs*, *throughputs* and *outputs*. Its fundamental equation is as follows:

fuel energy input = energy losses during combustion + energy losses during conversion + energy losses during distribution + energy losses during utilization + energy losses from utilization   (2.1)

2.3 PASSIVE ENERGY CASCADING—SUNDRY HEAT GAINS AND LOSSES

Energy losses are strictly those irrecoverable rejections to the external environment. Some energy losses which occur during combustion, conversion, distribution and utilization may become 'sundry gains' which offset some of the energy demands of a secondary utilization. For example, transmission heat losses from the jackets of furnaces and boilers, surface heat losses from internal heat distribution pipelines and heat losses from equipment, plant and processes, supply heat to the internal environment and reduce the demands for primary energy for space heating. Most electricity consumed within the internal environment (i.e. from lighting, electrical machines, electronics, computers, inlet ventilation fans, etc.) ends up as sundry gains. The few exceptions to this rule are the electricity supplied to extract fans, compressed air systems and refrigeration plant, where a sundry 'cold gain' may result in increased space-heating requirements. When the internal environment is cooled via air conditioning, sundry heat gains are disbenefits.

For a heated building the *energy audit equation* becomes:

heating fuel energy input + sundry gains from electricity + sundry gains from people (and/or livestock) + sundry gains from directly-fired process plant and equipment − sundry losses to cold plant

=

energy losses in flue gases + energy losses during conversion + energy losses from external distribution pipelines + energy losses through the external surfaces of the building via heat transmission through the fabric + energy losses in ventilating air + energy losses in process fluids or solids directly rejected to the external environment (2.2)

For a cooled building, the energy audit equation becomes:

electrical energy supplied for refrigeration = energy losses during conversion + energy gains to external distribution pipelines + sundry gains from electricity + sundry gains from people (and/or livestock) + sundry gains from directly-fired process plant and equipment − sundry losses to cold plant + energy gains through the external surfaces of the building via heat transmission through the fabric + energy in ventilating air + energy losses from input process fluids or solids (2.3)

2.4 THE AIM AND THE DETERMINATION

The aim of the energy audit is to obtain a simple, but comprehensive 'photograph' of the overall energy flow situation within a declared system boundary, which may be, for example, a building, or group of buildings, a factory, or a product line. This picture aids comprehension of the total overall system activity, reveals interrelations and allows priorities to be identified. It highlights major areas where inefficiencies or waste occurs and allows economic estimates, leading to fully-reasoned investment decisions, to be constructed. Diseconomic effects of certain energy-conserving investments will also emerge. Without the information contained in the energy audit, the energy manager operates blinkered and is prone to erratic, ill-conceived and non-optimal decisions.[1]

In constructing the energy audit, the energy manager must adopt the single-minded attitude of the financial accountant when constructing a financial audit. He must not be deflected from the purpose of constructing the balance sheet described by the energy audit equation. It is often easy to become absorbed in intricacies and, as a result, to spiral outwards from the objective.

2.5 FLOW CHART FOR THE CONSTRUCTION OF AN ENERGY AUDIT

The energy audit may be neatly subdivided into the *input* side, the *output* side, and the *throughput*.

The *input* side constitutes an analysis of fuel and electricity bills for a representative recent annual period.

The *output* side details the ultimate energy rejection to the external environment, mainly via heat transmission through the building fabric and ventilating air. The data is obtained from a site energy survey.

Analyses of the *throughputs* may require microaudits, or energy balances over individual items of plant and equipment, such as furnaces, boilers, refrigeration systems, steam autoclaves, compressors, etc., to ascertain operating efficiencies and to identify where sundry gains occur.

The systematic approach to an energy audit contains the following sequential steps:

1. Submit preliminary questionnaire.
2. Process responses from questionnaire.
3. Obtain fuel and electricity bills for a recent representative annual period.
4. Analyse fuel and electricity bills.
5. Conduct boiler house survey and efficiency measurements.
6. Investigate energy distribution systems.
7. Perform internal site survey.
8. Construct *input* side of the audit.
9. Obtain local climatic data.
10. Perform external site survey.
11. Quantify sundry gains.
12. Construct *output* side of the audit.
13. Construct the energy audit balance sheet.
14. Investigate any residual and iterate to balance the audit.
15. Analyse *throughputs*.

2.6 PRELIMINARY QUESTIONNAIRE

The purpose of the preliminary questionnaire is to extract preliminary information concerning the site, its function and the activities being conducted. It should include queries as to the purpose and function of the establishment, occupational patterns, air temperature and ventilation requirements. The questions might be asked during an initial meeting between client and consultant, or the questionnaire and responses might be conducted by post at the initial stages of the investigation.

A typical preliminary questionnaire might appear as in Table 2.1 overleaf.

2.7 THE PRELIMINARY REQUEST

A covering letter should request a set of copies of invoices for all fuel types and electricity consumed for a recent complete annual period.

2.8 FUELS AND MATERIALS SUPPLIED

Tables 2.2 and 2.3 show a returned preliminary questionnaire and a summary of information from the set of fuel bills supplied. Table 2.3 shows the fuels and electricity delivered for the period March 1991 to February 1992.

Using the conversion factors shown in Table 2.4 it is possible to produce Table 2.5 in common units.

Figures 2.1 and 2.2 show that the bulk of the energy (75 per cent) is consumed as gas and oil for factory heating. Fuel and electricity costs are shown in Table 2.6. Nightrate electricity is 55 per cent the cost of dayrate electricity, and oil is 245 per cent more expensive than gas. The gas supply for the site is on an interruptible tariff. Hence the need to substitute oil on occasions.

Table 2.1 Preliminary questionnaire for an energy audit

Energy Audit

Preliminary Questionnaire

Name of firm	
Nature of business	
Status (commercial, industrial or public sector)	
Address of premises to be surveyed	
Telephone number	
Fax number	
Contact	
Position	
Location	
Number of employees	
Hours of work Details of shift working Weekdays Weekends Annual shut-down	
Required inside temperature/s	
Required ventilation rates (if known)	
Working area (if known)	
Heated volume (if known)	
Annual energy bill	

Table 2.2 Questionnaire response

Name of firm	Bitusa Industries Ltd.
Nature of business	Mechanical Parts Manufacturer
Status (commercial, industrial or public sector)	Industrial
Address of premises to be surveyed	4 Back Street New Chesham Nr Whittleford UK
Telephone number	0234 750111
Fax number	0234 750112
Contact	J Jimmy
Position	Chief Works Engineer
Location	Boilerhouse (shed adjacent) The Merry Cockerel (eves and w-ends)
Number of employees	150
Hours of work Details of shift working Weekdays Weekends Annual shut-down	8760/annum 3×8 hours 3×8 hours Christmas
Required inside temperature/s	21 °C
Required ventilation rates (if known)	3 air changes/hour
Working area (if known)	8 000 m^2
Heated volume (if known)	40 000 m^3
Annual energy Bill	£280 000

Table 2.3 Fuels and energy delivered (March 91–February 92), raw units

Month	Coal tonnes	Oil litres	Gas therms	Electricity Dayrate kWh	Electricity Nightrate kWh
March	0	24 000	18 000	162 342	43 234
April	0	24 000	12 015	133 245	39 876
May	0	0	10 652	154 789	43 475
June	0	0	8 614	132 456	43 098
July	0	0	4 563	123 453	39 876
August	0	0	5 413	121 000	37 684
September	0	14 000	10 312	124 356	37 984
October	0	0	9 919	176 899	49 693
November	0	25 000	10 222	167 564	41 453
December	0	11 000	10 101	132 876	32 967
January	0	23 000	21 146	152 098	37 963
February	0	12 000	21 999	163 876	41 856
Totals	0	133 000	142 956	1 744 954	489 159

Table 2.4 Fuel conversion factors

Fuel type		Calorific value	
Coal	None used		
Oil	Light fuel oil	11.23	kWh per litre
Gas	Natural	29.3	kWh per therm

Table 2.5 Energy delivered (March 91–February 92), common units, kWh

Month	Electricity Dayrate	Electricity Nightrate	Oil	Gas
March	162 342	43 234	269 520	527 400
April	133 245	39 876	269 520	352 039
May	154 789	43 475	0	312 103
June	132 456	43 098	0	252 390
July	123 453	39 876	0	133 695
August	121 000	37 684	0	158 600
September	124 356	37 984	157 220	302 141
October	176 899	49 693	0	290 626
November	167 564	41 453	280 750	299 504
December	132 876	32 967	123 530	295 959
January	152 098	37 963	258 290	619 577
February	163 876	41 856	134 760	644 570
Totals	1 744 954	489 159	1 493 590	4 188 611

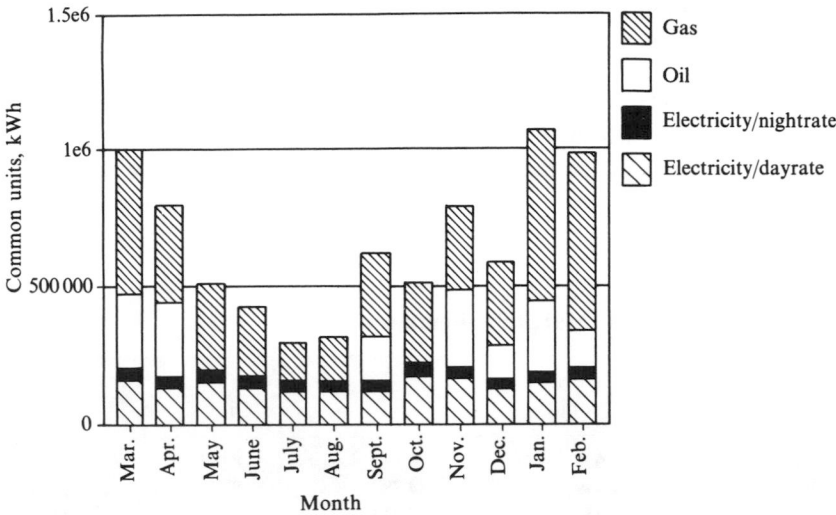

Figure 2.1 Energy delivered (March 91–February 92)—histogram.

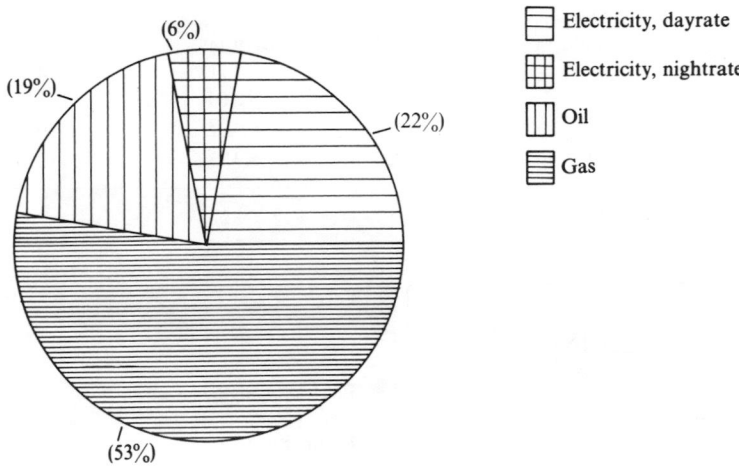

Figure 2.2 Energy delivered (March 91–February 92)—pie chart.

Table 2.6 Fuel and electricity costs, £

Coal	N/A	
Oil	0.375/litre	0.0333 kWh^{-1}
Gas	0.398/therm	0.0136 kWh^{-1}
Dayrate	Electricity	0.0683 kWh^{-1}
Nightrate	Electricity	0.0376 kWh^{-1}

During the accounting period, 1 744 954 kWh of dayrate electricity cost £119 180.4, 489 159 kWh of nightrate electricity cost £18 392.38, resulting in an average cost for electricity of £0.0616 per kWh.

In the same period, 1 493 590 kWh of oil cost £49 740 and 4 188 611 kWh of gas cost £56 965, giving an average cost for heating fuel of £0.0188 per kWh.

Table 2.7 and Figures 2.3 and 2.4 show the costs of energy delivered. These are dominated by the costs for electricity (57 per cent) even though they do not include charges for maximum demand or electrical supply capacity.

Table 2.7 Costs (£) of energy delivered (March 91–February 92)

Month	Electricity Dayrate	Electricity Nightrate	Gas	Oil	Totals
March	11 087	1 625	7 172	8 975	28 861
April	9 100	1 499	4 787	8 975	24 362
May	10 572	1 634	4 244		16 451
June	9 046	1 620	3 432		14 099
July	8 431	1 499	1 818		11 749
August	8 264	1 416	2 156		11 838
September	8 493	1 428	4 109	5 235	19 266
October	12 082	1 868	3 952		17 903
November	11 444	1 558	4 073	9 348	26 425
December	9 075	1 239	4 025	4 113	18 453
January	10 388	1 427	8 426	8 601	28 843
February	11 192	1 573	8 766	4 487	26 020
Totals	119 180	18 392	56 965	49 736	244 274

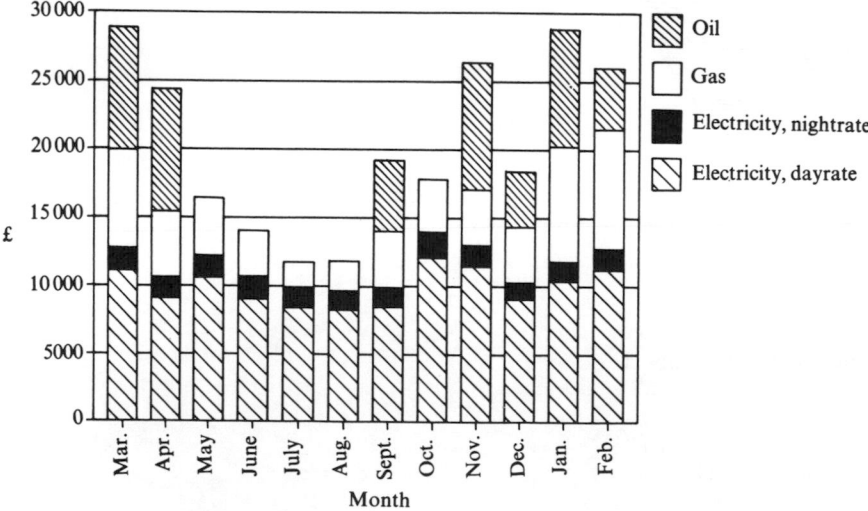

Figure 2.3 Costs of energy delivered (March 91–February 92)—Histogram.

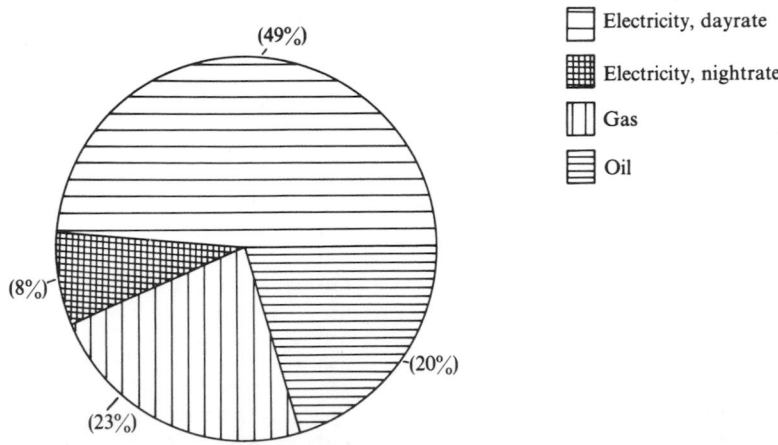

Electricity, dayrate

Electricity, nightrate

Gas

Oil

Figure 2.4 Costs of energy delivered (March 91–February 92)—pie chart.

Table 2.8 Electricity maximum demand (kVA) (March 91–February 92) supply capacity 1300 kVA

Month	Step 1	Step 2	Supply capacity
March	500	140	1300
April	500	50	1300
May	500	50	1300
June	500	60	1300
July	500	60	1300
August	500	40	1300
September	500	50	1300
October	500	60	1300
November	500	110	1300
December	500	100	1300
January	500	170	1300
February	500	190	1300

Electrical maximum demand is paid for in two steps as shown in Tables 2.8 and 2.9.

Figure 2.5 shows that the highest maximum demand over the accounting period was 690 kVA—much less than the supply capacity of 1300 kVA. This situation had come about because the company's business had contracted and personnel, equipment and activities had been relocated from various, now unused, outbuildings to a single central building. Whilst this space rationalization resulted in lower space heating demands, the management had neglected to renegotiate the electrical supply capacity. Reduction of the supply capacity to 800 kVA would save £3000 each year.

Recommendation 1 Renegotiate electrical supply capacity.

Table 2.9 Maximum demand charges (£) (March 91–February 92)

Month	Step 1, £/kVa		Step 2, £/kVA		Supply capacity, £/kVA		£	Totals
March	3.83	1 915	3.5	490	0.5		650	3 055
April	0.9	450	0.5	25	0.5		650	1 125
May	0.9	450	0.5	25	0.5		650	1 125
June	0.6	300	0.26	15.6	0.5		650	965.6
July	0.6	300	0.26	15.6	0.5		650	965.6
August	0.6	300	0.26	10.4	0.5		650	960.4
September	0.6	300	0.26	13	0.5		650	963
October	0.6	300	0.26	15.6	0.5		650	965.6
November	3.83	1 915	3.5	385	0.5		650	2 950
December	11.2	5 600	8.5	850	0.5		650	7 100
January	11.2	5 600	8.5	1445	0.5		650	7 695
February	11.2	5 600	8.5	1615	0.5		650	7 865
Totals		23 030		4905.2			7 800	35 735.2

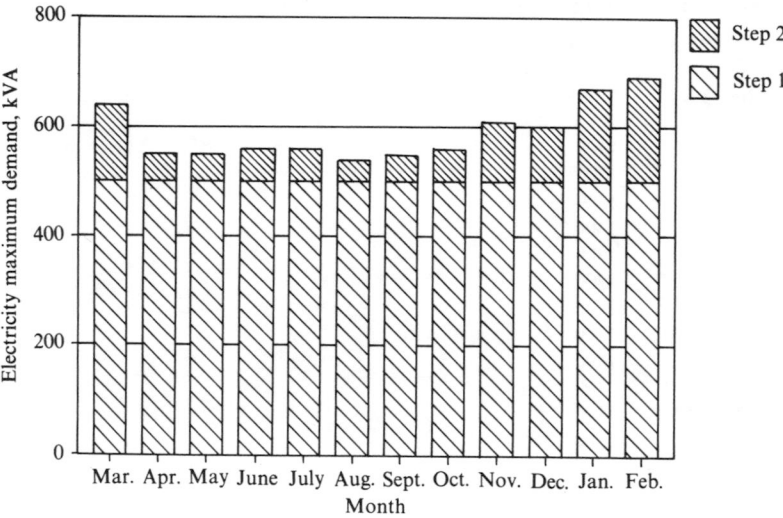

Figure 2.5 Electricity maximum demand (March 91–February 92).

Figure 2.5 also reveals a seasonal variation in maximum demand (kVA), resulting from the use of electricity for heating purposes. This should be investigated as the maximum demand premium is very high in the winter months (Fig. 2.6). As an example of the misuse of electricity. Suppose a 2 kW 'illegal' electric fire is turned on for the month of January. This will consume 2232 kWh, costing £137 and will result in a further £25.5 for kVA charges, making a total of £162.5. If the heat were supplied using gas, the cost would be only £30, saving £132.50 in the month of January alone.

Recommendation 2 Investigate the use of electricity for heating purposes.

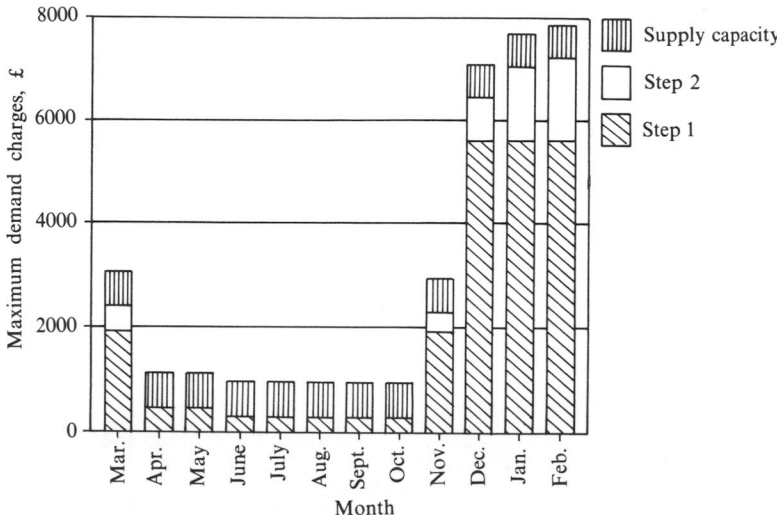

Figure 2.6 Electricity maximum demand charges (March 91–February 92).

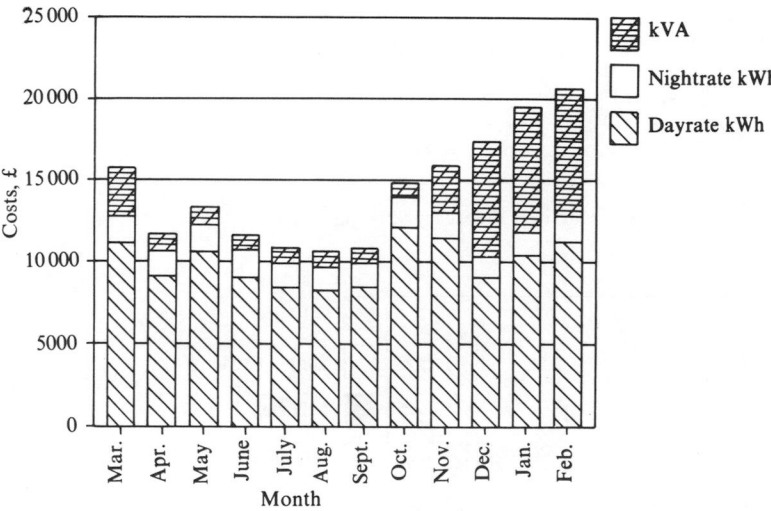

Figure 2.7 Total electricity costs (March 91–February 92)—Histogram.

This should result in evening out the monthly maximum demands over the year, and could save 240 kVA of maximum demand charges as well as kWh costs. This will be further investigated later.

When the maximum demand and supply capacity charges are included in the costs for electricity (Figs 2.7 to 2.11), the total costs for electricity, although only 25 per cent of the total energy used, becomes 62 per cent of the total annual bill £280 009.6 (Fig. 2.11).

Tables 2.10 and 2.11 give the annual breakdowns of energy (kWh) and costs (£).

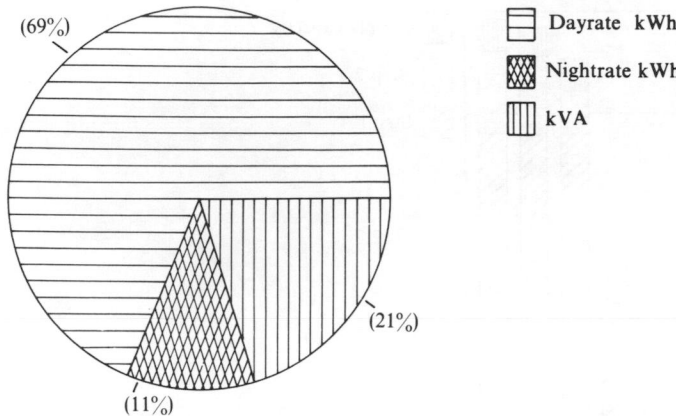

Figure 2.8 Total electricity costs (March 91–February 92)—pie chart.

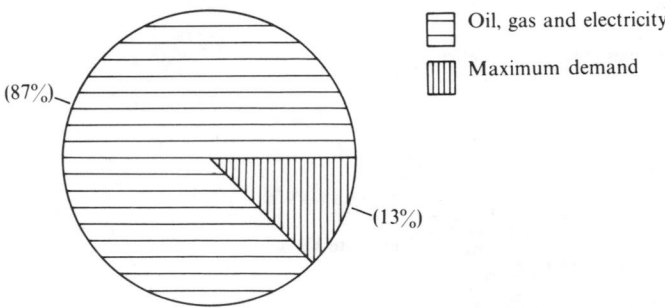

Figure 2.9 Energy and maximum demand costs (March 91–February 92)—pie chart.

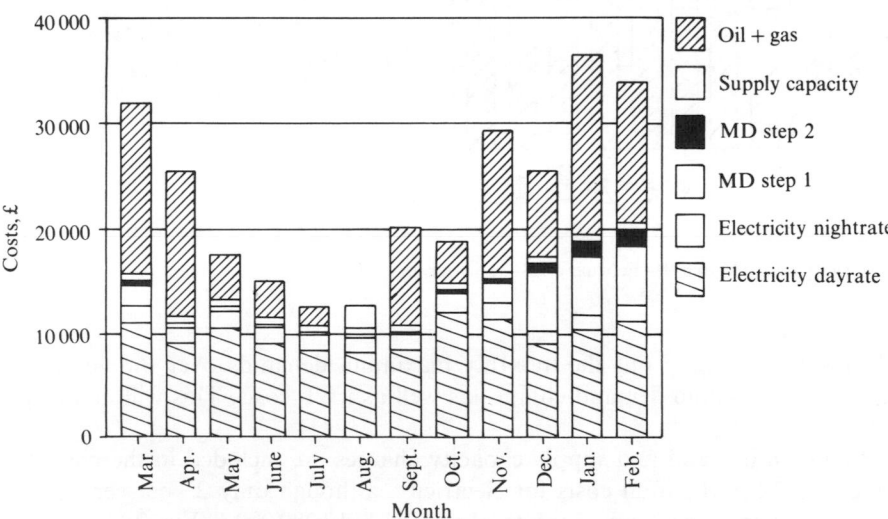

Figure 2.10 Energy and maximum demand costs (March 91–February 92)—histogram.

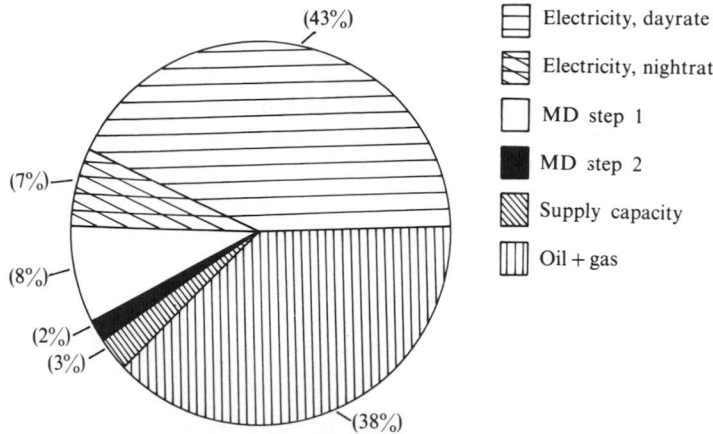

Electricity, dayrate

Electricity, nightrate

MD step 1

MD step 2

Supply capacity

Oil + gas

Figure 2.11 Breakdown of energy and maximum demand costs (March 91–February 92).

Table 2.10 Annual breakdown of energy consumption, kWh (March 91–February 92)

Oil	1 493 590	Total heating	5 682 201
Gas	4 188 611		
Dayrate electricity	1 744 954	Total electricity	2 234 113
Nightrate electricity	489 159		
Total	7 916 314	Heat-to-power ratio	2.54

Table 2.11 Annual breakdown of energy and MD costs, £ (March 91–February 92)

Oil	49 736		
Gas	56 965	Total heating	106 701
Dayrate electricity	119 180	Total electricity	173 307
Nightrate electricity	18 392		
MD step 1	23 030		
MD step 2	4 905		
Supply Capacity	7 800		
Total	280 008		

2.9 ENERGY INDICES

The energy figures may be normalized using the working floor area, the enclosed volume of the site and the number of personnel. The concept of the mean kW, which equals annual energy use divided by the number of working hours per annum can be employed to compare

Table 2.12 Annual energy consumption per unit area, kWh m^{-2}

Oil	186.6	Total heating	710.2
Gas	523.6		
Dayrate electricity	218.1	Total electricity	279.3
Nightrate electricity	61.1		
Total	989.5		

Table 2.13 Annual energy consumption per unit volume, kWh m^{-3}

Oil	37.3	Total heating	142.0
Gas	104.7		
Dayrate electricity	43.6	Total electricity	55.9
Nightrate electricity	12.2		
Total	197.9		

Table 2.14 Annual energy consumption per capita, kWh

Oil	9 946	Total heating	37 866
Gas	27 920		
Dayrate electricity	11 626	Total electricity	14 906
Nightrate electricity	3 253		
Total	52 720		

Table 2.15 Energy consumption per unit area, kW mean m^{-2}

Oil	0.021	Total heating	0.081
Gas	0.060		
Dayrate electricity	0.025	Total electricity	0.032
Nightrate electricity	0.007		
Total	0.113		

Table 2.16 Energy consumption per unit volume, kW mean m^{-3}

Oil	0.004	Total heating	0.061
Gas	0.012		
Dayrate electricity	0.005	Total electricity	0.006
Nightrate electricity	0.001		
Total	0.023		

Table 2.17 Energy consumption per capita, kW mean

Oil	1.14	Total heating	4.32
Gas	3.19		
Dayrate electricity	1.33	Total electricity	1.70
Nightrate electricity	0.37		
Total	6.02		

Table 2.18 Annual energy costs per unit area, £ m^{-2}

Oil	6.21	Total heating	13.33
Gas	7.12		
Dayrate electricity	14.90	Total electricity	17.20
Nightrate electricity	2.30		
MD step 1	2.88		
MD step 2	0.61		
Supply capacity	0.98		
Total	35.0		

Table 2.19 Annual energy costs per unit volume, £ m^{-3}

Oil	1.24	Total heating	2.66
Gas	1.42		
Dayrate electricity	2.98	Total electricity	3.44
Nightrate electricity	0.46		
MD step 1	0.58		
MD step 2	0.12		
Supply capacity	0.20		
Total	6.92		

Table 2.20 Annual energy costs per capita, £

Oil	331.22	Total heating	710.93
Gas	379.71		
Dayrate electricity	794.10	Total electricity	1 154.65
Nightrate electricity	122.32		
MD step 1	153.53		
MD step 2	32.70		
Supply capacity	52.00		
Total	1 865.58		

energy consumptions with those for other sites. The figures for this site are: floor area, $8000 \, \text{m}^2$; enclosed volume, $40\,000 \, \text{m}^3$; number of personnel, 150; and number of working hours per year, 8760.

Tables 2.12 to 2.20 give the resulting energy indices in terms of kWh and costs (£) using the fuel and electricity costs of Table 2.6.

Comparisons with other sites

The indices given in Table 2.21 have been obtained for a variety of audited sites. It can be seen that the present site ranks second best in energy use per unit area, probably as a result

Table 2.21 Comparisons with other sites (all energy)

	kW mean m^{-2}	kW mean m^{-3}	kW mean per capita
1	0.071	0.015	1.05
2	**0.113**		
3	0.114	0.0208	1.28
4	0.141	0.023	2.13
5		**0.023**	
6	0.16	0.0268	2.4
7	0.17	0.0274	2.72
8	0.18	0.0278	3.13
9	0.18	0.0282	3.17
10	0.195	0.0285	3.24
11	0.198	0.0312	4.21
12			**6.02**
13	0.2	0.0342	6.6
14	0.209	0.036	6.87
15	0.21	0.0369	10.95
16	0.24	0.0379	11
17	0.377	0.0418	13.5
18	0.42	0.044	28

(Present site in **bold** type)

of the space rationalization exercise, fifth best in energy use per unit volume, and twelfth in energy use per capita. This latter index may reflect the rational use of the workforce. On the whole the site could be classified as a relatively frugal energy user.

2.10 CORRELATIONS

Degree-days

Degree-days are a measure of the variation of outside air temperature which enables building designers and users to determine how the energy consumption of a building is related to the weather.[3] Heating degree-day tables commonly published are based upon a base outside air temperature of 15.5 °C. It is assumed that when the outside air temperature is 15.5 °C then no heating will be required, as sundry internal heat gains due to people, lights, power and processes would raise the inside air temperatures to the required comfort level. The calculation of degree-days is based upon maximum and minimum diurnal temperatures and explained in reference 3. The degree-days used here are named *heating degree-days* in this book and defined as the mean number of degrees by which the outside temperature on a given day is less than the base, or control, temperature for the degree-days, added up for all the days in the period, usually one month. For that part of the winter when the outside temperature is consistently below the control temperature, the number of heating degree-days over a period is given by the difference between the average temperature and the control temperature multiplied by the number of days in the period.

Then, reversing this procedure,

mean monthly temperature = control temperature − (degree-days per month)/(days per month)
= control temperature − mean degree-days per day
= control temperature − mean heating degrees per day

Thus, if January (31 days) has 350 degree-days to a base of 15.5 °C,

$$\text{mean monthly temperature} = 15.5 - 350/31$$
$$= 15.5 - 11.29$$
$$= 4.21 \,°C$$

However, in the autumn and early spring, it is common for night temperatures to drop below the control temperature, whilst the daytime temperatures remain above it. On such a day, heating will not be required for part of the day. This is allowed for in calculating the degree-days by working, not with the average daily temperatures, but with a formula based upon the daily maximum and minimum temperatures, which are easier to record, and assuming that the outside air temperature follows a sine wave through the 24 hours of the day. Thus the method used here for converting published monthly degree-days to mean monthly temperatures becomes less reliable as the mean monthly temperatures approach the control temperature.

To be more accurate, one would use more continuous recordings of temperature to build up daily profiles and compute the monthly degree-days and mean monthly temperatures directly. Electronic building energy management systems can provide measured values of temperature every 15 minutes and so produce degree-day data which are more accurate than degree-days based upon the data provided by the Energy Efficiency Office.[4]

Table 2.22 lists the degree-days for the site and Fig. 2.12(a) shows the correlation between all energy consumption and degree-days for each month of the year. On average, the site consumes 3830 kWh per degree-day.

For monitoring and targeting purposes, and to make true comparison among different sites, the indices in Tables 2.12 to 2.20 can be further normalized using the degree-days per year.

Table 2.22 All energy versus degree-days, common units, kWh

Month	Electricity	Oil + gas	Totals	Degree-days
December	165 843	419 489	585 332	335
November	209 017	580 254	789 271	295
January	190 061	877 867	1 067 929	286
April	173 121	621 559	794 680	241
February	205 732	779 330	985 062	227
March	205 576	796 920	1 002 496	217
October	226 592	290 626	517 218	132
May	198 264	312 103	510 367	115
June	175 554	252 390	427 944	88
September	162 340	459 361	621 701	64
August	158 684	158 600	317 284	40
July	163 329	133 695	297 024	27
Totals	2 234 113	5 682 201	7 916 314	2 067

On average, the site consumes 3830 kWh per degree-day.

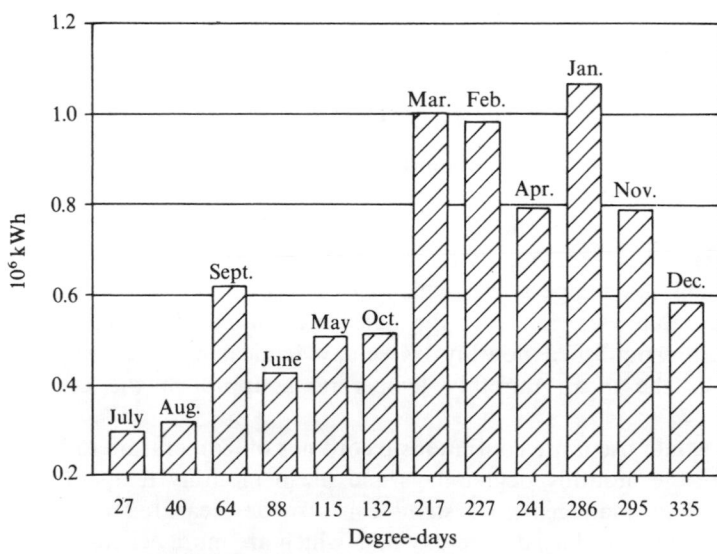

Figure 2.12a All energy versus degree-days—histogram.

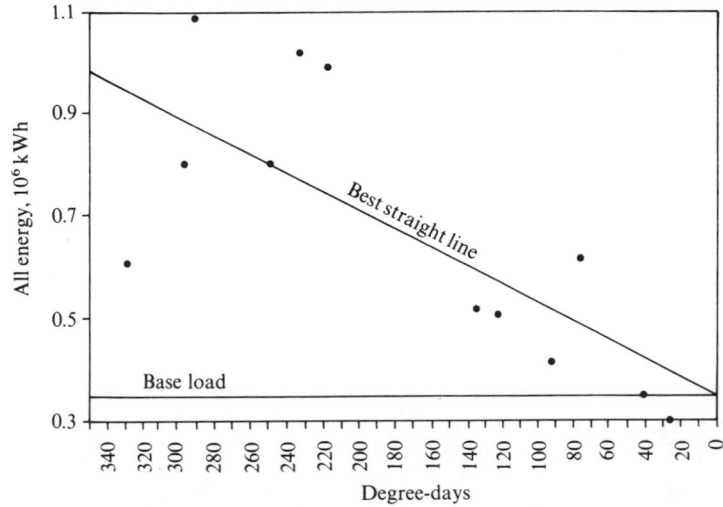

Figure 2.12b All energy versus degree-days—best fit.

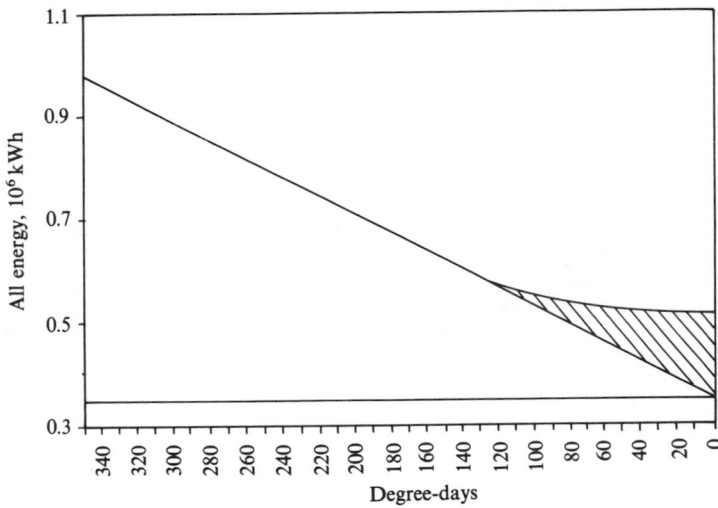

Figure 2.12c Hockeystick curve.

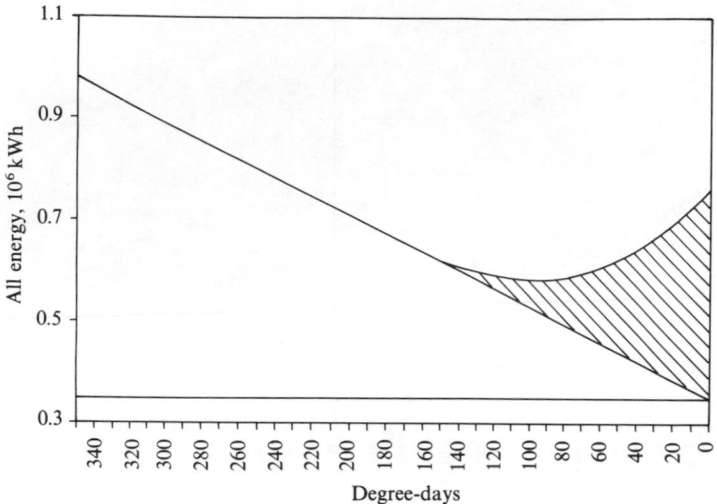

Figure 2.12d Effect of air conditioning.

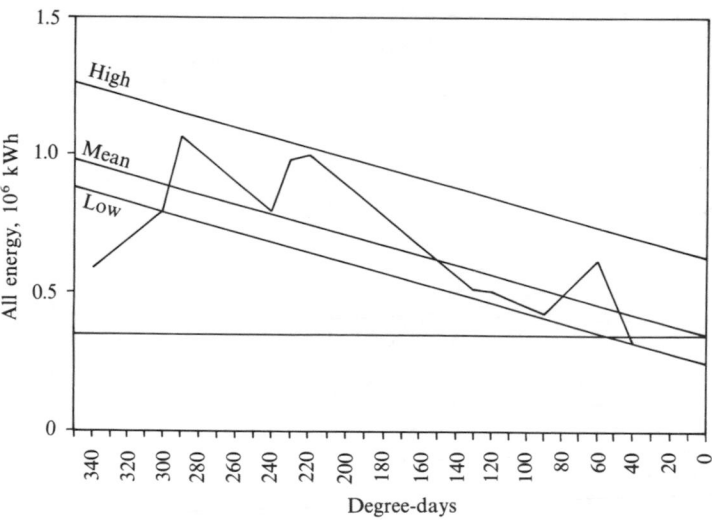

Figure 2.12e Scatter in the data.

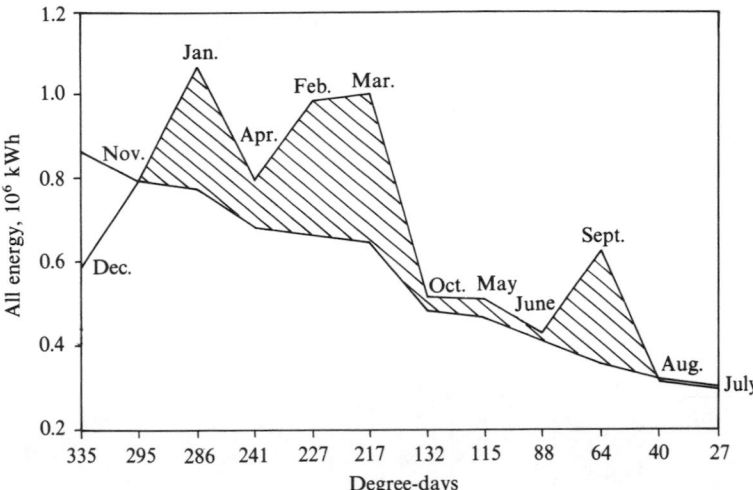

Figure 2.12f Low use base.

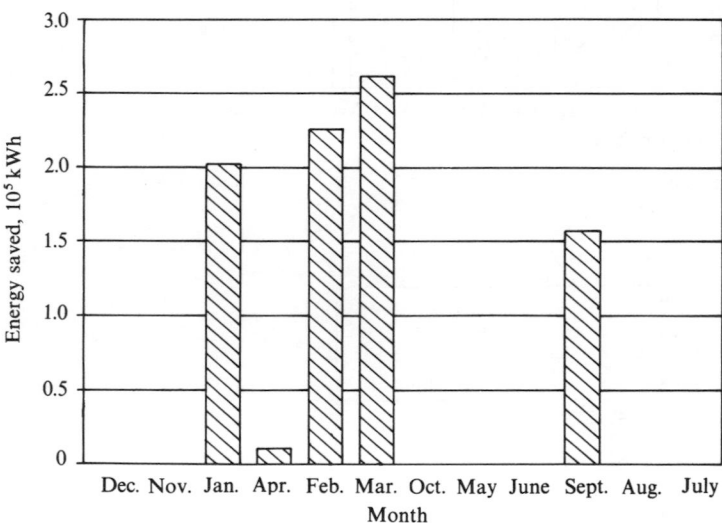

Figure 2.12g Savings with controls.

Figure 2.12(b) shows the results of a linear regression analysis on the data. When the total number of degree-days is zero, no heating should be required. The energy used then must correspond to the base-load for the site, in this case 348 530 kWh. Although not exhibited in the present data, such curves often resemble 'hockey-sticks' as depicted in Fig. 2.12(c). In the moderate months, the boiler system may be oversized for supplying small loads. Frequent on–off cycles result in a greater proportion of casing heat transmission and stack losses from

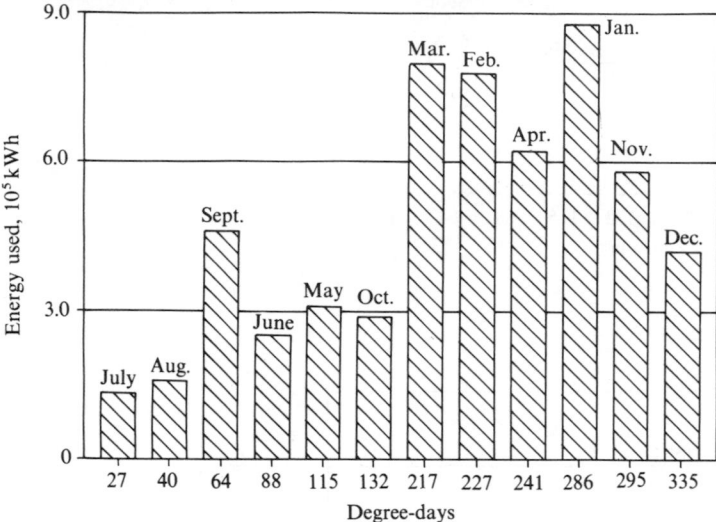

Figure 2.13 Oil and gas use with degree-days.

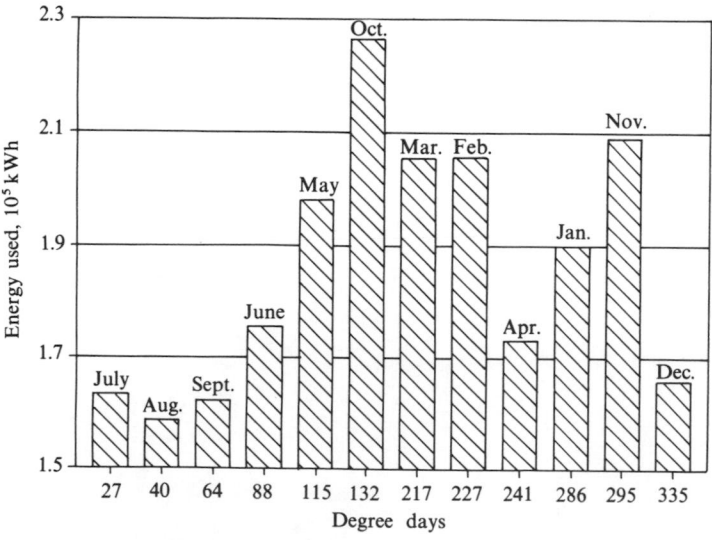

Figure 2.14 Electricity use with degree-days.

the boiler. The 'wasted' energy is shaded in Fig. 2.12(c). This may be regained by proper boilerhouse control, the use of thermal storage or by the provision of supplementary heating at the point of use. Figure 2.12(d) shows a similar effect, with an increase in energy consumption as the number of degree-days decrease, indicating the presence of refrigeration and/or air-conditioning. The scatter in the data is high (Fig. 2.12(e)) telling of a lack of feedback heating controls. The high and low bands bound the possible energy wastage resulting.

Recommendation 3   Investigate the heating control arrangements for year-round heating.

If the heating energy supplied were controlled to the lower bound of Fig. 2.12(e), then the total energy saved over the year is shown by the shaded area in Fig. 2.12(f), amounting to 860 000 kWh or £11 500 of gas (Fig. 2.12(g)).

Figures 2.13, 2.14 and 2.15 show how the oil + gas consumption, the electricity consumption and the monthly maximum demand vary with degree-days. There is clearly a correlation between electricity usage and degree-days. Figure 2.15 shows that the monthly maximum demand varies from 540 kVA in August to 690 kVA in February, a difference of 150 kVA, indicating the 'clandestine' use of electrical heating systems in the winter months. The probable electrical baseload for the site is 560 kVA. Thus, as Table 2.23 shows, £2782 might be saved by eliminating electrical heating, or by ensuring that electrical heating equipment is turned off during times of peak electrical demand.

Recommendation 4   Investigate the use of electrical heating equipment, assess the efficacy of space heating arrangements in winter months and seek to eliminate or control electrically-heated equipment.

Figure 2.14 shows that the electrical baseload (kWh) is probably 164 000 kWh. Thus Table 2.24 shows that 273 760 kWh will also be saved by adopting this recommendation.

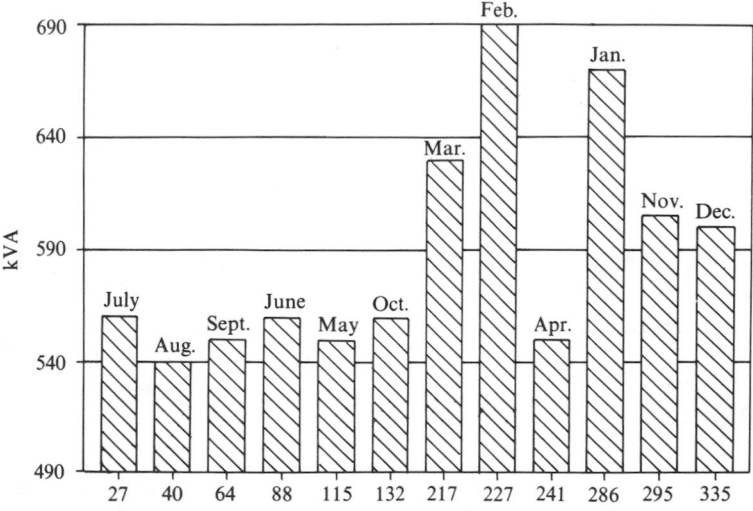

Figure 2.15 Maximum demand variation with degree-days.

Table 2.23 Maximum demand cost of electrical heating

Month	Maximum demand, kVA	Baseload	Step 2 Excess	Cost, £ per kVA	Clandestine cost
December	600	560	40	8.5	340
November	605	560	45	3.5	157
January	670	560	110	8.5	935
April	550	560	0	0.5	0
February	690	560	130	8.5	1105
March	630	560	70	3.5	245
October	560	560	0	0.26	0
May	550	560	0	0.5	0
June	560	560	0	0.26	0
September	550	560	0	0.26	0
August	540	560	0	0.26	0
July	560	560	0	0.26	0
Total					2782

Table 2.24 Electricity used for heating

Month	Base load electricity	'Heating' electricity
December	164 000	1 843
November	164 000	45 017
January	164 000	26 061
April	164 000	9 121
February	164 000	41 732
March	164 000	41 576
October	164 000	62 592
May	164 000	34 264
June	164 000	11 554
September	164 000	0
August	164 000	0
July	164 000	0
Total		273 760

The annual average cost per kWh is £0.062, thus the available savings from cutting electrical heating is £16 858.

The total annual savings from *recommendation 4* then becomes £19 641.

2.11 THE BASE TEMPERATURE

Because the sundry gain is not in the same ratio to the heating energy for all sites,[5] the base temperature approach has been developed as an improved analytical technique. In this, the degree-day concept is abandoned and the energy utilization correlated with mean monthly

Table 2.25 Degree-days to mean monthly temperatures

Month	Monthly degree-days	Days per month	Degree-days per day	Mean monthly temperature, °C (15.5–DD/D)
January	286	31	9.23	6.27
February	227	29	7.83	7.67
March	217	31	7.00	8.50
April	241	30	8.03	7.47
May	115	31	3.71	11.79
June	88	30	2.93	12.57
July	27	31	0.87	14.63
August	40	31	1.29	14.21
September	64	30	2.13	13.37
October	132	31	4.26	11.24
November	295	30	9.83	5.67
December	335	31	10.81	4.69

outside air temperatures. Table 2.25 converts the degree-days per month to mean monthly temperatures, using the procedure outlined earlier.

Figure 2.16 shows the relationship between maximum demand and mean monthly outside air temperature, T_o. A regression analysis indicated that the data can be described by

$$kVA = 677.3 - 9 \times T_o \tag{2.4}$$

and this 'best straight line' is plotted in Fig. 2.17, along with the monthly data points.

The striking feature revealed is the random scatter of the data about the mean line, up to 150 kVA, indicating a lack of maximum demand control. With an outside air temperature

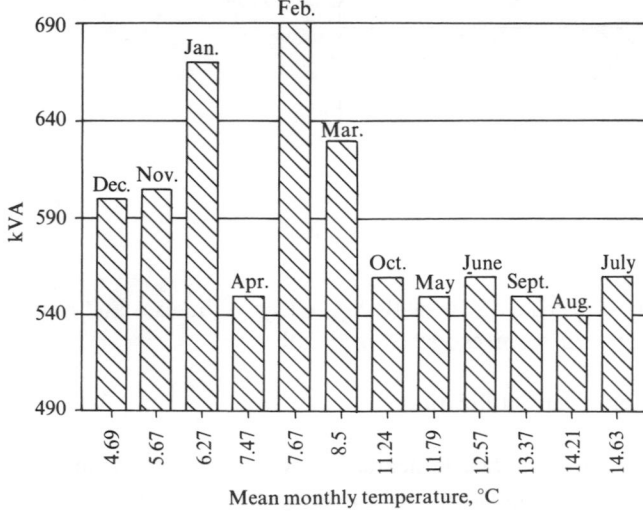

Figure 2.16 Maximum demand versus mean monthly outside air temperatures—histogram.

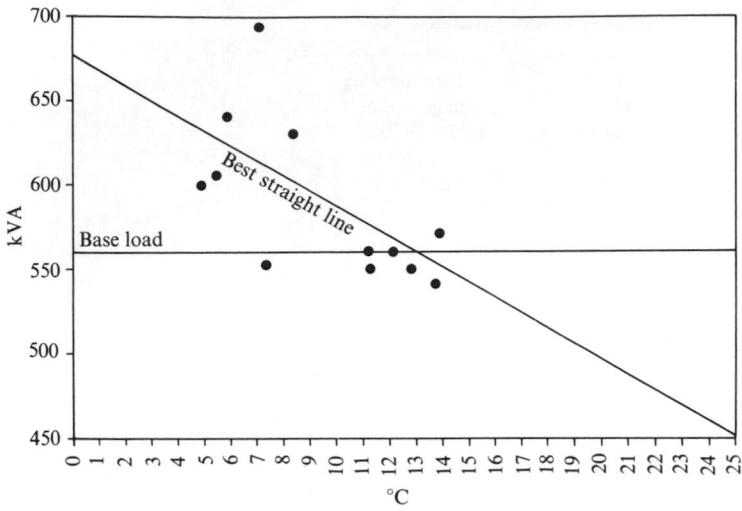

Figure 2.17 Maximum demand versus mean monthly outside air temperatures—best fit.

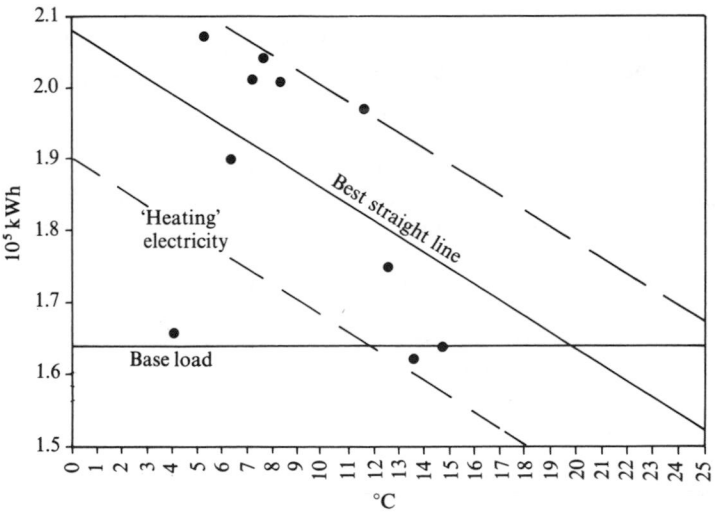

Figure 2.18 Electrical kWh versus mean monthly outside air temperatures—best fit.

of 7 °C, recorded maximum demands vary from 550 to 690 kVA. Clearly, the uses and abuses of electricity (annual cost £173 307) should be investigated in detail.

Recommendation 5   Investigate electricity utilization.

Again, considerable random scatter in the data is indicated in Fig. 2.18, where the monthly electricity consumptions (kWh) are plotted against outside air temperatures. Whilst the best straight line, given by

$$kWh = 208\,124 - 2230 \times T_o \qquad (2.5)$$

yields the electricity energy used for heating purposes (i.e. a function of outside air temperature), the scatter in the data encompasses 50 000 kWh per month, corresponding to £36 800 per year. If the electricity used for heating purposes were replaced by gas, the scatter in the baseload for electricity may be reduced. It should nevertheless be also eliminated by monitoring and controlling usage. The boundary lines in Fig. 2.18 show the range of possible savings, 30 000 kWh, or £2200 per year.

2.12 HEATING ENERGY CHARACTERISTIC— THE 'ENERGY SIGNATURE'

In order to correlate energy for space heating with outside air temperature, the electricity used for space heating must be added to the kWhs for oil and gas. This results in the data given in Table 2.26.

This data may be represented by

$$kWh = 1\,041\,244 - 55\,377 \times T_o \qquad (2.5a)$$

Again, considerable scatter in the data is exhibited. In November and January, for example, the mean outside air temperature was approximately the same, although the heating energy delivered in January was 150 per cent of that in November. The scatter corresponds to 400 000 kWh of heating energy, of which 100 000 kWh (£1880) might be saved by improved heating controls.

Table 2.26 Heating energy dependence on outside air temperatures

Month	Heating energy, kWh	Mean monthly outside air temperature, °C
December	421 332.3	4.69
November	625 271.6	5.67
January	903 928.8	6.27
April	630 680.5	7.47
February	821 062.7	7.67
March	838 496.0	8.5
October	353 218.7	11.24
May	346 367.6	11.79
June	263 944.2	12.57
September	459 361.6	13.37
August	158 600.9	14.21
July	133 695.9	14.63
Total	595 596.1	

Annual mean outside air temperature: 9.84 °C

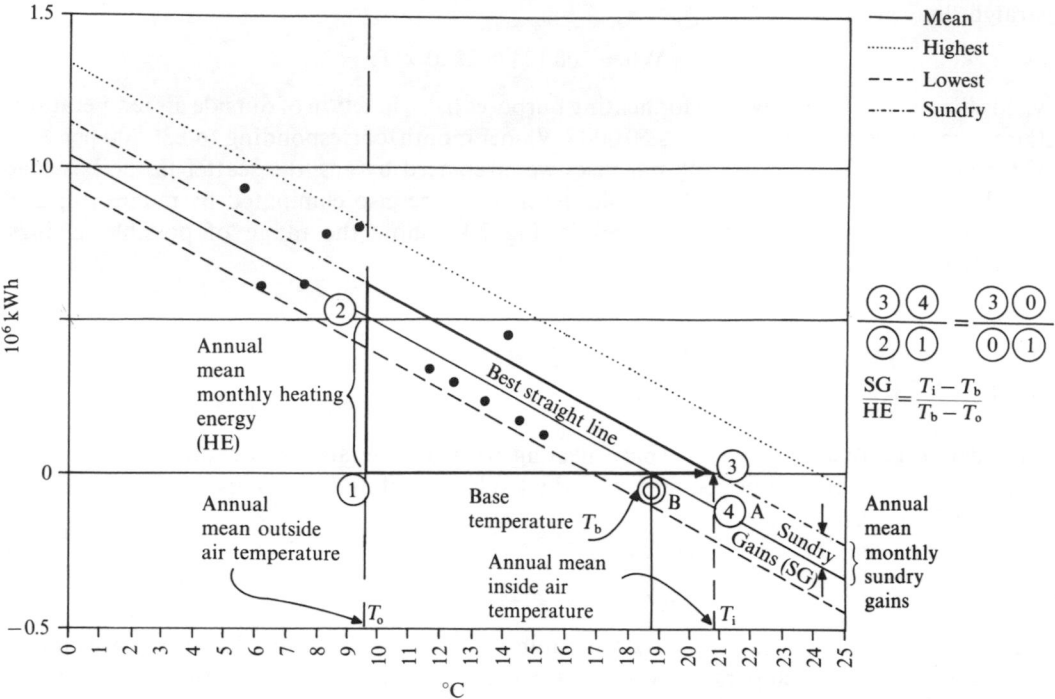

Figure 2.19 The energy signature—heating energy versus mean outside air temperature.

Figure 2.19 shows that, when the outside air temperature is 19 °C (the base temperature, T_b) or above, no heating is required. Thus the ratio of sundry gains to total heating energy for the site is considerably smaller than that implied using degree-days to a base of 15.5 °C. The mean inside air temperature is 21 °C, and so the internal sundry gains, A, are sufficient to raise internal temperatures by 2 °C, B. These sundry gains, emanating from electrical equipment, people and process losses to the internal environment, offset fabric transmission losses and heat losses accompanying ventilating air according to the *sundry gain equation*:

$$\text{sundry gains (kWh/annum)} = \text{transmission heat losses} + \text{ventilation losses}$$
$$= (UA + mc_p) \times (T_i - T_b) \times \text{hours/annum}/1000 \qquad (2.6)$$

where U = overall transmission heat loss coefficient for the building fabric, described in detail later, $W\,m^{-2}\,K^{-1}$
A = exposed area of the building, m^2
m = rate of ventilating air throughput, $kg\,s^{-1}$
c_p = specific heat of the ventilating air, $J\,kg^{-1}\,K^{-1}$
T_i = mean inside air temperature, °C
T_b = base temperature.

If n is the number of air changes per hour, then

$$m = \rho V n/3600 \qquad (2.7)$$

where ρ = density of the air, kg m^{-3}

V = enclosed volume of the building, m^3.

The density and specific heat of air at room temperature are $1.2\,\text{kg m}^{-3}$ and $1000\,\text{J kg}^{-1}\,\text{K}^{-1}$ respectively. Substituting these values,

$$m = 1.2\,Vn/3600 \tag{2.8}$$

and

$$mc_{\text{p}} = 1200\,Vn/3600 = nV/3 \tag{2.9}$$

The sundry gain equation thus becomes

$$\text{sundry gains (kWh per year)} = (UA + nV/3) \times (T_{\text{i}} - T_{\text{b}}) \times (\text{hours per year})/1000 \tag{2.10}$$

The areas and the volume of the building are known. The U-value must be obtained from a detailed site survey (later). The mean inside temperature is specified and the base temperature, T_{b}, for the building is obtained from Fig. 2.19, i.e. when no heating energy is required.

The *heating energy equation* is expressed similarly as

$$\text{heating energy (kWh per year)} = (UA + nV/3) \times (T_{\text{b}} - T_{\text{o}}) \times (\text{hours per year})/1000 \tag{2.11}$$

where T_{o} (°C) is the mean annual outside air temperature.

From the sundry gain equation and the heating energy equation:

$$(UA + nV/3) \times (\text{hours per year})/1000 = \frac{\text{sundry gains (kWh per year)}}{(T_{\text{i}} - T_{\text{b}})}$$

$$= \frac{\text{heating energy (kWh per year)}}{(T_{\text{b}} - T_{\text{o}})}$$

or

$$\frac{\text{sundry gains (kWh per year)}}{\text{heating energy (kWh per year)}} = \frac{(T_{\text{i}} - T_{\text{b}})}{(T_{\text{b}} - T_{\text{o}})} \tag{2.12}$$

For the present site, $T_{\text{i}} = 21\,°\text{C}$, $T_{\text{b}} = 19\,°\text{C}$ and $T_{\text{o}} = 9.84\,°\text{C}$, and so

$$\frac{\text{sundry gains (kWh per year)}}{\text{heating energy (kWh per year)}} = \frac{(T_{\text{i}} - T_{\text{b}})}{(T_{\text{b}} - T_{\text{o}})}$$

$$= \frac{21 - 19}{19 - 9.84}$$

$$= 0.22$$

The annual heating energy is $5\,955\,961\,\text{kWh}$. In order to satisfy the heating energy equation, this must be modified for boilerhouse and distribution losses. Assuming, for the moment, a boiler/distribution efficiency of 75 per cent:

$$\text{heating energy supplied} = 5\,955\,961$$
$$= 273\,760\,(\text{electricity}) + 5\,682\,201\,(\text{oil} + \text{gas})$$

heating energy delivered to the building
$$= 273\,760 + 0.75 \times 5\,682\,201$$
$$= 4\,535\,411\,kWh$$

and the total electricity supplied for purposes other than heating is $(164\,000 \times 12) = 1\,968\,000\,kWh$.

From the above equation, the sundry gains are $(0.22 \times 4\,535\,411) = 997\,790\,kWh$, or 45 per cent of the electricity supplied.

The exposed area, A, of the building is $9820\,m^2$, the enclosed volume, V, $40\,000\,m^3$ and the heating hours per year 8760. Thus, from the sundry gain equation:

$$\text{sundry gains (kWh per year)} = (UA + nV/3) \times (T_i - T_b) \times \text{(hours per year)}/1000 \qquad (2.10)$$

i.e.

$$997\,790 = (9820U + 13\,333n) \times 2 \times 8760/1000$$

or

$$997\,790 = (9820U + 13\,333n) \times 17.52.$$

Transposing,

$$9820U + 13\,333n = 56\,952$$

from which

$$n = 4.27 - 0.74U$$

Alternatively, using the heating energy equation:

$$\text{heating energy (kWh per year)} = (UA + nV/3) \times (T_b - T_o) \times \text{(hours per year)}/1000 \qquad (2.11)$$

i.e

$$4\,535\,411 = (9820U + 13\,333n) \times (19 - 9.84) \times 8760/1000$$
$$4\,535\,411 = (9820U + 13\,333n \times 80.24$$
$$56\,523 = 9820U + 13\,333n$$
$$n = 4.27 - 0.74U$$

yields the same result.

The equation relating the number of air changes per hour to the U-value is extremely important. Whilst, the overall U-value may be estimated from a detailed site survey, the mean number of air changes over the annual period is most difficult to obtain from measurements. Methods available will be discussed on page 232.

If the overall U-value can be estimated accurately, then the mean number of air changes per hour can be calculated from the above equation.

For the present, it is possible to examine the possible range of combinations of values for U and n

$$n = 4.27 - 0.74U \qquad (2.13)$$

Overall U-values (see later) for buildings vary from $0.5\,W\,m^{-2}\,K^{-1}$ (very highly insulated) to $6\,W\,m^{-2}\,K^{-1}$ (i.e. for a glasshouse). The number of hourly air changes can vary from 0 to 40.

Substituting the range of U-values, Table 2.27 lists the range of likely combinations of U and n. It can be seen that the overall U-value can lie in the range 2.0 to $3.5\,W\,m^{-2}\,K^{-1}$ and the mean annual number of air changes per hour between 1.0 and 3.0.

Alternatively, boilerhouse records might be used to obtain the heating energy delivered.[5]

Table 2.27 Combinations of U and n

	U	n	
	0	4.27	
	0.5	3.90	
	1.0	3.53	
unlikely	1.5	3.16	
	2.0	2.79	
	2.5	2.42	
	3.0	2.05	
	3.5	1.68	
	4.0	1.31	
	4.5	0.94	unlikely
	5.0	0.57	
	5.5	0.20	

2.13 ENERGY OUTPUT—REJECTION TO THE EXTERNAL ENVIRONMENT

The heating energy delivered is lost to the external environment via fabric transmission through the walls, roof and base, according to the heating energy equation. The energy auditor is now in a position to visit the site to conduct the *external energy survey* (Table 2.28).

Table 2.28 External survey

Dimensions:	Roof details					
Length, m	Pitched corrugated aluminium sheet					
U-value $W\,m^{-2}\,K^{-1}$	with foil-backed 100 mm plasterboard					
Facade	A	B	C	D	Roof	Base
Facing	S	W	N	E		
Description	single Glazing	335 mm Brick	335 mm Brick	335 mm Brick	See above	Concrete
Length, m	114	70	114	70		
Height, m	5	5	5	5		
Area, m^2	570	350	570	350	7 980	7 980
U-value, $W\,m^{-2}\,K^{-1}$	5.6	1.7	1.7	1.7	1.9	0.06
UA	3 192	595	969	595	15 162	478.8

Total UA (minus base)	$20\,513\,W\,K^{-1}$
Total exposed area (minus base)	$9\,820\,m^2$
Overall U-value	$2.09\,W\,m^{-2}\,K^{-1}$
Total enclosed volume	$V = 39\,900\,m^3$
from equation for U versus n	$n = 2.72$ per hour

Values for the component U-values are listed in reference 6. Commonly encountered materials are included in the database (Appendix 1).

2.14 PRELIMINARY ENERGY AUDIT

Having examined the *input* and *output* sides of energy use at the site, it is now possible to construct an *input/output* energy audit for the site, according to:

Input energy, kWh per year
  fuel energy input + electricity for heating + other electrical energy input

=

Output energy
  fuel energy losses during combustion, conversion and distribution
    + electrical directly rejected during utilization
    + fabric transmission heat losses + ventilation heat losses. (2.14)

  Neglecting the small proportion of internal sundry gains from personnel (this would be significant in, for example, theatres and restaurants):

Input energy kWh per year
$$5\,682\,201 + 273\,760 + 1\,968\,000$$

=

Output energy
  $0.25 \times 5\,682\,201 + 1\,968\,000 - 997\,790$ (sundry gains) $+ UA(T_i - T_o) \times$ hours per year/1000
    $+ 0.33nV(T_i - T_o) \times$ hours per year/1000

$$7\,923\,961 = 1\,420\,550 + 970\,210 + 2.09 \times 9820 \times (21 - 9.84) \times 8760/1000$$
$$+ 0.33 \times 2.72 \times 40\,000 \times (21 - 9.84) \times 8760/1000$$

$$7\,923\,961 = 1\,420\,550 + 970\,310 + 2\,006\,439 + 3\,510\,033$$

Finally,

Input energy, kWh per year $7\,923\,961$

=

Output energy, $7\,907\,332$

  This balanced audit has been obtained by equating the fuel and electricity inputs from an annual set of fuel bills to the energy losses occurring during combustion, conversion and distribution (assumed here to be 25 per cent), that part of the electricity utilisation which does not result in sundry gains to the internal environment, the heat losses by transmission through the building fabric, and the heat losses associated with ventilating air.

  The furnace/boiler efficiency (assumed here at 75 per cent) should be measured and an estimate of annual mean efficiency should be made by examining boilerhouse records.

  The fraction of annual electricity utilization which does not result in sundry gains to the internal environment was estimated by plotting the heating energy supplied each month against mean monthly external air temperatures. Although the mean inside air temperature was maintained constant at 21 °C throughout the year, no heating energy would be required when the outside air temperature exceeded 19 °C, the *base temperature* for the site. Thus the internal sundry gains from electrical equipment and personnel (neglected here) are sufficient to raise the internal temperature by 2 °C with respect to the outside air temperature. This results in a linear relationship between internal sundry gains and the overall U-value and mean number of air changes for the building.

A second linear relationship, derived from the same graph, relates total heating energy delivered per mean annual temperature difference between internal and external environments to the overall U-value and the mean number of air changes for the building. These two linear relationships contain three unknowns, the internal sundry gains, the overall U-value and the mean number of air changes per hour. A combination of the latter two variables was eliminated, resulting in a value for the internal sundry gains which can be compared with the overall electricity consumption.

A relationship between the overall U-value and the annual mean number of air changes per hour was then obtained. The former was estimated from a site survey and thence the annual mean number of air changes could be estimated.

Heating energy balance

Heating energy input, kWh per year

$$\text{Fuel energy input} + \text{electrical heating} + \text{sundry gains}$$

=

Heating energy output
fuel energy losses during combustion, conversion and distribution
$+$ fabric transmission heat losses $+$ ventilation heat losses   (2.15)

Substituting the numbers from above,

$$5\,682\,201 + 273\,760 + 997\,790 = 1\,420\,550 + 2\,006\,439 + 3\,510\,033$$

Heating energy input, kWh per year $6\,953\,751$

=

Heating energy output $6\,937\,022$

2.15 EFFECTS OF ENERGY CONSERVATION ON THE ENERGY CHARACTERISTIC

Energy conservation

From the heating energy equation, the slope of the regression line relating heating energy to outside air temperature is proportional to $(UA + nV/3)$. Thus Fig. 2.20 shows the effects of adding or reducing insulation levels (or the amount of reject heat recovered from ventilating air).

Increasing energy conservation measures results in lowering the base temperature from T_b to T_{bM} in the diagram, as the constant sundry gains offset reduced heating energy losses, and also decrease the negative slope of the graph. The ensuing annual savings are 'indicated' by the area O E T_b minus the area O E_M T_{bM}. The overall annual savings depend upon the annual outside temperature profile (i.e. the number of hours at given temperatures). The frequencies of occurrence may be deduced from the degree-days/month. Alternatively, these savings are

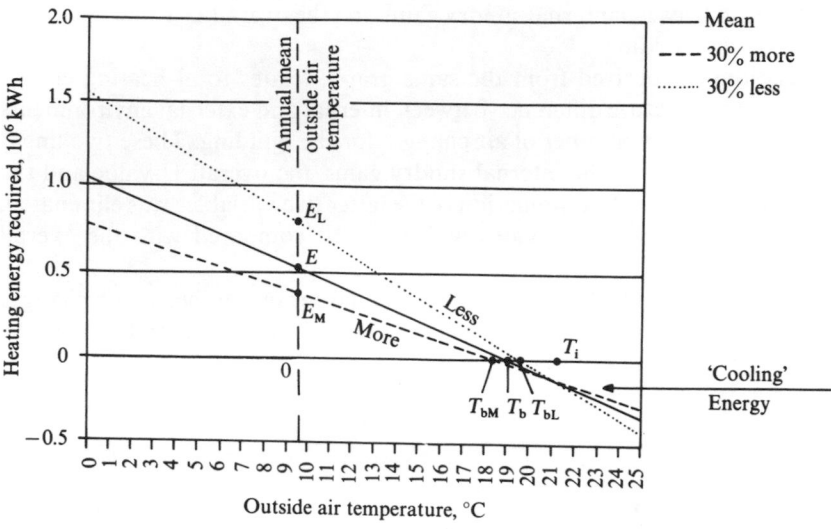

Figure 2.20 Effects of insulating or waste heat recovery on the energy signature.

more easily quantified using the sundry gain and heating energy equations. Note that the amount of 'cooling energy' required, when the outside air temperature is higher than the base temperature, also changes.

Reducing energy conservation measures results in raising the base temperature from T_b to T_{bL} in the diagram, as the constant sundry gains offset increased heating energy losses, and also increase the negative slope of the graph. The ensuing annual lossess are indicated by the area $O\,E_L\,T_{bL}$ minus the area $O\,E\,T_b$.

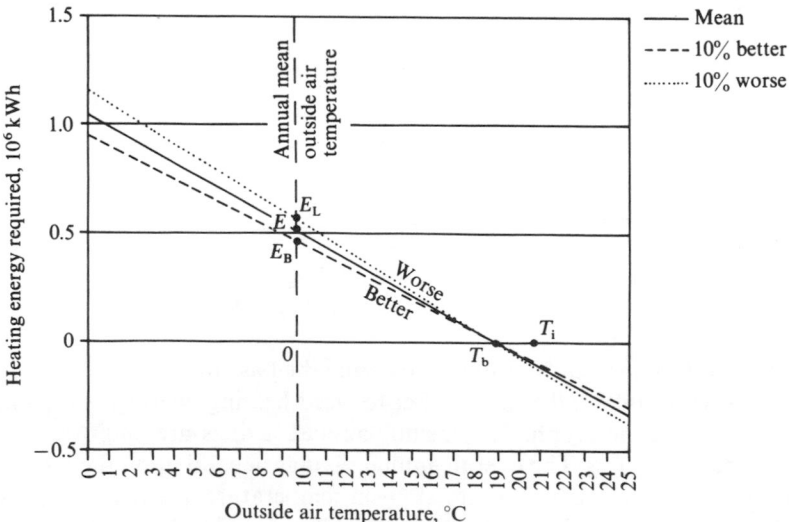

Figure 2.21 Effects of boiler/distribution efficiency changes on the energy signature.

Boiler/distribution efficiencies

Increasing the boiler/distribution efficiency (Fig. 2.21) results in no change in base temperature but decreases the negative slope of the graph. The ensuing annual savings are indicated by the area $O E T_b$ minus the area $O E_B T_b$.

Decreasing the boiler/distribution efficiency has the opposite effect.

Inside air temperature

Figure 2.22 shows the substantial effects of changes in inside air temperature T_i. Considering that all the heating energy consumed in a year serves to raise this temperature by only 8.16 °C above the mean annual outside temperature, a change of 1 °C in the mean annual inside temperature, results in a change of 12.25 per cent in the annual energy bill for heating.

If the mean inside air temperature were reduced by 3 °C to 18 °C, the resulting savings are indicated by the area $O E T_{b21}$ minus the area $O E_L T_{b18}$.

If the mean inside air temperature increased by 3 °C to 24 °C, the resulting increased energy consumption is indicated by the area $O E_H T_{b24}$ minus the area $O E T_{b21}$.

Recommendation 6   Question the need for 21 °C mean inside air temperature.

Sundry gains

Reducing the amount of sundry gains, resulting from, for example, internal energy conservation measures, such as insulating internal pipelines and process equipment, or replacing lighting systems with more efficient luminaires, raises the base temperature while the slope of the graph remains the same, Fig. 2.23. Thus, although energy may be saved internally, more heating energy will be needed to offset fabric transmission and ventilation losses. The amount

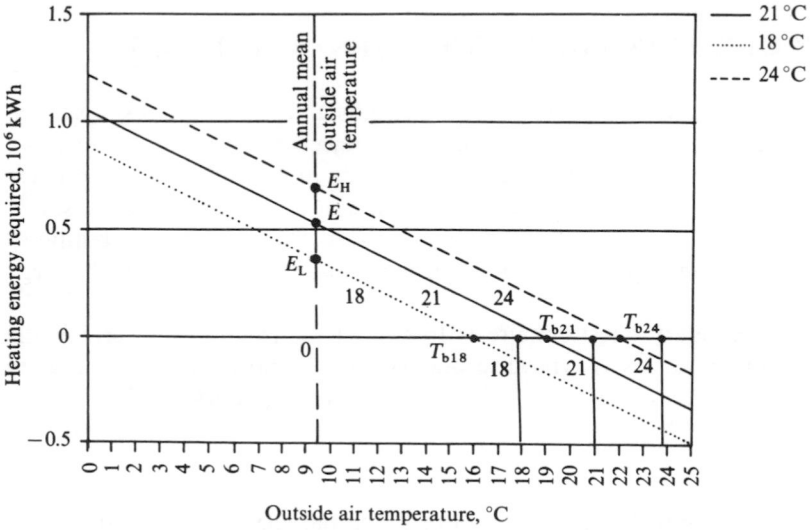

Figure 2.22 Effects of inside air temperature changes on the energy signature.

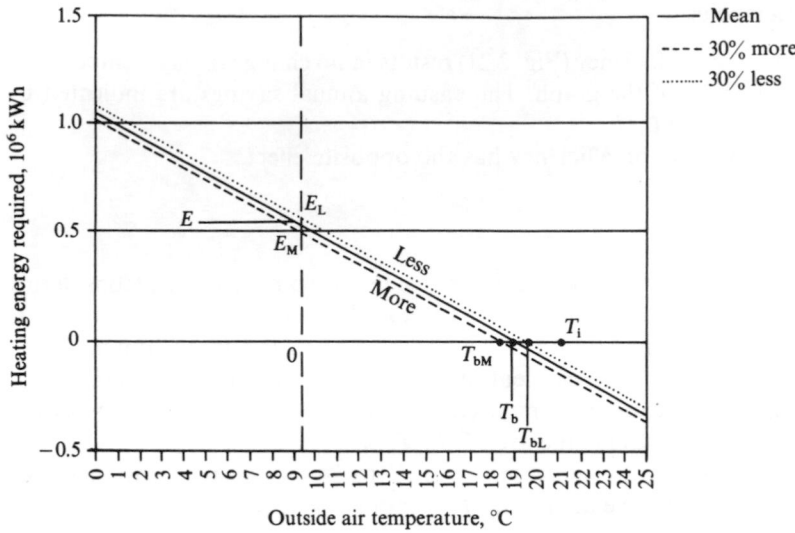

Figure 2.23 Effects of internal sundry gains on the energy signature.

is indicated by the area $O E_L T_{bL}$ minus the area $O E T_b$. If the energy 'saved' (i.e. from becoming a sundry gain) internally was supplied from electricity, then, whilst no net energy savings will be achieved, financial savings will result from the price difference between gas and electricity.

Conversely, increased sundry gains result in lowering the base temperature and reducing heating energy requirements. Economic analyses of 'internal' energy conservation measures should be discounted by the need for extra heating energy to counter the loss in sundry gains.

2.16 SECONDARY QUESTIONNAIRE—THE CLIENT INTERVIEW

Having obtain an *input/output* audit from the annual fuel bills and the external site survey, energy *throughputs* should be investigated to ascertain direct rejects of energy to the external environment other than those losses via fabric heat transmission and ventilation (e.g. process fluid effluents, flue gases from internal furnaces, etc.). The aim is to quantify the uses of electricity and process fuels which do not result in sundry gains. The energy contributing to sundry gains should also be disaggregated to facilitate economic appraisals of 'internal' energy conservation retrofit measures.

The purpose of the secondary questionnaire is to extract information concerning the internal activities, plant and processes. A typical secondary questionnaire might appear as in Table 2.29 and should include questions extracted from the following pages.

Energy management practices

The interview should commence with a presentation of the initial energy audit and recommendations and discussions.

Table 2.29 Secondary questionnaire for an energy audit

Energy Audit
Secondary Questionnaire

Name of firm	Bitusa Industries Ltd
Address of premises surveyed	4 Back Street New Chesham Nr Whittleford USA
Telephone number	0234 750111
Fax number	0234 750112
Interviewer	I Probe
Interviewee	J Jimmy
Position	Chief Works Engineer
Location	Boilerhouse

Summary of initial *recommendations*

1. Renegotiate electrical supply capacity.
2. Investigate the use of electricity for heating purposes.
3. Investigate the heating control arrangements for year-round heating.
4. Investigate the use of electrical heating equipment, assess the efficacy of space heating arrangements in winter months and seek to eliminate or control electrically-heated equipment.
5. Investigate electricity utilization.
6. Question the need for 21 °C mean inside temperature.

Information should be then sought under the following headings:

Energy management procedures

Who is responsible for energy management?
Position in the organization?
Reporting to?
Full or part-time?
Qualifications and experience?
What is done?
Has an energy flow diagram been prepared?
What has been achieved?

Financial practices

Who controls the capital spending budget?
Who controls the recurrent spending budget?

Upon what financial criteria should cost-effectiveness calculations be performed?
What is the period available to complete this exercise?
Is there a list of energy-saving investments under review, ranked in order of priority, with detailed costing and cost-benefit calculations?
If not, why not?

Comments on energy consumption

Is energy consumption, about right, too high, too low?
What are the areas of high energy consumption?
What tariffs are used?
Why these?
When were these last reviewed?
Can off-peak tariffs be used?
Can maximum demand be cut?
Can power factors be improved cost-effectively?

Monitoring and recording practices

How is energy consumption reviewed?
By whom?
When was the last review?
How is energy consumption analysed?
Does the analysis normalize the data with level of activity?
– by building, by product, by month, by year, by cost, by use activity, by sector or section?
What units of measurement are used?
What are the metering control arrangements?
Is there an energy consumption forecast/budget?
Have standards been set (i.e. for a given task or product or building)?
Is consumption compared with previous periods, other locations, other companies or other industries?
What are the monitoring and targeting procedures?
Should an energy management system be installed?
What should be the nature of this system (computational or manual)?

Personnel energy awareness

Are details of energy consumption made known to employees?
Are employees made aware of the need for energy conservation?
What steps have been made to promote energy awareness via education and training, posters, etc.?

Current energy conservation measures

What steps have been taken to reduce energy consumption?
What steps have been taken to cascade or recycle energy (e.g. incineration or sale of combustible of recyclable scrap or refuse, waste heat recovery from air, water or hot products)?

Comments of energy inefficiencies

Are there obvious incidents of energy wastage?

Conditions of buildings, plant and equipment

Is insulation and draughtproofing adequate and in good repair (roof, walls, floors, interzones, doors, windows)?

Furnaces

Is plant operating efficiently?
Are furnaces efficiency tested?
What are the maintenance procedures?
What are the control arrangements?

Boilers

Is plant operating efficiently?
Are boilers efficiency tested?
Is optimum blowdown maintained?
What are the maintenance procedures?
What are the control arrangements?

Boilerhouse auxiliaries

Is plant operating efficiently?
How is it examined?
What are the maintenance procedures?

Heat distribution systems

Are there leaks of hot water or steam?
Are pipes and ducts adequately lagged?
What are the maintenance procedures?

Major items

What are the major energy-consuming items of plant and equipment?

Energy storage systems

How are storage tanks heated?
What are their temperatures?
Why these temperatures?
Are storage tanks adequately insulated?

What are the maintenance procedures?
What are the control arrangements?

Process plant

Is plant operating efficiently?
Are process temperatures and pressures at the lowest essential levels?
What are the maintenance procedures?
What are the control arrangements?

Space heating services

For what periods are the buildings heated?
How is heating controlled?
Could the temperature be reduced?
Does the temperature vary in different zones?
Are parts of the building being heated unnecessarily?
What are the maintenance procedures?

Lighting systems

Are parts of the building being lit unnecessarily?
How is lighting controlled?
What are the mantenance procedures?

Power and electrical services

Is plant operating efficiently?
What are the maintenance procedures?

Mechanical ventilation

What is the minimum ventilation rate?
Why this value?
How is the building ventilated?
How does air get in?
How does air get out?
Do ventilation rates vary in different zones?
Are parts of the building being ventilated unnecessarily?
Is there any evidence of vertical stratification?
What are the maintenance procedures for fansets?
What are the control arrangements?

Air conditioning systems

Is plant operating efficiently?
What are the maintenance procedures?
What are the control arrangements?

Domestic hot water systems

Are there leaks of hot water?
What are the maintenance procedures?
What are the control arrangements?

Compressed air services

What is the compressed air requirement?
Where does inlet air to the compressor come from?
Where is the inlet air duct sited?
What is compressed air used for?
What are the delivery temperature and pressure?
What are the pressures at the points of use?
Is there any evidence of unauthorized use or leakage?
Is plant operating efficiently?
What are the maintenance procedures?
What are the control arrangements?

Refrigeration plant

Is plant operating efficiently?
What are the maintenance procedures?
What are the control arrangements?

Chilled water distribution systems

Is their evidence of leakage?
Is insulation adequate?
What are the maintenance procedures?

Steam plant

Is plant operating efficiently?
Is there obvious leakage of steam?
Is condensate recovered?
Are steam traps cleaned and unclogged?
What are the maintenance procedures?
What are the control arrangements?

Other services

Other plant

Special equipment and processes

Energy throughputs

Information should be sought concerning the major energy-consuming items of plant and equipment (Table 2.30).

Table 2.30 Major energy-consuming items—work parameters

Item	Description	Power rating, kW	Efficiency	Operating hours and loading patterns	Control arrangements	Maintenance procedures
Furnaces						
Boilers						
Boilerhouse auxiliaries						
Heat distribution systems						
Energy storage systems						
Process plant						
Space heating services						
Lighting systems						
Power and electrical services						
Mechanical ventilation						
Air conditioning systems						
Domestic hot water systems						
Compressed air services						
Refrigeration plant						
Chilled water distribution systems						
Steam plant						
Kitchens and catering						
Other services						
Other plant						
Special equipment and processes						

2.17 THE SECONDARY REQUEST

A request should be made for the following information:

- Plan drawings of the buildings
- Elevations of the buildings
- Details of the structural components of the buildings
- Specifications of the boilers and associated plant
- Plans of the layout of process equipment and piping
- Details of air handling units and other space heating arrangements
- Details and layout of the inlet and extract ventilation fan sets

2.18 INTERNAL ENERGY AUDIT CHECKLIST

During the internal site survey, the energy auditor should start at the boilerhouse and track energy flows to the final point of use and rejection to the external environment. An energy flow chart or a Sankey diagram should be constructed during the quest.

The energy management checklist given in Appendix 2 has been constructed for guidance. This covers:

- Fuels
   coal
   oil
   gas
- Electricity
- Energy release—furnaces
- Energy conversion—boilers, autoclaves and liquid heaters
- Heat distribution systems
- Heating systems
- Energy storage systems
- Plant and equipment
- Lighting
- Thermal insulation
- Ventilation
- Air conditioning systems
- Domestic hot water systems
- Compressed air services
- Refrigeration plant and chilled water distribution systems
- Steam plant
- Waste heat and materials reclamation
- Controls

Table 2.31 Savings made by reducing compressed air delivery pressures

Delivery pressure, lb in^{-2} (10^5 Nm^{-2})	Adiabatic delivery temperature, K	Total work done, MJ kg^{-1}	Savings by reduction, MJ kg^{-1}	Saving, %	Saving per £1000 bill
100(6.90)	489	0.020	0	0	0
90(6.21)	477	0.019	0.012	6	60
80(5.52)	461	0.018	0.020	10	100
70(4.83)	443	0.016	0.046	22	220
60(4.14)	407	0.013	0.082	40	400
50(3.45)	403	0.012	0.086	42	420
40(2.76)	377	0.009	0.112	54	540
30(2.07)	348	0.007	0.141	68	600
20(1.38)	310	0.003	0.179	87	870
14.7(1.013)	283	0	0.02	100	1000

The energy manager should consider only those major energy-consuming items of plant and equipment and should remember to track back energy saved to that fuel (money) saved at the boilerhouse.

Table 2.31 shows the *minimum* savings resulting from reduced compressed air delivery pressures.

2.19 ENERGY THROUGHPUTS—ENERGY FLOW CHARTS—ENERGY AUDIT

The final energy audit relates all *inputs, throughputs* and *outputs* so that the effects of introducing energy saving measures for one activity on other activities may be quantified. *Energy saved should be tracked back to that fuel (money) saved at the boilerhouse.* Diseconomic effects arising from system changes should also be identified.

The object of the overall energy audit is to complete the charts shown in Figs 2.24(a) to (d), in terms of kWh per year (Fig. 2.24b), £ per year (Fig. 2.24c), and kW mean per year (Fig. 2.24d), so that energy-consuming sectors and loss centres may be viewed in proper perspective.

The summary data for Bitusa Industries Ltd is given in Table 2.32.

An analysis of compressed air usage indicated that, over the year, 470 000 kWh of the electricity delivered was used to supply compressed air.

Table 2.32 Summary data—Bitusa Industries Ltd

Number of employees	150
Working area	$8\,000\,m^2$
Enclosed volume	$40\,000\,m^3$
Hours/annum	8 760
Recommended air changes	3 per hour
Calculated air changes	2.72
Overall U-value	2.09

Table 2.33 Percentage breakdown of fabric transmission heat losses

Component	UA, W K^{-1}	%
Walls	2 159	10.3
Glazing	3 192	15.2
Roof	15 162	72.2
Base	479	2.3
Total	20 992	100

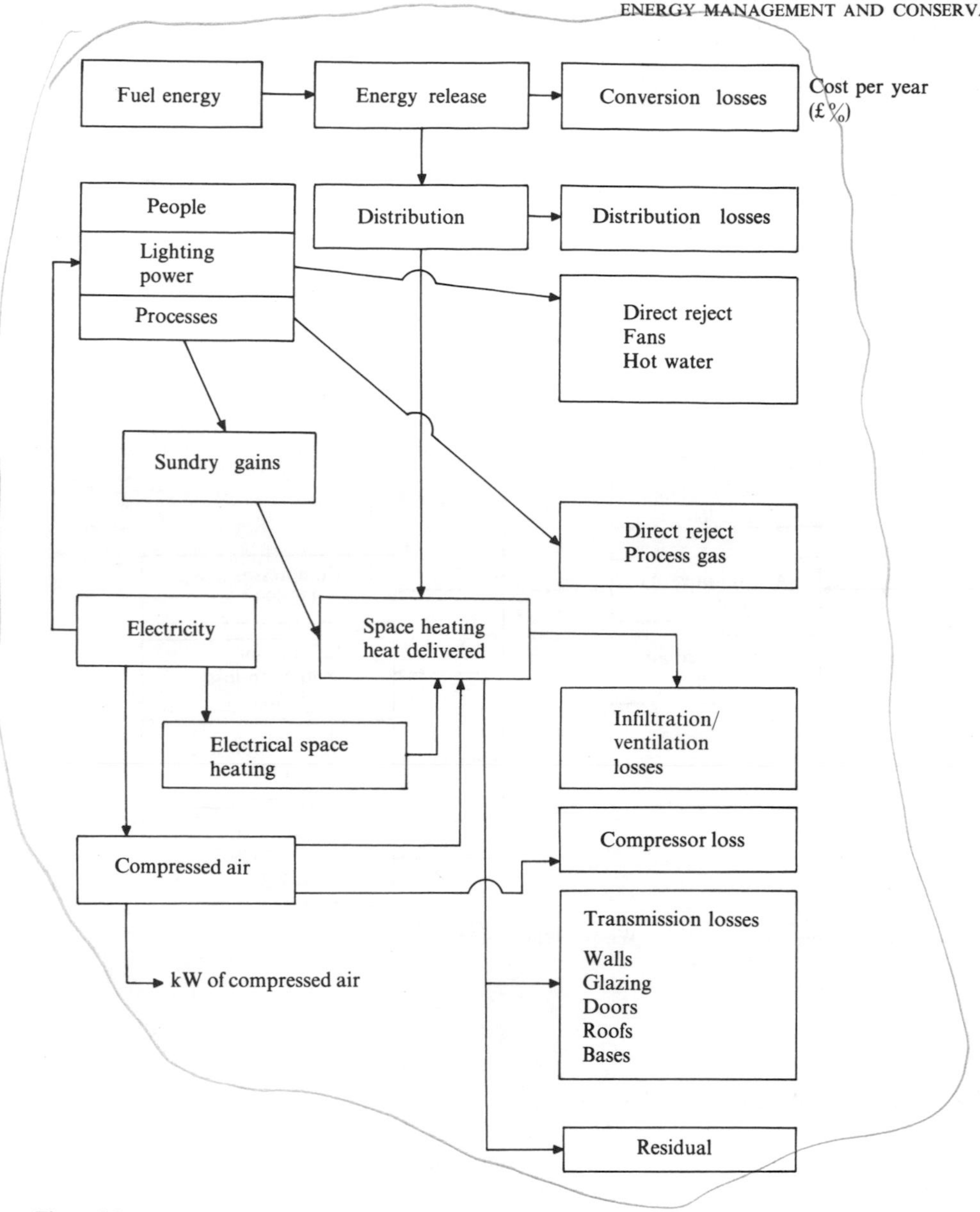

Figure 2.24a Energy audit chart.

Thus the chart (Table 2.34) gives the breakdown of energy use over the period considered, and this data is used to construct energy and monergy audit charts.

Final energy audit—Bitusa Industries Ltd

The costs, in terms of energy and money, over the year considered are shown in Table 2.35.

In addition, £36 000 was paid in electrical maximum demand and supply capacity charges.

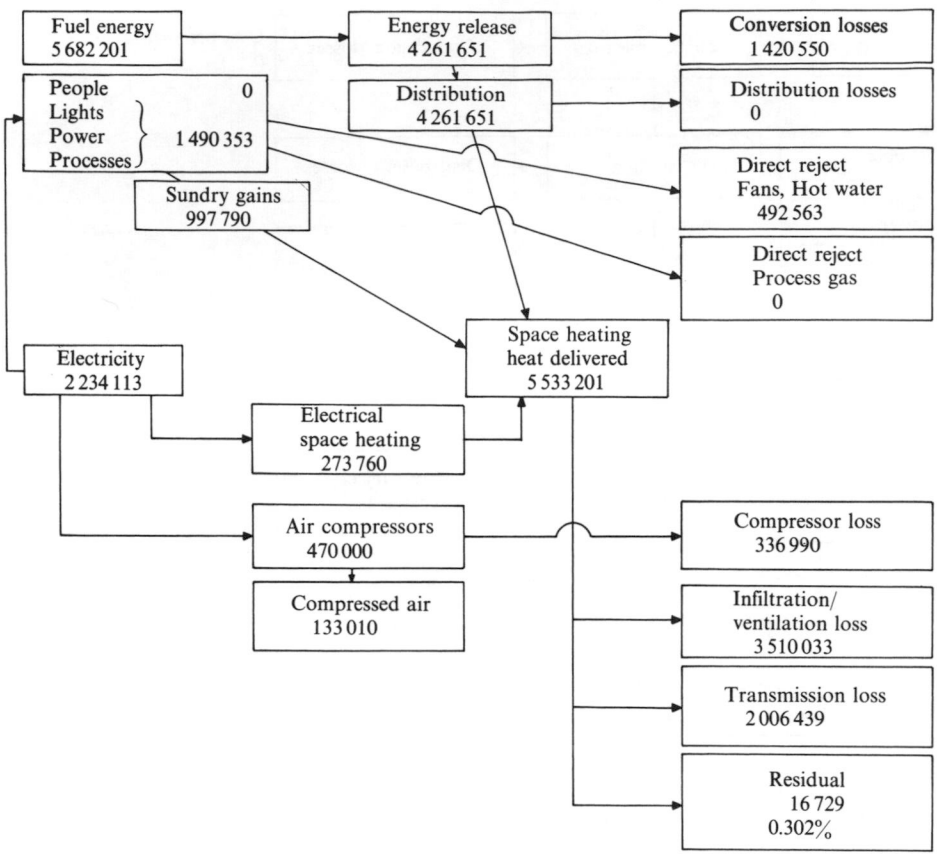

Figure 2.24b Bitusa Industries energy audit—kWh per year.

Table 2.34 Breakdown of energy use, kWh per year

	Total	Heating	Compressed air	Other
Electricity	2 234 113	273 760	470 000	492 563
(Sundry gain)		997 790		
Oil + gas	5 682 201	5 682 201		
Totals	7 916 314	6 953 751	470 000	492 563
Combustion losses		1 420 550		
Heat delivered		5 533 201		
Ventilation losses		3 510 033		
Fabric transmission losses		2 006 439		
Compressor losses			336 667	
Compressed air delivered			133 333	
Direct reject				49 563

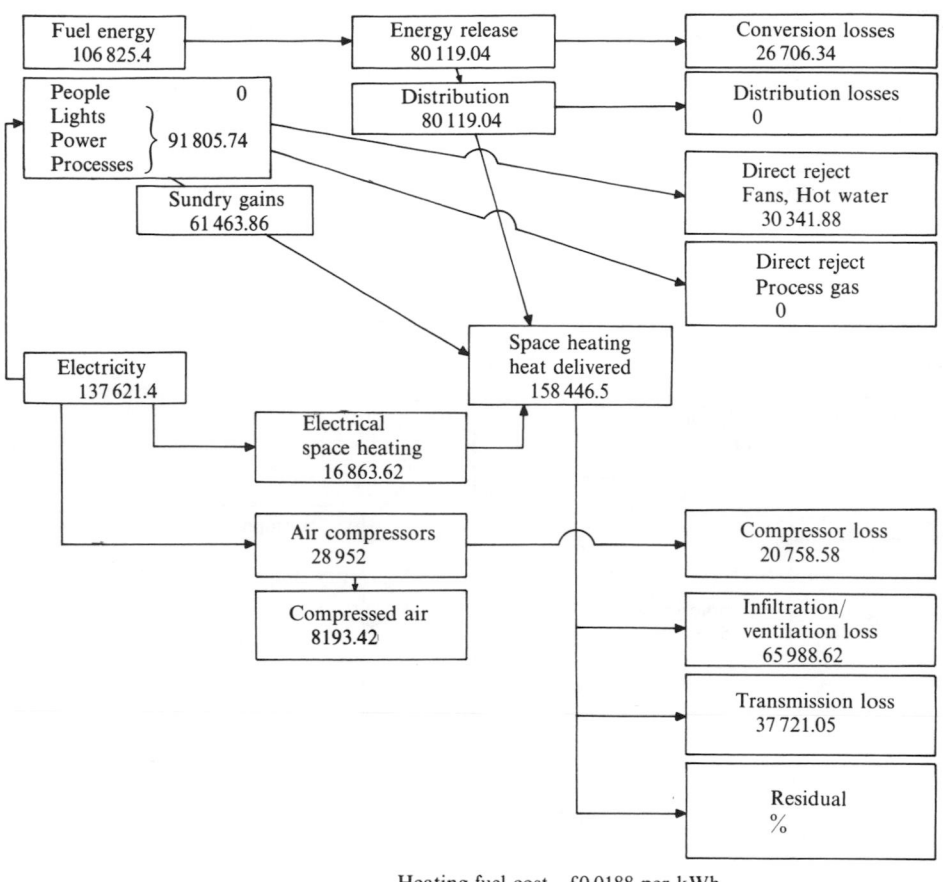

Heating fuel cost £0.0188 per kWh
Electricity cost £0.0616 per kWh

Figure 2.24c £ per year.

Table 2.35 Breakdown of costs

Factor	kWh	%	£	%
Ventilation losses	3 510 033	44	65 989	27
Fabric transmission losses	2 006 439	25	37 721	15
Directly rejected energy	492 563	6	30 342	12
Fuel conversion losses	1 420 550	18	26 706	11
Air compressor losses	336 990	4	20 759	9
Total losses	7 766 575	98	181 517	74
in addition				
Electrical 'added' cost of sundry gains	(997 790)	(13)	42 705	18
'Additional' cost for electrical heating	(273 760)	(3)	11 717	5
Compressed air produced	133 010	2	8 194	3
Totals	7 900 000	100	244 000	100

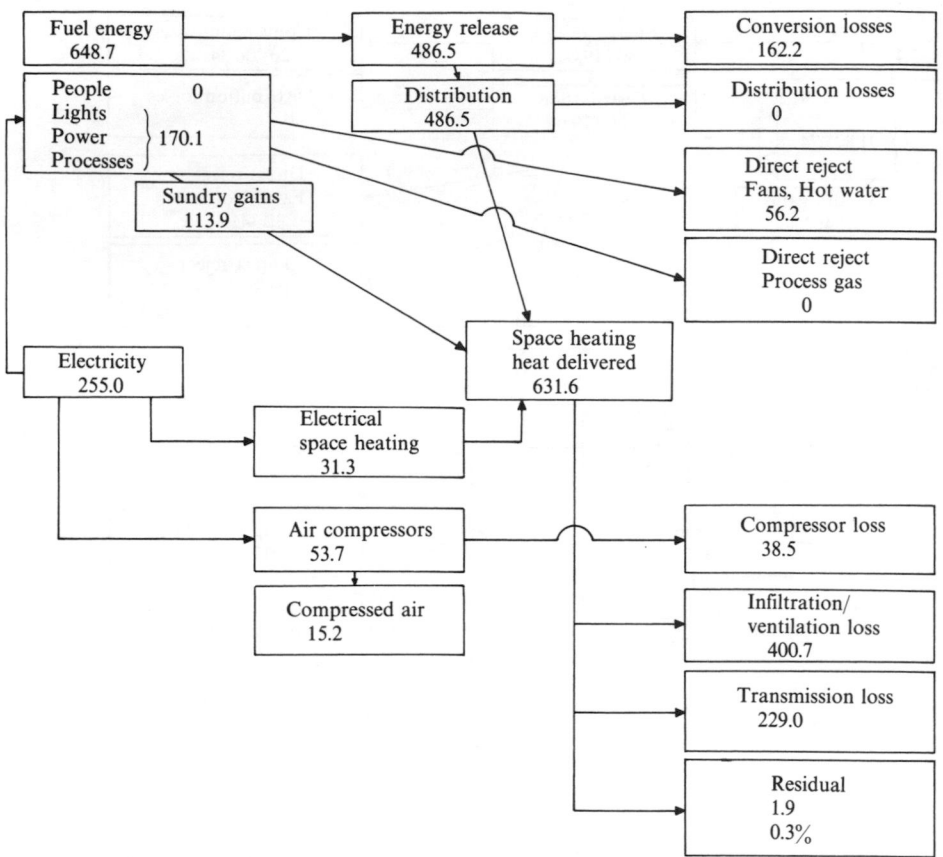

Figure 2.24d kW mean per year.

2.20 ENERGY-SAVING OPTIONS

Ventilation losses may be reduced by minimizing the rate of fresh air change. Since this is already below the recommended value, it is unlikely that a reduction could be accomplished.

A reduction in mean inside air temperature from 21 °C to 18 °C would however save $(3/(21 - 9.84)) \times £65\,989 = £17\,700$, which becomes £23 600 fuel saved at the boilerhouse. This would require close temperature control.

Fabric transmission losses may be reduced by improving insulation or by reducing the mean inside air temperature.

The overall building U-value is dominated by that of the roof, the heat losses through which cost £27 000 per year. The addition of a layer 50–100 mm of fibreglass insulant could halve this loss, giving an annual saving of £13 500 (£18 000 at the boilerhouse), but may be an expensive retrofit measure. Quotations from contractors should be requested.

A reduction in mean inside air temperature from 21 °C to 18 °C would however save $(3/(21 - 9.84)) \times £37\,721 = £10\,140$, or £13 520 at the boilerhouse.

Directly rejected energy accounted for £30 342. A further, more detailed, study of these losses should be made.

Fuel conversion losses may be reduced by proper boiler control, which could increase the annual mean conversion efficiency to 80 per cent and so save £5340 per year on current usage.

The introduction of a condensing boiler could increase overall efficiency by a further 10 per cent, saving £10 680 per year on current usage. Estimates from contractors should be sought.

Air compressor losses accounted for £20 759 at electricity prices. If the 'cooling' compressor air were redirected to aid heating the building, £6335 of heating fuel could be saved, resulting in savings of £8447 at the boilerhouse. The uses for compressed air should be further investigated. Each £-worth of compressed air saves £3.5 of electricity (see Appendix 2—checklists).

Electrical heating should be abolished and replaced with gas-fired systems, saving £11 717 per year, which becomes £8788 at the boilerhouse due to boiler inefficiency. To this must be added savings on maximum demand premiums of £2783, making a total of £12 571.

An overall summary of the potential savings is shown at Table 2.36.

All uses for electricity should be investigated and reductions sought. Any resulting reduction in sundry gains must then be made up by the increased (but less expensive) use of gas for heating.

Some of the annual £36 000 for electrical maximum demand and supply capacity charges might also be saved from reductions in compressed air utilization and other electrical uses.

The heat to power ratio, 2.5, for the site indicates that the installation of a combined

Table 2.36 Summary of potential savings

Factor	Saving, £
Inside air temperature reduction	37 120
Roof insulation	18 000
Boiler control	5 340
Use of compressor cooling air	8 447
Abolish electrical heating	12 571
Reduce electrical supply capacity	3 000
Total savings	84 478
Initial energy bill	244 000
% saving	35
Final energy bill	159 522

heat and power system may be suitable. This should be further investigated, but only after the above less-expensive measures have been introduced.

Thus savings in excess of £80 000 are possible for the site. It now remains to obtain estimates for the various retrofit measures and to conduct economic cost-effectiveness calculations.

2.21 INVESTMENT OPPORTUNITIES AND PROJECT PLAN

A summary of the investment opportunities (ranked in order of cost-effectiveness), investment plans and projected future cost advantages is shown at Tables 2.37 to 2.39.

Table 2.37 Summary of investment opportunities

Retrofit measure	Annual savings, £	Capital costs, £	Straight rate of return, %
A. Reduce electrical supply capacity	3 000	0	infinity
B. Inside air temperature reduction	37 120	1 000	3712
C. Abolish electrical heating	12 571	2 000	629
D. Boiler control	5 340	2 000	267
E. Use of compressor cooling air	8 447	10 000	85
F. Roof insulation	18 000	40 000	45
Totals	84 478	55 000	154

Table 2.38 Investment plan

Year	Retrofit measure	Capital cost, £	Savings year on year, £	Accumulated savings, £
1	A	0	3 000	3 000
2	B and C	3 000	52 691	52 691
3	D, E and F	52 000	84 478	85 169

Table 2.39 Projected savings over three years

Initial energy bill	£244 000
Gross invested	£55 000
Net invested	£zero
Final energy bill	£160 000
Residual capital	£85 169

APPENDIX 2A POWER FACTOR CORRECTION
ELECTRICAL CIRCUITS AND POWER FACTOR CORRECTION

Resistive circuits

The simplest type of electrical circuit is a resistive circuit, Fig. 2A.1, where an alternating current is applied to a resistive load.

If $V = 100$ volts and $I = 100$ amps, since $V = IR$, $R = V/I = 1$ ohm, in the system shown.

Voltage and current are in phase, and, at any part of the alternating cycle

$$\text{power} = \text{voltage} \times \text{current} = VI \tag{2A.1}$$

Figure 2A.2 shows the response of such a circuit, having a maximum voltage and a maximum current of 100 volts and 100 amps respectively—all input electrical energy is converted to power.

Such are the characteristics of sine waves that the root mean square values for the current and voltage are $(1/2)^{1/2} = 0.707$ times the maximum values. The average power is then $0.707 \times 100 \times 0.707 \times 100 = 5000\,\text{W}$.

Example
Degrees lead/lag: $\phi = 0$
Maximum voltage: 100 volts
Maximum current: 100 amps
Maximum power: 10 kW

Inductive circuits

Figure 2A.3 shows a solenoid coil being supplied with an alternating supply of current. In the coil, there is set up an alternating magnetic field, which induces an electromotive force, emf, in the coil. This induced emf is at a maximum when the current and magnetic field are passing through zero and it lags behind the current and magnetic field by one quarter of a cycle (i.e. 90 degrees after the current).

The induced voltage lags the current by 90 degrees.
The circuit current lags the applied voltage by 90 degrees (Fig. 2A.4).

The induced emf opposes the applied emf and the flow of current and is somewhat like a resistance. It is known as the *inductive reactance*.

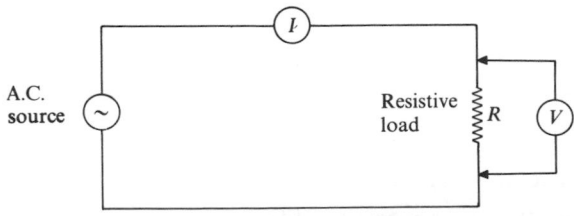

Current and voltage in phase

Figure 2A.1 Resistive circuit.

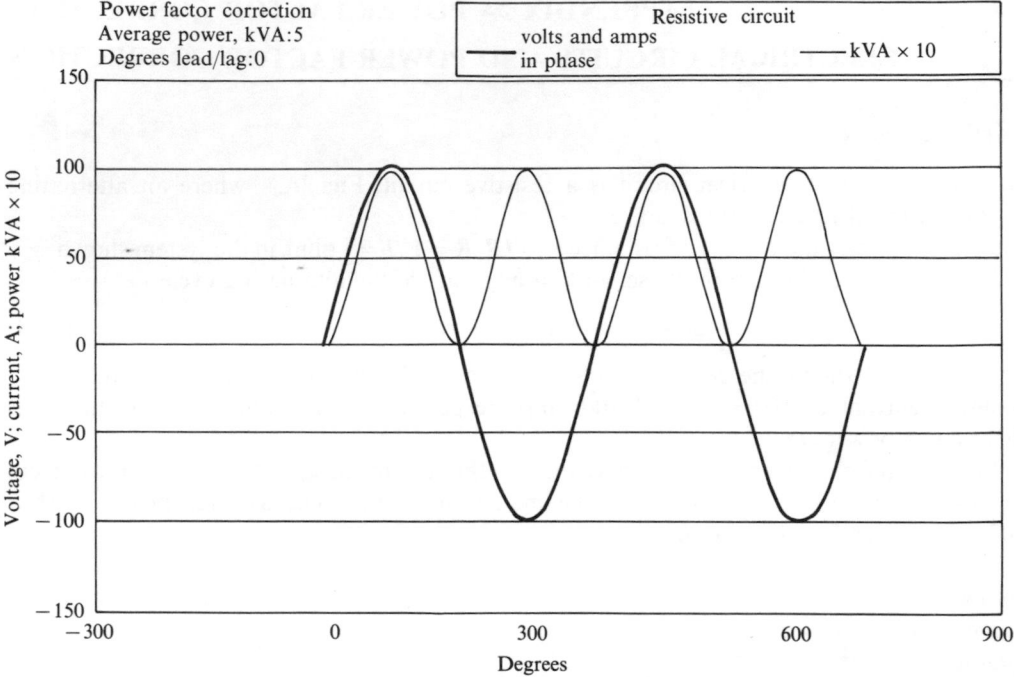

Figure 2A.2 Phase coincidence in a resistive circuit.

RMS power: $0.707 \times 100 \times 0.707 \times 100 = 5000\,\text{W} = 5\,\text{kW}$

degrees	Volts	Amps	kVA
0	0	0	0
30	50	50	2.5
60	86.6	86.6	7.5
90	100	100	10.0
120	86.6	86.6	7.5
150	50	50	2.5
180	0	0	0
210	−50	−50	2.5
240	−86.6	−86.6	7.5
270	−100	−100	10.0
300	−86.6	−86.6	7.5
330	−50	−50	2.5
360	0	0	0
RMS voltage V	70.7		
RMS current A		70.7	
kVA			5.0

$$\text{Power factor} = \text{Average kVA/Average power}$$
$$= 5/5 = 1.0 = \cos \phi = \cos(0) = 1.0 \qquad (2\text{A}.2)$$

Inductive reactance is given by

$$X_L = 2\pi f L \qquad (2A.3)$$

where X_L is the inductive reactance in ohms, f is the frequency of the alternating current source in Hertz (Hz), and L is the inductance of the circuit in henrys (H).

No power is converted in a purely inductive circuit. All circuits, however, have some resistance (Fig. 2A.5), and Fig. 2A.6 shows the response of a typical circuit where voltage leads the current by 30 degrees.

The angular separation between voltage and current is known as the *phase angle*, ϕ.

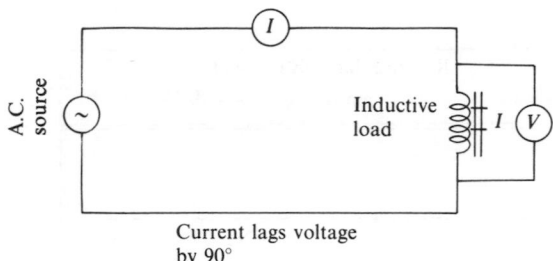

Current lags voltage
by 90°

Figure 2A.3 Inductive circuit.

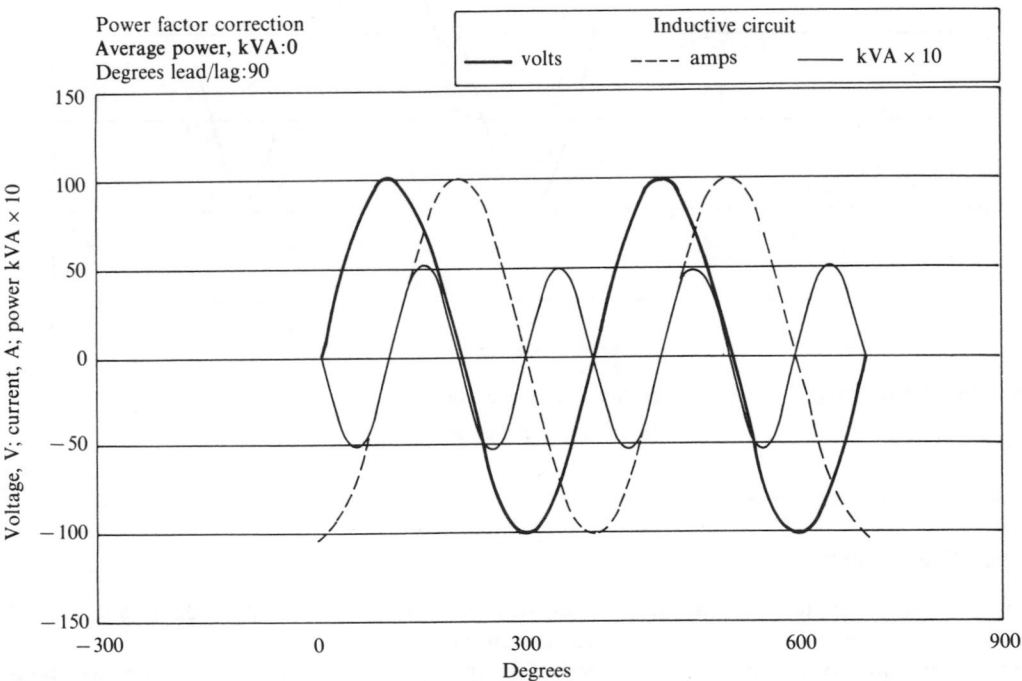

Figure 2A.4 Current lagging voltage in an inductive circuit.

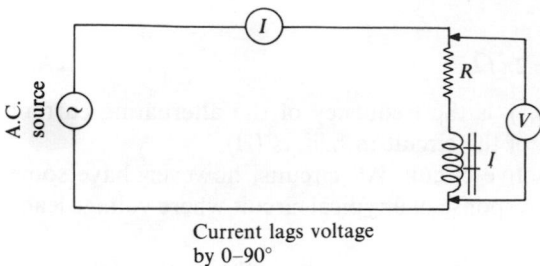

Current lags voltage
by 0–90°

Figure 2A.5 Resistive–inductive circuit.

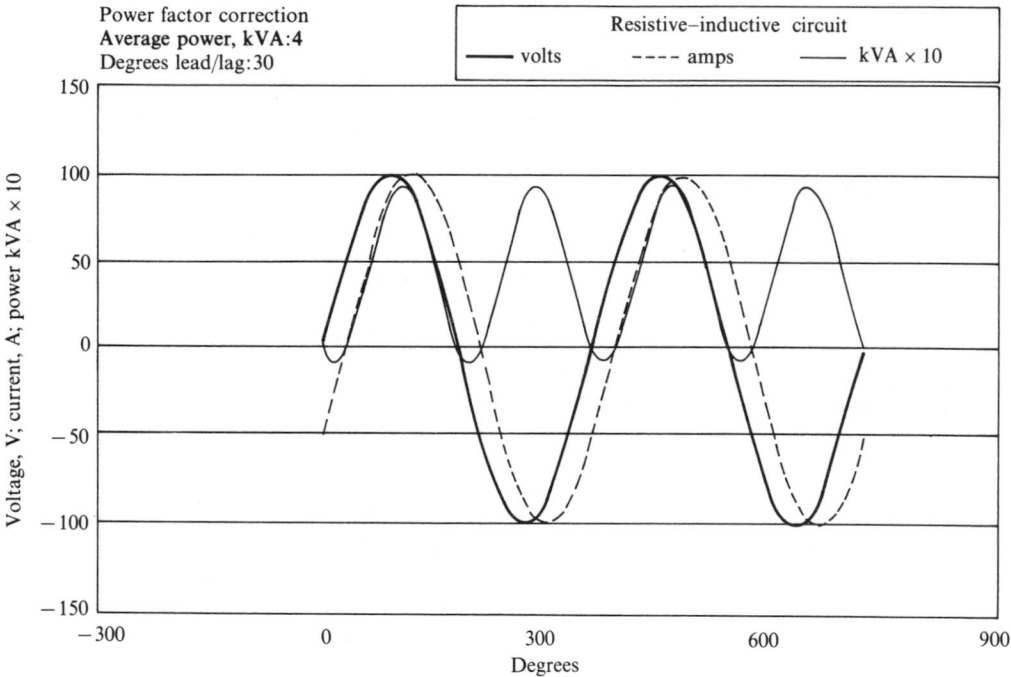

Figure 2A.6 Current lagging voltage in a resistive–inductive circuit.

Capacitive circuits

When two metal plates are separated by an insulator, the effect is to provide a *capacitor*. If such a capacitor is placed in an alternating current circuit (Fig. 2A.7) and the circuit is closed, current will start to flow before a voltage builds up across the plates. The current will be a

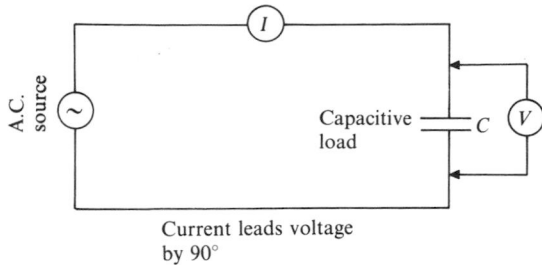

Current leads voltage
by 90°

Figure 2A.7 Capacitive circuit.

maximum when the voltage across the capacitor is a minimum (Fig. 2A.8). In other words:

The current leads the voltage by 90 degrees.
The capacitive voltage lags the current by 90 degrees.

No power is converted in a purely capacitive circuit. All circuits, however, have some

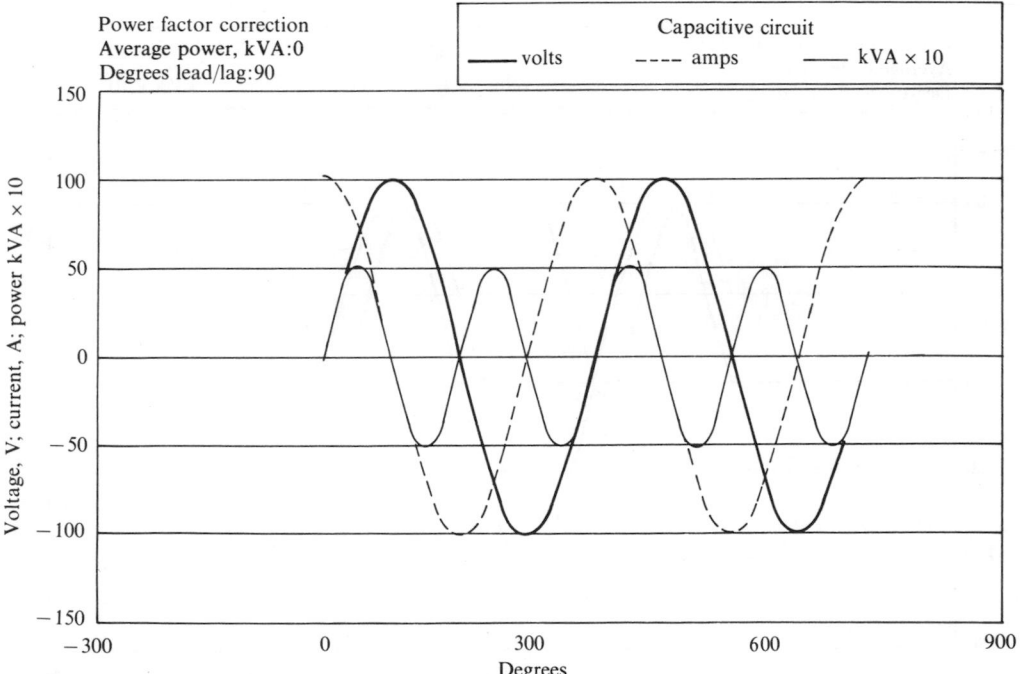

Figure 2A.8 Current leading voltage in a capacitive circuit.

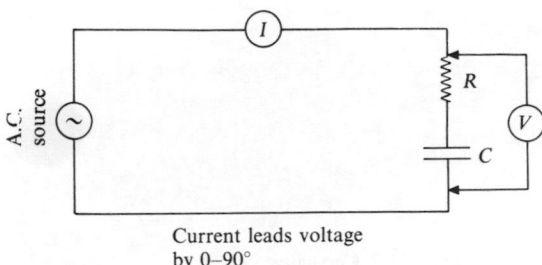

Current leads voltage
by 0–90°

Figure 2A.9 Resistive–capacitive circuit.

resistance (Fig. 2A.9), and Fig. 2A.10 shows the response of a typical circuit where current leads the voltage by 30 degrees.

The reciprocal of the capacitor current is similar to a resistance and is known as the *capacitive reactance* of the condenser.

Capacitive reactance is given by

$$X_C = 1/2\pi f C \qquad (2A.4)$$

where X_C is the capacitive reactance in ohms, f is the frequency of the alternating current source in Hertz (Hz), and C is the capacitance of the circuit in farads (F).

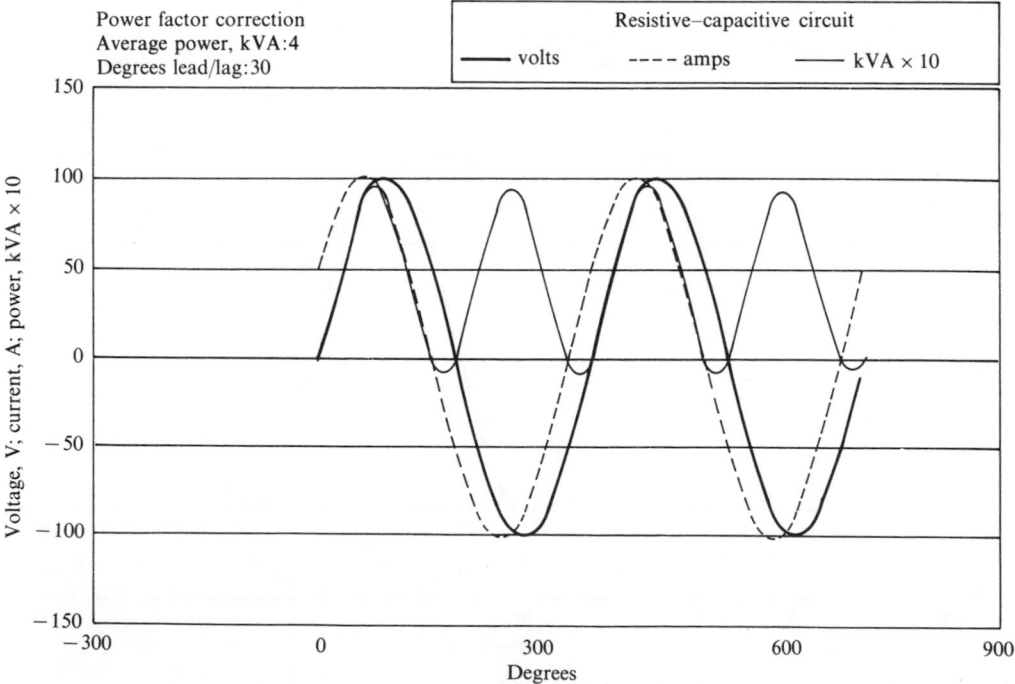

Figure 2A.10 Current leading voltage in a resistive–capacitive circuit.

Reactance

As the capacitive and inductive currents differ in phase by half a cycle (180 degrees), they act in opposition to one another. The total *reactance* of a series circuit containing an inductance and a capacitance is the difference between the inductive and capacitive reactances.

Vectors and phasor diagrams

Vectors are straight lines which have a specific direction and length. They may be used to represent the direction and magnitude of an alternating current or voltage.

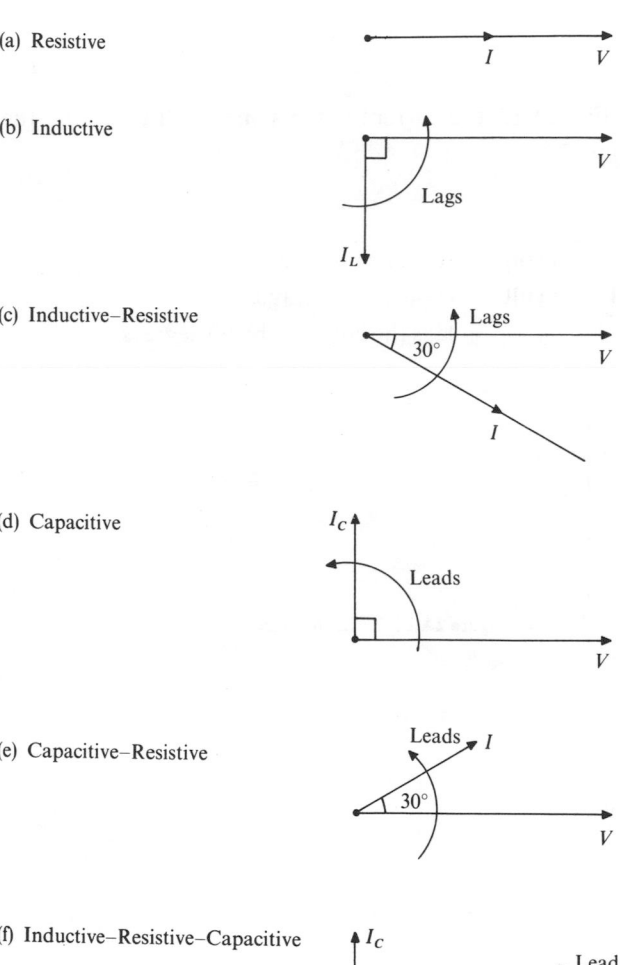

(a) Resistive

(b) Inductive

(c) Inductive–Resistive

(d) Capacitive

(e) Capacitive–Resistive

(f) Inductive–Resistive–Capacitive

Figure 2A.11 Phasor diagrams for (a) resistive; (b) inductive; (c) inductive–resistive; (e) capacitive–resistive; and (f) inductive–resistive–capacitive circuits.

When starting a diagram showing a set of vectors, a *phasor* diagram, a horizontal line is drawn is drawn in the *x*-axis. Rotation in a counter-clockwise direction from this reference line is considered to be in a positive direction. Figures 2A.11(a) to (f) show the phasor diagrams for the following configurations:

- A purely resistive circuit
- A purely inductive circuit
- An inductive/resistive circuit
- A purely capacitive circuit
- A capacitive/resistive circuit
- An inductive/resistive/capacitive circuit

Series circuit

In a series alternating current circuit, Fig. 2A.12, the current is the same in all parts and the voltages must be added using vector addition.

Example

| Voltage across resistor | V_R | 4 volts | reference vector |
| Voltage across capacitor | V_C | 7 volts | capacitive voltage, lags the current* by 90 degrees |

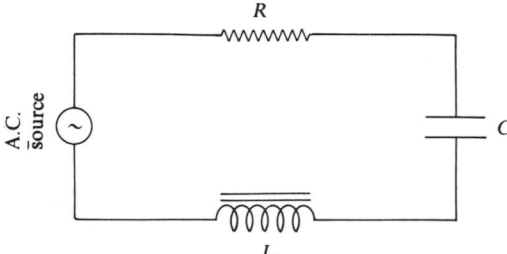

Figure 2A.12 Series AC circuit.

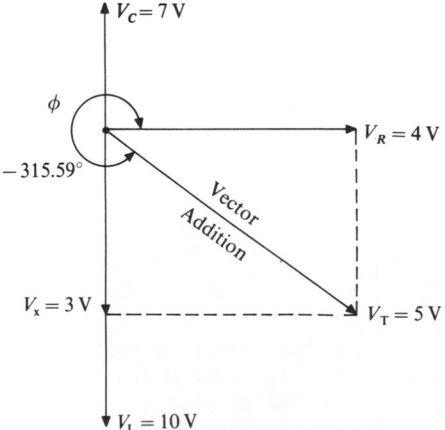

Figure 2A.13 Vector diagram for a series AC circuit.

Voltage across inductor V_L 10 volts   Induced voltage, lags the
current* by 90 degrees.

* and hence the resistive voltage, as the current is the same through each component in a series circuit.

Referring to Fig. 2A.13,

$$V_L = 10 \text{ volts}$$
$$V_X = V_L - V_C = 3 \text{ volts}$$

The resultant voltage is given by

$$V_T = (V_{R^2} - V_{X^2})^{0.5} = 5 \text{ volts}$$

and acts at an angle given from $\cos(\phi) = V_X/V_R = 3/4$, or $\phi = 41.41$ (318.59 degrees anti-clockwise).

Parallel circuit

The procedure for parallel circuits (Fig. 2A.14) is essentially similar, but then the voltage across each component is equal and the currents through each component are summed by vector addition.

Example

Current through resistor I_R 6 amps   reference vector
Current through capacitor I_C 10 amps   capacitive current,
   leads the current by 90 degrees
Current through inductor I_L 8 amps   induced current lags, the
   current by 90 degrees.

Referring to Fig. 2A.15,

$$I_C = 10 \text{ amps}$$
$$I_L = 8 \text{ amps}$$
$$I_R = 6 \text{ amps}$$

$I_X = 2$ amps, $I_T = 6.32$ amps at $\cos(\phi) = 6/6.32$, or $\phi = 18.3$ degrees anticlockwise.

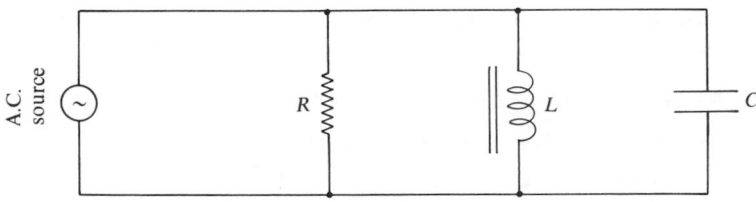

Figure 2A.14 Parallel AC circuit.

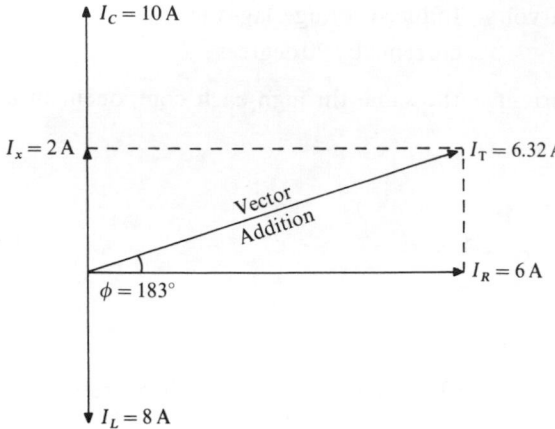

Figure 2A.15 Vector diagram for a parallel AC circuit.

Impedance

When a circuit containing resistance, inductance and capacitance is connected to an alternating current supply, the current is less than that which would flow were the circuit purely resistive.

The combination of resistance, inductance and capacitance is termed the *impedance*, Z (ohms) of a circuit. From Ohm's Law,

$$I = V/R \tag{2A.5}$$

or

$$I = V/Z \tag{2A.6}$$

or

$$\text{Impedance, } Z = V/I \tag{2A.7}$$

If an electrical motor takes 10 amps from a 240 volt supply,

$$Z = 240/10 = 24 \text{ ohms}$$

In general, for a series circuit,

$$Z = (R^2 + \text{abs}(X_L - X_C)^2)^{0.5} \tag{2A.8}$$

Power

When voltage and current are in phase, the average power in an alternating current circuit is obtained by multiplying the rms values of voltage and current. If, however, there is a difference in phase between the voltage and the current, the power will be less. The non-useful component of the 'power' is known as the *kilovolt-amps reactive*, kVA_r.

When the *power factor* is unity as for the purely resistive circuit,

$$kVA_r = \text{zero}, kW = kVA, \text{ i.e. no losses}$$

When the power factor is zero as for purely inductive or capacitive circuits,

$$kVA_r = kVA, \text{ i.e. total loss}$$

Power factor = average true power/apparent power = watts(W)/(volts(V) × current(A)).

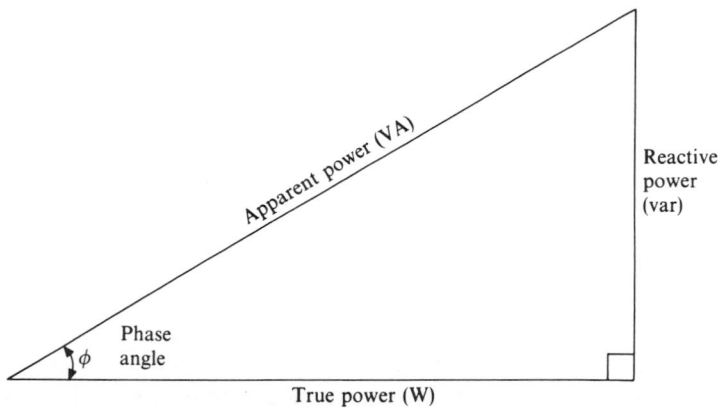

Figure 2A.16 Power triangle.

From the *power triangle* (Fig. 2A.16), it can be seen that power factor $= \cos(\phi)$ and

$$\text{watts(W)} = \text{power factor(p.f.)} \times \text{volts(V)} \times \text{current(A)} \qquad (2A.9)$$

In electrical motors, an inductive reactance is set up, causing the current to lag the applied voltage. Power factors of 0.7 are common.

Example

Power supplied, V	240 V
Motor input, W	3 kW
Power factor, p.f.	0.7
Current	$W/(\text{p.f.} \times V)$
	$3000/(0.7 \times 240)$
	17.86 amps

If the power factor were unity, the current would be given by

$$3000/(1.00 \times 240) = 12.5 \text{ amps}$$

Thus the low power factor of 0.7 results in 43 per cent extra current, and hence power, having to be supplied to the system.

Power factor correction

For a series circuit, $X_L = X_C$ to balance inductive and capacitive reactances, and so $2\pi fL = 1/2\pi fC$. The size of the capacitor needed is then given by

$$C = \frac{1}{(2\pi f)^2 L}$$

Similar calculations can be carried out for three phase supplies, and series/parallel circuits.

Why raise the power factor? The more the power factor departs from unity the greater the current required to provide the power. The 'non-power' current is dissipated as waste heat, resulting in inefficient operation. In addition, the sizes of switchgear, fusegear, cables and transformers have to be larger. Heating losses in the transmission of higher currents are also greater $(=I^2 R)$.

To discourage low power factor loads, most electricity suppliers impose penalty kVA charges for low power factors.

Practical power factors

Aggregate industrial loads often have power factors as low as 0.6. Capacitors may be connected at motor terminals to achieve the best effect, at the main terminal for a multimotor machine, or at the distribution centre or branch feeder. In the latter two cases the diversity of loads result in complex voltage and current characteristics. It is then not economically possible to increase the overall power factor to unity. A value of 0.9 is often the best that can be achieved.

REFERENCES

1. O'Callaghan, P. W.: *Design and Management for Energy Conservation*, Pergamon Press, Oxford, 1981.
2. O'Callaghan, P. W.: *Building for Energy Conservation*, Pergamon Press, Oxford, 1978.
3. UK Energy Efficiency Office, Fuel Efficiency Booklets Nos. 1–18, 1988.
4. Energy Management, J Energy Efficiency Office, UK Department of Energy, March/April, 30, 1992.
5. Adderley, E. A.: The Management and Implementation of Energy Thrift in Hospitals, PhD Thesis, Cranfield Institute of Technology, July 1989.
6. CIBSE Guide Books, Chartered Institution of Building Services Engineers, London, 1991.

FURTHER READING

Andreas, J. C.: *Energy-Efficient Electric Motors*, Marcel Dekker, New York, 1982.
Fardo, S. W., and Patrick, D. R.: *Electrical Power Systems Technology*, Prentice-Hall, Englewood Cliffs, New Jersey, 1985.
Morley, A., and Hughes, E.: *Electrical Engineering Science*, Longmans, London, 1949.
UK Energy Efficiency Office, Fuel Efficiency Booklet No. 9—Economic Use of Electricity.

THREE

ENERGY IN MANUFACTURE

3.1 INTRODUCTION

In manufacturing a product, energy is consumed during the exploration, excavation, transportation, refining and utilization of raw materials, such as fossil fuels, metals, plastics, refractories, paper and chemicals, as well as during manufacture itself. Buildings and building services equipment, such as heating and ventilating systems, lighting and air conditioning consume energy and materials. Further commodities are expended during marketing and distribution. All along this chain, energy and materials are consumed by infrastructures and services.

In the examination of an energy flow system, the energy manager (Fig. 3.1) must take great care to define the boundaries of the system under scrutiny, and to identify commodity flows across these boundaries.

The energy audit may concern:

- The energy 'content' of a product (Fig. 3.2)
- The energy consumed by a manufacturing procedure
- The energy requirements of a built facility (Fig. 3.3) or service, such as a transportation system

In each case, the audit should include an environmental audit of solid, liquid effluent and gaseous wastes.

3.2 ENERGY AND ENVIRONMENTAL ANALYSES OF PRODUCTS

In assessing the historical energy content and environmental impact of a product, its input materials and rejected pollution should be listed and the production process fully outlined. A *cradle-to-grave* approach should be adopted.

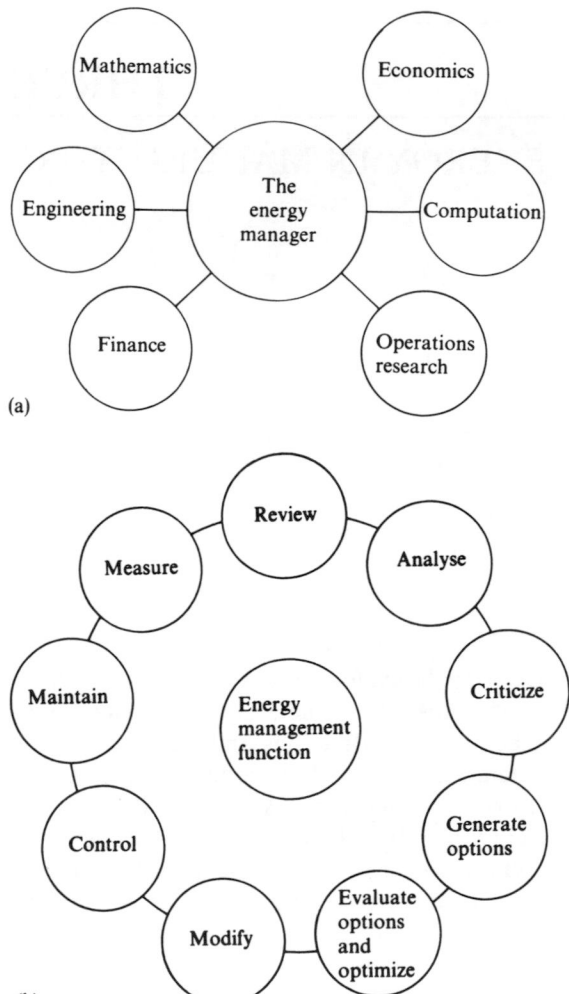

(a)

(b)

Figure 3.1 (a) The energy manager; (b) The energy management function.

Following method study practices, the following symbols may be useful for overlays of production processes.

= energy/material storage
* energy release
⌀ conversion of energy/materials
○ utilisation of energy and materials
⸗ flow resistance
⊥ rejection to the environment

Figure 3.3 shows, for example, a diagram for the steady-state commodity flows through a building and also indicates where efficiency improvements might be made. Different colours

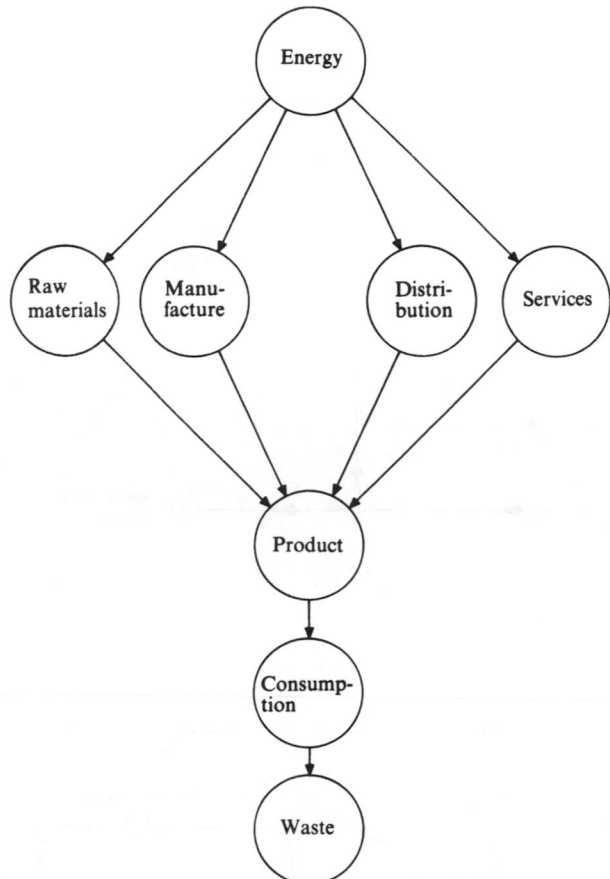

Figure 3.2 Energy in products.

may be used for different materials, or forms of energy. String diagrams may be constructed to examine product paths.

Once the flow chart has been established (Fig. 3.4) high demand operations should be critically examined (Fig. 3.5) with a view towards greater energy efficiency and the minimization of pollution.

Where fossil fuels, electricity, materials or products enter the system boundary, they bring with them their historical energy, materials and 'pollution' contents.

Energy costs for energy commodities

In delivering fossil fuels and electricity, energy and materials are consumed and pollution takes place. In order to quantify these, the energy manager should obtain data[1] concerning deliveries of energy resources and materials to the energy industries over a recent representative period, together with the amounts of the fuels exported over the same period. (See Table 3.1.)

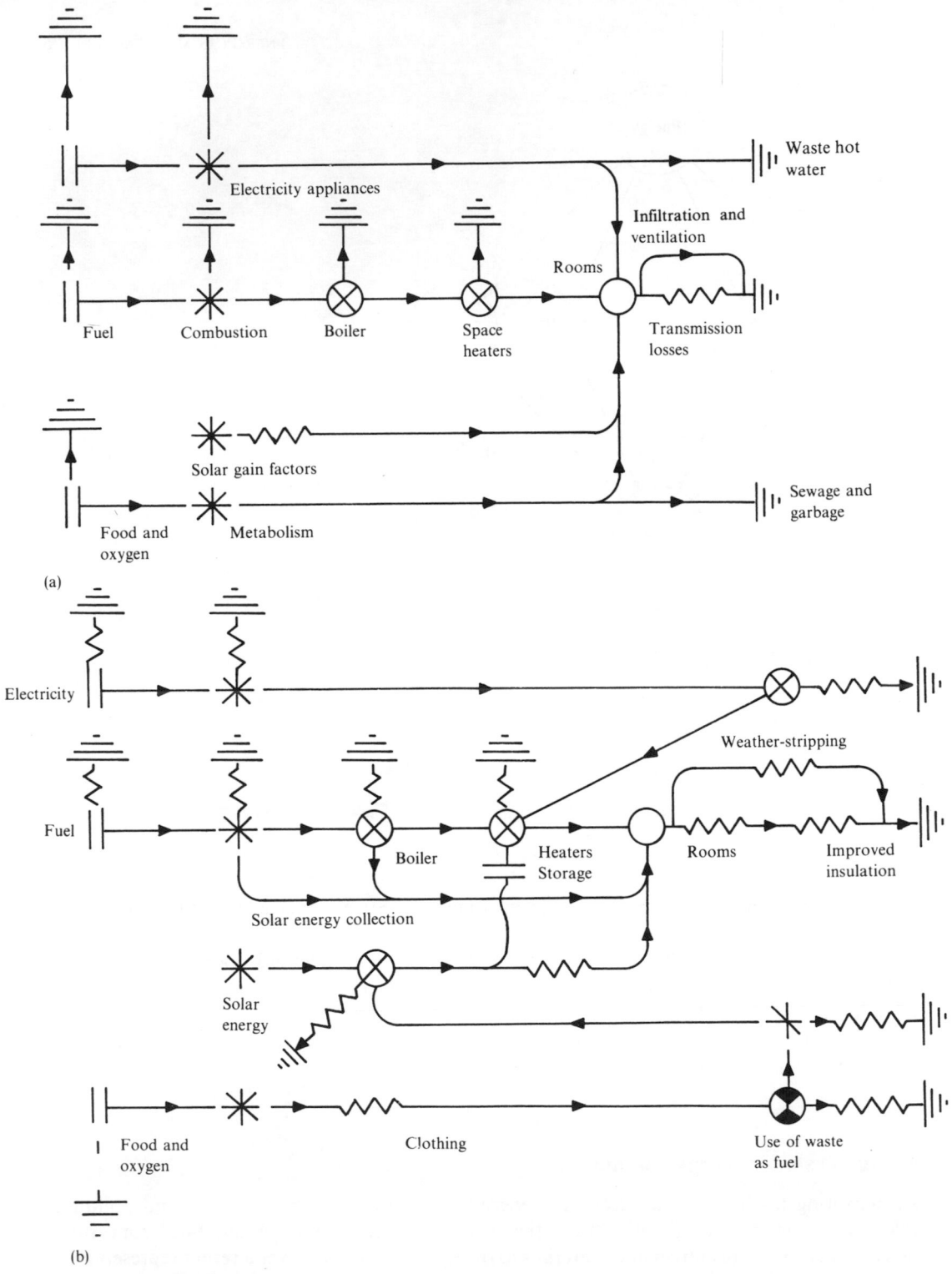

Figure 3.3 Energy flows through a building: (a) Common system; (b) Improved system.

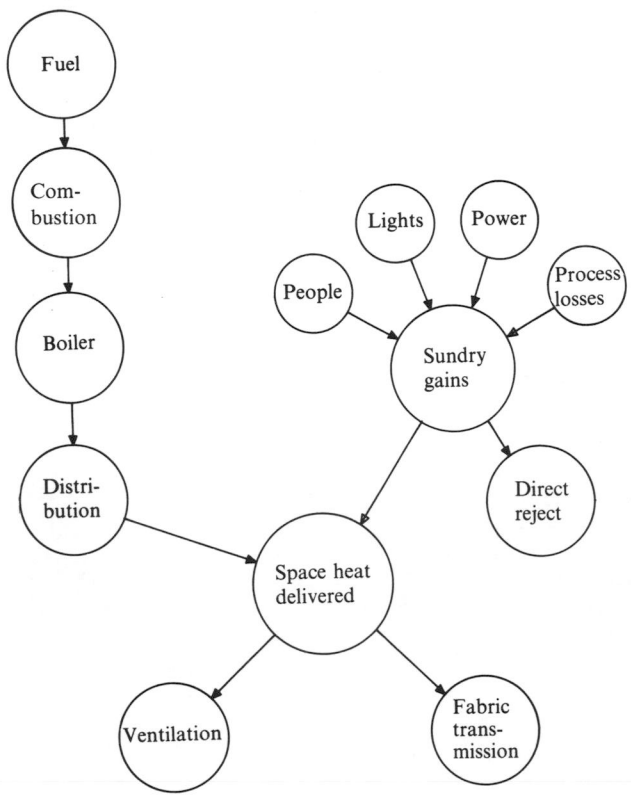

Figure 3.4 The energy audit.

	Primary question	Present facts	Alternatives	Selected alternative
Purpose	What is done?	Is it necessary? Why?	What else could be done?	What should be done?
Means	How is it done?	Why like this?	How else could it be done?	How should it be done?
Place	Where is this done?	Why there?	Where else could it be done?	Where should it be done?
Time	When is this done?	Why then?	When else could it be done?	When should it be done?
Source (grade)	Where does the input resource come from?	Why this source?	What other source could be used?	What source should be used?
Sink (grade)	Where does the rejected resource go?	Why there?	Where could it be redirected?	Where should it be redirected?

Figure 3.5 Critical assessment chart: description of operation.

Table 3.1 Energy costs and contents of fuels and electricity

Type	Historical energy cost, MJ kg^{-1}	Energy content, MJ kg^{-1}
Coal	1.2	28.3
Town gas	13.3	32.0
Methane[†]	0	53.4
Petrol	5.9	43.7
Fuel oil	3.9	44.7
Uranium	1200	??
Electricity	3.2 MJ MJ^{-1}	1.0 MJ MJ^{-1}

Type	Historical energy cost, MJ m^{-3}	Energy content, MJ m^{-3}
Coal	1900.0	45 000
Town gas	16.0	38.4
Methane[†]	0	35.5
Petrol	4700.0	34 800
Fuel oil	3330.0	37 000

[†] Principal constituent of natural gas

Natural gas production has a high energy use efficiency, approaching 100 per cent, coal production has an energy efficiency of about 95 per cent, oil has 90 per cent, whilst the production of electricity is only 25 per cent efficient in energy terms.[2]

The approximate densities of fossil fuels at room temperature and pressure (kg m^{-3}) are as below:

Coal	1590
Town gas	1.2
Methane	0.665
Petrol	796
Fuel oil	828

From Chapter 1, in 1987, 27×10^{12} kg of CO_2 was released globally accompanying the release of 312×10^{12} MJ from fossil fuels. This averages out to 0.0866 kg of CO_2 per MJ.

Energy contents of raw materials

Similarly, raw materials, e.g. metals, plastics, glass, have historical energy and 'pollution' contents. These can again be quantified by analysing industrial energy and materials deliveries and exports of materials. Table 3.2 shows the results of such an analysis.

Table 3.2 Energy and pollution costs of raw materials[2]

Material	Energy cost, $MJ\,kg^{-1}$	Environmental cost, kg of $CO_2\,kg^{-1}$
Steel	50–65	4.3–6.7
Aluminium	60–270	5.2–2.3
Copper	25–85	2.2–7.4
Zinc	60–70	5.2–6
Lead	25–50	2.2–4.3
Cement	8–9	0.7–0.8[†]
Plaster	3	0.26
Plastic	10	0.87[‡]
Glass	20–50	1.7–4.3
Brick	2	0.17
Paper	25	2.2
Wood	4–6	0.4–0.5
Rubber	150	13
Sand, gravel	0.1	0.009

[†] Plus the kg of CO_2 produced in the process itself
[‡] Fossil fuel tied up in plastics 140–150 MJ

Products

Having built the building blocks, quantity surveying data may be used to assess the energy and environmental costs of products (Table 3.3).[3,4]

3.3 ENERGY CONSUMPTION IN MANUFACTURING

The construction of an energy and materials audit, or input/output balance, is appropriate for any resource-consuming activity. The energy costs of fuels, raw materials and products must be estimated before any realistic attempt can be made to reduce the historical resource content of products. A detailed knowledge of all inputs, throughputs and outputs occurring during manufacture must be obtained before the resource utilization efficiencies of processes can be improved most cost-effectively. The optimal cost-effectiveness resulting from the application of any resource saving technique can never be achieved by restricting examination to individual components of a system. Each system must be studied as a whole to identify major waste centres and to compare the cost-effectivenesses and environmental impacts of the many alternative or retrofit actions possible.

All leaks of fuels, materials and energy should be prevented. Energy may be saved in the careful storage of fuels (Appendix 2).

3.4 ENERGY CONSERVATION

The consumption of energy implies four basic processes:

- Combustion
- Transformation

Table 3.3 Energy and environmental costs of products

Product	Energy cost, MJ	Environmental cost, kg of CO_2
Coal-fired power station	10^{10}	$8.66 \times 10^{8\dagger}$
Nuclear power station	3.5×10^{10}	$3 \times 10^{9\dagger}$
North sea oil rig	10^{10}	$8.66 \times 10^{8\dagger}$
Semi-detached house	2.5×10^5	$2 \times 10^{4\dagger}$
1000 cc motor car	8×10^4	$7 \times 10^{3\ddagger}$
Bus	6×10^5	5×10^4
Locomotive	5×10^6	4.33×10^5
Ship	2×10^9	1.7×10^6
Jumbo jet	7×10^7	6×10^6
Bicycle	6×10^3	500
Washing machine	8×10^3	700
Refrigerator	5×10^3	433
Vacuum cleaner	2.4×10^3	210
Colour TV	2.5×10^4	22×10^3
Radio	1.5×10^3	130
Record player	2×10^3	173
3-piece suite	1.8×10^4	1.56×10^3
Table	3.5×10^3	300
Bed	3×10^3	260
Newspaper	2	0.2
Magazine	10	0.9
Book	15	1.3
Milk bottle	8	0.7
Aluminium can	5	0.4
Plastic bottle	80	7.0
$1 m^2$ double glazing	6×10^3	520.0
$1 m^2$ solar energy collector	3.5×10^3	300
1 kg insulant	10–200	0.9–17

$\dagger$ plus CO_2 released in cement making
$\ddagger$ one litre of petrol releases 40 MJ and 3.5 kg of CO_2, thus the car would have to travel 10 000 miles, or for about 1 year, to match its historical energy cost.

- Utilization
- Rejection to the environment

There is much scope for improvements in all four activities. The efficiencies of combustion processes may be maximized, conversion efficiencies can be increased, activities may be performed more economically, and waste heat or materials may be reclaimed (e.g. by employing heat exchangers, incinerating waste, or recycling waste materials). Energy and material flows should be redirected, inhibited or enhanced so as to achieve the maximum overall efficiency of resource utilization in space and time. All resource consuming systems should be designed on the basis of least resource running costs, coupled with long-life and low-maintenance requirements, as well as upon initial capital cost. Often, the greatest immediate savings can be made by questioning the purpose of an activity, closely specifying its resource requirements, and ensuring tight controls so that the specification is adhered to.

That which is not measured cannot be controlled. Energy or materials cannot readily be conserved unless accurate and comprehensive measurements in consistent units are first obtained for all consuming activities within the system boundary.

Financial rewards may be gained at little cost by employing 'good-housekeeping' (e.g. structural repairs, draughtproofing and ventilation control, trimming control systems, turning down thermostats, turning off lights, and switching off plant and equipment when not required).

Plant and equipment should be matched to the required purpose and should be selected on sensible extreme conditions. All systems should be operated at rates corresponding to maximum efficiency (normally fully loaded in continuous operation). Intermittent operations and fluctuations should be avoided. Efficiency checks should be carried out frequently using standardized procedures. The use of energy accumulation should be considered to match intermittent supplies with variable demands (i.e. to balance load factors), to peak lop and use off-peak electricity, and to increase overall energy efficiencies of boilers and distribution systems.

Optimal gas and electricity tariffs should be chosen. Electricity should be metered to all sectors and equipment contributing to peak electrical demand should be identified. Attempts should be made to balance electrical load factors by peak demand lopping (i.e. by staggering start-up times, rescheduling peak activities). The use of standby generators for peak lopping might be considered. Electrical motors should be selected so that they run at near full load. The introduction of a 'total energy' system utilizing combined heat and power should be considered. Electricity is the second most expensive commodity (the most expensive is compressed air and should be used only as a last resort).

The installation of thermal insulation and 'resource conserving' capital plant and equipment should follow only after obvious wastages have been identified and eliminated. Insulation should be applied to high temperature surfaces before attempts are made to insulate lower temperature surfaces. The 'law of diminishing returns' applies in so far as a second increment of insulation applied to a surface results in less *extra* energy savings than the first increment. Doubling the insulation thickness does not double the financial savings. There occurs a financial break-even point when the lifetime savings resulting from an additional increment of insulation equals the cost of the increment. Thereafter, further additions result in net financial losses unless the ratio of the costs for energy to the costs for insulant alters.

Greater energy efficiency always requires an expenditure of materials and vice versa (e.g. the greater the area of a heat exchanger, the more effective the transfer of heat).

High-grade energy (which may be 'hot' or 'cold' with respect to the environmental datum) should not be allowed to be dissipated directly to the environment. The energy rejected from a high-grade process should be collected and redirected via heat exchangers (or simply fans or pumps) to be employed at another place, collected and stored to be employed at another time, or concentrated for another higher-grade purpose using a heat pump or other thermal transformer, as long as these operations are economically justifiable.

Attempts should be made to introduce feedback from energy loss centres to higher-grade stations in the energy flow sequence (e.g. by recycling materials, heat pumping or incinerating waste). Attempts to reduce or reuse waste should be made before any attempts at recycling or recovery. Waste energy and materials should be reused wherever economically possible ensuring that practical grade, time and space-matched uses have been found for the reclaimed amounts. The value of the savings must clearly exceed the cost of recovery.

Solar gains, lighting dissipations and high-temperature thermal loads, emanating from electronics and electrical systems, should be extracted by cooling windows, louvers, shutters, luminaires or equipment, using air or water at outside environmental temperatures. This avoids the wasteful practice in air conditioning systems of allowing such energy to infiltrate into and so disturb the thermal equilibrium of a room, for which it is necessary to use high grade chilled water or refrigerant to remove the excess heat via a large heat transfer surface in order to regain comfort conditions.

Heat distribution systems and hot/chilled water services

All pipework, storage vessels, pumps and valves should be insulated to economic and effective levels. Fabric transmission heat loss calculations should be made and the ratio of heat loss per unit area to heat throughput should be evaluated. The total heat loss associated with energy distribution systems should be estimated for the energy audit. The effective operation of pumps, values and air vents should be ensured. The settings and operations of all control systems should be checked. The integrities of vapour seals on cold pipes and equipment should be maintained.

Space heating systems

Maintenance and operating procedures should be reviewed. Particular attention should be paid to start-up, shut-down times and schedules and control procedures. Heaters and cooling equipment should be correctly sized and positioned. Checks should be made for vertical temperature stratification. Unoccupied zones should be identified. The use of waste heat for space heating should be considered. Combustion efficiency checks should be made for direct-fired heaters. Checks should be made for air leaks into casings. All heat transfer surfaces should be kept clean and free from obstructions. Steam traps should operate efficiently and should not leak. Condensate should be recovered and used for another heating purpose.

Lighting systems

High-efficiency lighting systems should be installed and kept clean. Optimal balances between natural daylighting and artificial lighting in different zones should be sought. In general, the smaller a window, the less transmission losses through the glazing and the greater the amount of artificial lighting required. Different levels of lighting are appropriate for different zones and activities (i.e. design studios versus corridors). Lighting controls should be investigated and reviewed. Lighting should not be provided where it is not needed. Wall colours should be light and reflecting surfaces should be kept clean. Heat recovery from luminaires might be considered.

3.5 TRANSPORTATION SYSTEMS

The direct use of fuels for the transportation sector in the United Kingdom has risen from 15 per cent in 1980 to 25 per cent in 1990. Integrated international networks for production and distribution have increased multifold. These transportation networks have resulted in the dispersal of production systems across Europe, to areas where labour, land, raw materials and local facilities are cheaper, whilst distribution outlets and marketing activities remain in urbanized areas. Industrial, commercial and domestic centres have become increasingly separated. A vicious spiral has developed—more suburbs—more traffic—more roads—declining public transport and rail services—more traffic—more suburbs....

An energy-conscious transportation network arranges that least energy is expended in travel along the production and distribution lines. Routes and load schedules of conveyers, trucks and other vehicles both inside and outside the factory should be critically examined to ascertain whether each journey is absolutely necessary? Carrier sizes should match the loads carried and vehicles and equipment should be regularly serviced and lubricated.

3.6 WATER CONSERVATION

Water and effluent management often fall within the remit of the energy manager. From an analysis of water bills and water usage patterns, a water audit can be constructed in the same manner as the energy audit. Loss and wastage centres can be identified and rectified. Monitoring and targeting exercises, in comparing current water usage with previous periods, can reveal the onset of new leaks or wastages. Discharge rates of liquid effluent should be correlated with water usage. Possibilities for water cascading, reuse or reclaim should be investigated. Required water qualities for different processes should be specified. Water charges may be based upon rateable value or meter readings. From the water audit, possible financial savings arising from installing a water meter may be estimated. Water consumption in like establishments (e.g. schools, hospitals) can be compared by normalizing monthly usage by working areas, number of personnel or amount of product. Automatic flushing schedules should be optimized and flushing should not occur when not necessary (i.e. when buildings are unoccupied). Infra-red detectors might be employed to detect personnel or flushings, and lights might be controlled by door openings.

3.7 RULES FOR THE EFFICIENT CONSERVATION OF ENERGY AND MATERIALS

- The purposes for which expenditures of energy or materials are required should be critically examined.
- As much useful work, heat, or other purpose fulfilment, should be extracted from a degrading energy or materials chain, as is compatible with economic and other considerations.
- The quality, not the quantity, of energy (materials) is the subject of conservation.
- Each energy operation should be examined critically and systematically in isolation and in relation to all other events occurring within the system boundary.
- The manner and extent of all energy and materials use should be challenged, including the appropriateness of the process method and the size of the plant involved.

Sources

- Fuels and materials should be used only when and where required.
- Space or time delays inevitably incur losses.

- Stocks should be maintained at minimum levels plus emergency reserves.
- Attention should be paid to the delivery, storage and handling systems. The financial, energy and materials costs of these should be assessed.
- Comprehensive and accurate monitoring and metering of all energy and materials inputs, throughputs and outputs should be accomplished.
- A continuous fuels log should be maintained.
- Procedures should be standardized.
- Qualities should be checked.
- Information should be easily accessible, comprehensible and disaggregable.
- Attempts should be made to account for all inputs in terms of outputs.
- Storage areas should be made secure against loss or theft.

Plant, equipment, systems, products

- All hardware should be matched to the purpose for which it is required.
- All systems should be operated at rates corresponding to maximum efficiency (normally fully loaded in continuous operation).
- Intermittent operations and fluctuations should be avoided.
- Efficiency checks should be carried out frequently using standardized procedures.
- Plant should be selected on sensible extreme conditions.
- Energy or materials cannot readily be conserved unless accurate and comprehensive measurements in consistent units are first obtained for all activities within the system boundary.
- Greater overall efficiency can always be achieved at the cost of additional complexity.
- Greater energy efficiency always requires an expenditure of materials and vice versa (e.g. the greater the area of a heat exchanger, the more effective the transfer of heat).
- The 'law of diminishing returns' applies.
- Side benefits and diseconomies: incidental benefits or penalties arising from each consuming activity should be identified and carefully evaluated.
- Only the most efficient component branches in energy or materials utilization chains should be adopted.
- Overall efficiencies are always lower than that of the most inefficient link in the chain.
- Product designs should maximize lifetime, promote easy maintenance and repair, require little additional energy or materials inputs during active life, and should facilitate re-use, recycling, easy disposal and natural degradation and recycling.

Improvements

- Systems should be modelled and evaluated accurately so that the cost-effectivenesses of conservation options can be compared realistically.
- Careful assessments of real savings should be made, including maintenance costs.
- Full audits in common units should be carried out before and after improvements.
- Evaluations should be obtained with respect to quantities of energy and materials, financial costs of energy and materials, energy costs of energy and materials, exergy degradation and overall exergetic efficiencies.
- Representative periods should be adopted for these analyses.

- A continuous *monitoring and targeting* procedure should be implemented.
- The selection of new plant, processes, or energy (or material) conserving measures should be made, not on least capital cost criterion alone, but upon the basis of least total cost over the lifetime of the system.
- Random factors should be eliminated—the system should be isolated from its external environment.
- All leaks of fuels, materials and energy should be prevented.
- Attempts should always be made to reduce demand before increasing energy or materials supplies.
- It should be ensured that modifications have no hidden diseconomic effects and that they comply with safety, fire and statutory regulations and codes of practice.

Energy and exergy

- Energy (materials) grade and availability should be matched to the purpose for which it is required in terms of temperature, pressure, heat flux, and the qualities of materials.

The choice of an energy (materials) form to suit a particular application should not be made arbitrarily. Forms of different qualities are suitable for specific applications. High-grade energy should be used only for high grade purposes, such as producing work, fuels, materials or electrical potential.

Whenever, or whenever, energy quality (exergy) must be degraded, attempts should be made to:

- Obtain useful heat (i.e. from a temperature reduction)
- Obtain useful work (i.e. from a pressure reduction)

A degrading energy chain should be made to do as much work and other useful activity as possible during the process.

- The input and output grades of energy supplied to and rejected from any particular component activity should be such that the exergetic efficiency of the activity is a maximum.
- Energy flows should be redirected within the overall system such that the overall exergetic efficiency of the system is a maximum.
- If additional energy has to be added to a degrading energy chain, it is more efficient to introduce low-grade energy, preferably 'waste' heat, rejected from a higher grade activity, at the lower end of the chain.

For example, fuel should not be burnt at 1000 degrees centigrade in order to provide space heating for a room whose temperature needs to be raised only a few degrees above that of the outside air.

- If energy is to be removed from a degrading energy chain, it is more efficient to withdraw this energy before it is degraded, using the external environment as a sink, at the higher end of the energy chain.

Reclamation

- High-grade energy (which may be 'hot' or 'cold' with respect to the environmental datum) should not be allowed to dissipate directly to the environment. The energy rejected from a high-grade process should be collected and redirected via heat exchangers (or simply fans or pumps) to be employed at another place, collected and stored to be employed at another time, or concentrated for another higher grade purpose using a heat pump or other thermal transformer, as long as these operations are economically justifiable.
- The effects on the desired purpose of by-passing, deleting, or moving back up degrading energy chains should be examined.
- Attempts should be made to introduce feedback from energy loss centres to higher-grade stations in the energy flow sequence (e.g. by recycling materials, heat pumping or incinerating waste).
- Attempts to reduce or reuse waste should be made before any attempts at recycling or recovery.
- Waste energy and materials should be reused wherever economically possible ensuring that practical grade, time and space-matched uses have been found for the reclaimed amounts.
- The value of the savings must clearly exceed the cost of recovery.

Waste heat, materials and ambient energy

- The use of external waste heat, materials or ambient energy (i.e. solar, wind, waves, tides, external air temperatures) should be considered before increasing the rate of usage of fuels.
- Solar energy is the most pollution-free power source.

Energy storage

- The use of energy accumulation should be considered to:

(1) Balance load factors
(2) Peak lop and use off-peak electricity
(3) Increase overall energy efficiencies of boilers and distribution systems
(4) Harness ambient energy

Management

Great care should be taken to ensure the cleanliness, correct operation and planned systematic maintenance of all storage, release, distribution, utilization, insulation, heat recovery, instrumentation and control systems.

Pollution

- Waste and pollution should be closely monitored and minimized.
- Waste in all forms not only squanders human effort, energy, time and materials, but also damages the external environment and disrupts ecological harmony.
- The reduction of waste is especially desirable where materials have intense availability contents or where their historical availability costs are high. Metals, glass, plastics, paper and refractories are examples of such materials.

- Design improvements, which prolong the lifespan, or promote the reuse or easy recycling of these energy-intensive materials are highly desirable.
- Methods of waste collection, sorting and reclamation should be developed.
- Improved recycling techniques are needed.

Education

- All personnel should be made fully aware of the energy and materials implications of their activities and decisions.
- All associated personnel should be availed information, demonstrations of achievements and reports of failures of 'improvement' activities.

3.8 LAWS OF ENERGY AND MATERIALS FLOWS

- Energy or matter can be neither created nor destroyed.
- Energy or matter are always conserved, although transductions may occur.
- Exergy degrades via equilibrium processes.
- All energy and materials tend to degrade to entropic disorder by being dispersed over a greater volume.
- Human consumerism accelerates this process.
- As a result of utilization, energy and materials are eventually downgraded to an equilibrium state having zero availability corresponding to the environmental datum.
- The environmental potential should be regarded as the reference datum.
- The availability of either energy or materials requires a source or sink at a potential different to that of the environment.
- The efficiency of an energy or materials conversion is always less than 100 per cent.
- Decreasing entropy is a futile process in the long term but may be of temporary use in the short term.
- Energy must be lost in reducing entropy (work production).
- The supply of work can reduce entropy (heat pumping).
- Activities, if completely described, all lead to energy or materials rejection to the environmental datum, i.e. the total entropy of the considered control volume through which the energy or materials flow, rises.
- Energy reclamation involves the expenditure of materials.
- Materials reclamation involves the expenditure of energy.
- Reclamation of energy or materials reduces pollution.

REFERENCES

1. DTI, Digest of United Kingdom Energy Statistics, Department of Trade and Industry, HMSO, London, 1992.
2. O'Callaghan, P. W.: *Design and Management for Energy Conservation*, Pergamon Press, Oxford, 1981.
3. Chapman, P.: *Fuels Paradise*, Penguin, Harmondsworth, 1975.
4. Lenihan, J. and Fletcher, W. W.: 'Energy Resources and Environment', Vol. I, Blackie, London, 1975.

FUNDAMENTAL CONCEPTS

4.1 THERMOPHYSICAL TRANSPORT PROPERTIES

Transport properties are those of solids, liquids and gases which dictate the way in which heat or momentum are transported through these materials.

Density

The *density*, $\rho\,(\text{kg m}^{-3})$, of a substance is its mass per unit volume. Its value determines the weight of a body of given volume, or its momentum when travelling at speed.

Thermal conductivity

The *thermal conductivity*, $k\,(\text{W m}^{-1}\,\text{K}^{-1})$, of a substance determines the rate at which heat is transported though it in the steady-state (i.e. not dependent upon time).

Specific heat

The *specific heat*, $c\,(\text{J kg}^{-1}\,\text{K}^{-1})$, of a substance determines how quickly (i.e. time-dependent heat transfer) a body will heat up or cool down.

The *specific heat at constant pressure*, $c_p\,(\text{J kg}^{-1}\,\text{K}^{-1})$, of a gas or vapour is larger than the *specific heat at constant volume*, $c_v\,(\text{J kg}^{-1}\,\text{K}^{-1})$, because some of the heat supplied at constant pressure produces work against an attempted volume change.

For a perfect gas, *Boyle's Law* states that, at constant temperature, $T\,(\text{K})$, the product of the pressure on a gas, $P\,(\text{N m}^{-2})$, and the volume occupied by that gas, $V\,(\text{m}^3)$, is constant; i.e.

$$PV = \text{constant} \tag{4.1}$$

Also for a perfect gas, *Charles' Law* states that, under a constant pressure, $P\,(\text{N m}^{-2})$, the quotient of the volume, $V(m^3)$, occupied by the gas and the absolute temperature, $T(K)$, of that gas is also constant; i.e.

$$V/T = \text{constant} \tag{4.2}$$

The combination of these two laws leads to the *characteristic gas equation*

$$PV = mRT \tag{4.3}$$

where m is a mass of a perfect gas (kg) and R is the *characteristic gas constant* $(J\,kg^{-1}\,K^{-1})$. Thus, the density of a gas, $\rho = m/V = P/RT$.

The *universal gas constant*, $R_o = 8.3143\,kJ\,kmol^{-1}\,K^{-1}$. For any gas, its characteristic gas constant may be calculated from

$$R = R_o/M \tag{4.4}$$

where M is the molecular weight of the gas. For air, $M = 28.87\,kg$, $R = 8314.3/28.87 = 288\,J\,kg^{-1}\,K^{-1}$.

From the *First Law of Thermodynamics*, energy is conserved. Thus, when heat is supplied to a confined gas or vapour, its *internal energy* increases, manifested by a rise in temperature, and some work is performed. The rise in internal energy is given by $mc_v\Delta T(J)$. The *total energy*, or *enthalpy*, which must be supplied to accomplish this, is given by $mc_p\Delta T(J)$. The difference between the energy supplied and the rise in internal energy is the *work performed*. Thus

$$mc_p\Delta T(J) - mc_v\Delta T(J) = \text{work done} \tag{4.5}$$

The work done by a gas or vapour in expanding at constant pressure is given by $P\Delta V$ and from the characteristic gas equation $P\Delta V = mR\Delta T$. Substituting

$$mc_p\Delta T - mc_v\Delta T = mR\Delta T$$

and cancelling common terms

$$c_p - c_v = R \tag{4.6}$$

The ratio of the specific heats, c_p/c_v, is known as the *adiabatic index*, γ, for a gas or vapour.

Expansion

The *coefficient of linear expansion*, $\alpha\,(m\,m^{-1}\,K^{-1})$, of a substance determines how much it will expand with increasing temperature.

The *coefficient of volumetric expansion*, $\beta\,(m^3\,m^{-3}\,K^{-1})$, determines the additional volume by which a volume of gas will expand with increasing temperature.

From Charles' Law, the *coefficient of volumetric expansion*, $\beta\,(m^3\,m^{-3}\,K^{-1})$ is equal to the reciprocal of the absolute temperature, $T(K)$, for a perfect gas; i.e.

$$\beta = 1/T \tag{4.7}$$

Viscosity

The *dynamic viscosity*, $\mu\,(kg\,m^{-1}\,s^{-1})$, of a fluid (i.e. liquid or gas) is the property which resists flow, when the fluid is subject to a pressure difference. It is caused by forces acting between the molecules of a fluid. It is measured by timing the rate at which a fluid, subject to a pressure difference, passes through an orifice.

Pressure (altitude) dependence

Transport properties, *except density*, do not vary significantly with pressure (i.e. associated with altitude). The pressure dependence of gas densities may be calculated from the characteristic gas equation.

4.2 THERMODYNAMICS AND EXERGY

The *Zeroth Law of Thermodynamics* states that if two substances are each in equilibrium with a third substance, then the two substances are in equilibrium with each other. All substances seek this general equilibrium state. Hence the tendency of materials to degrade by dispersion, decay or oxidation to a state in equilibrium with the environmental datum.

The release of stored solar energy in the combustion of fossil fuels converts energy from a relatively stable form to the unstable form of thermal energy, producing waste and pollution in the process.

Entropy may be considered as the amount of unavailable energy within a given system. *Thermodynamic availability* cannot be recycled. The energy engineer must extract as much useful work and heat from a degrading energy chain as possible before the energy in transit assumes the mean properties of the environment. Energy may be *cascaded*, e.g. the waste heat from one process may be supplied to another, heat from people and lights become sundry gains to buildings, and so on. The energy manager must seek these cascading opportunities.

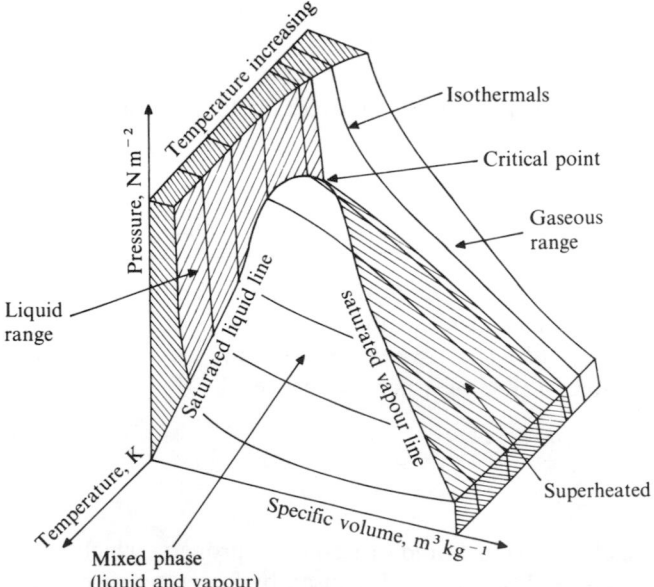

Figure 4.1 The pure substance.

Table 4.1 Forms of energy

Mechanical	Potential	(Stored energy)	
	Kinetic	• Translation	
		• Rotation	(Energy in motion)
		• Vibration	
	Pressure	(Strain)	Energy
Electrical	Electromagnetic	(Solar energy)	Magnetic
	Electrostatic		
Chemical	Fuels		
Nuclear	Fission	Fusion	
Thermal	Heat		

The *First Law of Thermodynamics* states that 'energy and materials are always conserved' they can be neither created nor destroyed, only converted from one state (Fig. 4.1) or form (Table 4.1) to another.

Thus the energy manager is not attempting to *conserve energy*. His task is to conserve the *quality* of energy and its ability to perform work and render useful heat, or cold.

Thermodynamics is concerned with the ways in which fluids behave as they are heated, cooled, expanded and compressed. Work and heat appear at the boundary of a system undergoing a thermodynamic process (Fig. 4.2). When a system is heated, its temperature and *internal energy* rises and some work is performed; i.e.

$$\text{heat} = \text{work} + \text{rise in internal energy} = \text{constant}$$

Conversely, when work is done on a system, its temperature and internal energy rises and some heat is released to the environment.

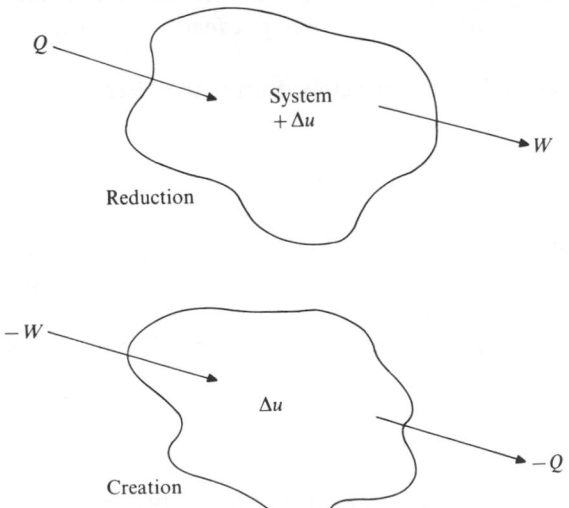

Figure 4.2 Entropy reduction and creation.

Table 4.2 Specific heats of gases

Gas type	c_p/R	c_v/R	c_p/c_v
Monatomic	5/2	3/2	5/3
Diatomic	7/2	5/2	7/5
n degrees freedom	$(n+2)/2$	$n/2$	$1+2/n$

The ratio of these specific heats is known as the *adiabatic index* for a gas ($\gamma = c_p/c_v$).

Table 4.3 Gas data at room temperatures

Material	Chemical formula	Molecular weight, kg	R, $\mathrm{J\,kg^{-1}\,K^{-1}}$	c_p, $\mathrm{J\,kg^{-1}\,K^{-1}}$	c_v, $\mathrm{J\,kg^{-1}\,K^{-1}}$	γ
Hydrogen	H_2	2	4157	14 550	10 393	1.4
Oxygen	O_2	32	260	910	650	1.4
Nitrogen	N_2	28	297	1 040	743	1.4
Carbon	C	12				
Carbon monoxide	CO	28	298	1 043	745	1.4
Carbon dioxide	CO_2	44	189	819	630	1.3
Water (steam)	H_2O	18	462	1 860	1 398	1.33
Methane	CH_4	16	520	2 190	1 670	1.31
Sulphur	S	32				
Sulphur dioxide	SO_2	64	130	1 365	1 235	1.11
Nitrous oxide	N_2O	44	189	877	688	1.27
Air	71% N_2 29% O_2	29	287	1 005	7 18	1.4

Molecules of *monatomic* gases, such as helium, argon and other inert gases are composed of single atoms. *Diatomic* gases, such as hydrogen, oxygen, nitrogen and carbon monoxide, have two atoms in a molecule. Other gases are said to have *n degrees of freedom*—translational, rotational, vibrational, etc.

Data for a number of gas types and gases of interest, developed using the *kinetic theory of gases*, are shown in Tables 4.2 and 4.3.

Work

When a gas expands, its pressure falls, its volume increases and it performs work in displacing its surroundings. The amount of work done is given by the area under the pressure versus volume curve, as shown in Fig. 4.3, the integral of $P\,dV$.

The relationship between pressure and volume can follow one of five possible processes (Fig. 4.4):

- A *constant volume* process, $V = $ constant
- A *constant pressure* process, $P = $ constant
- A *constant temperature* process, $T = $ constant, $PV = $ constant or *isothermal* process

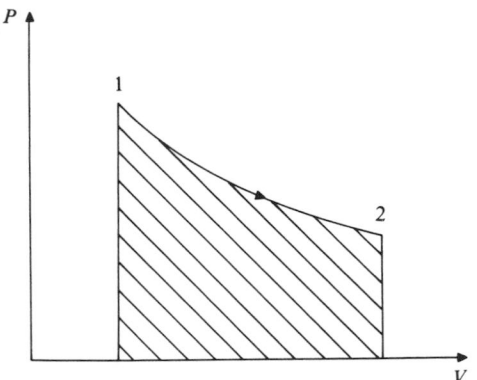

Figure 4.3 Expansion producing work.

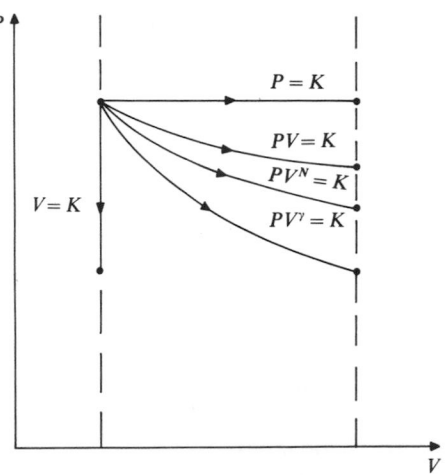

Figure 4.4 Expansion processes.

Table 4.4 Work done, heat transferred and state changes for a perfect gas undergoing various expansion processes

Process	W	Q	ΔU	ΔH	ΔS
Constant volume $V = K$	0	$\Delta U = mc_v\Delta T$	$mc_v\Delta T$	$mc_p\Delta T$	$mc_v \ln(T_2/T_1)$
Constant pressure $P = K$	$P(V_2 - V_1)$	$\Delta U = mc_p\Delta T$	$mc_v\Delta T$	$mc_p\Delta T$	$mc_p \ln(T_2/T_1)$
Isothermal $T = K,\ PV = K$	$P_1 V_1 \ln(V_2/V_1)$ $= P_2 V_2 \ln(P_1/P_2)$	$= W$	0	0	$mR \ln(V_2/V_1)$
Adiabatic $Q = 0$ $PV^\gamma = K,\ S = K$	$\dfrac{P_1 V_1 - P_2 V_2}{(\gamma - 1)}$	0	$mc_v\Delta T = W$	$mc_p\Delta T$	0
Polytropic $PV^n = K$	$\dfrac{P_1 V_1 - P_2 V_2}{(n - 1)}$	$\dfrac{P_1 V_1 - P_2 V_2}{(n - 1)}$ $- mc_v\Delta T$ $= \dfrac{(\gamma - n)}{(\gamma - 1)} W$	$mc_v\Delta T$	$mc_p\Delta T$	$mc_v \ln(T_2/T_1)$ $+ R \ln(V_2/V_1)$

$K \equiv$ constant
Isentropic means reversible adiabatic, i.e. that no heat is transferred to or from the system during the process, and that the process is fully reversible.

- An *adiabatic* process, where no net heat enters or leaves the system $PV^\gamma = \text{constant}$
- A *polytropic* process, $PV^n = \text{constant}$

Table 4.4 lists the work done, heat transferred and state changes for a perfect gas undergoing these processes. The changes in total heat and entropy content involved are denoted ΔH and ΔS, respectively.

Definitions

1st Law	$Q = \Delta U + W$
For a perfect gas	$PV = mRT$
Work done	$W = \int P\, dV$
Enthalpy	$H = U + PV$
Adiabatic index	$\gamma = c_p/c_v$
Gas constant	$R = c_p - c_v$
Entropy change	$\Delta s = mc_v \ln(T_2/T_1) + R \ln(V_2/V_1)$
Internal energy change	$\Delta U = mc_v \Delta T$
Total energy change	$\Delta H = mc_p \Delta T$

A compression process is the reverse of an expansion process, but then the work is supplied to the system to compress the gas.

Table 4.5 lists the overall efficiencies of various engines and other heat-to-work energy converters.

All engines are cyclic devices. This means that they must pressurize a fluid, such as air or steam, originally at environmental temperature and pressure, then allow the fluid to expand, performing work, and then return the fluid to its initial state in equilibrium with the environment, ready to start the cycle again. This is achieved in internal combustion engines by mixing air with oil or gas, igniting the mixture to raise its temperature and confining the mixture so that it pressurizes. Following the resisted expansion process, the low pressure mixture is

Table 4.5 Approximate overall efficiencies, η%, of some important energy transducers

System	Nuclear	Chemical	Thermal	Mechanical	Electrical	Electro-magnetic	η
Resistance heating			o ←		→ o		100
Electrical generator				o →	→ o		98
Muscular action		o →		→ o			95
Large electrical motor				o ←	→ o		93
Dry battery		o →			→ o		90
Hydroelectric power station				o →	→ o		88
Power station boiler		o →	→ o				88
Domestic gas boiler		o →	→ o				85
Gearbox and transmission				o			80

Table 4.5 (*continued*)

System	Nuclear	Chemical	Thermal	Mechanical	Electrical	Electro-magnetic	η
Pumped liquid storage				o			76
Car battery		o———————————→			o		74
Fuel cell		o———————————→			o		70
Ideal heat engine		o———→	o———→	o			68
Industrial oil boilers		o———→	o				
5% of full load							50
52% of full load							70
50% of full load							75
100% of full load							77
120% of full load							75
Domestic oil boiler		o———→	o				66
Small electric motor				o ←———	o		64
Space rocket		o———→	o———→	o			50
Steam turbine			o———→	o			40–46
Stirling engine			o———→	o			44
Thermal power plant		o———→	o———→	o———→	o		40
Gas laser					o———→	o	40
Diesel engine		o———→	o———→	o			38
Joule (Brayton) cycle		o———→	o———→	o			36
Aircraft gas turbine		o———→	o———→	o			36
Industrial gas turbine		o———→	o———→	o			35
Oil-fired power station		o———→	o———→	o———→	o		33
Nuclear power plant	o———————→		o———→	o———→	o		30
Petrol car engine (Otto)		o———→	o———→	o			26
Coal-fired power station		o———→	o———→	o———→	o		25
Fluorescent lamp					o———→	o	20
Wankel engine		o———→	o———→	o			18
Solar cell					o ←———	o	10
Thermocouple			o———————————→		o		8
First steam train		o———→	o———→	o			8
Nuclear battery	o———————————————→				o		6
Incandescent lamp					o———→	o	5
Watt engine		o———→	o———→	o			2.7–4.5
Smeaton engine		o———→	o———→	o			0.8–1.4
Newcomen engine		o———→	o———→	o			0.5

Diesel engines run hotter than petrol engines and so have higher carnot efficiencies.
o denotes the form of energy conversion used in the system.

rejected to the environment where it disperses and cools to environmental conditions, completing the cycle.

In the steam engine, water at environmental conditions is boiled via the combustion of fossil fuels, the high pressure steam is expanded to perform work and the low pressure steam is then condensed and cooled to the original state, ready for the next cycle.

Because heat must be rejected to the environment to complete an engine cycle, not all of the heat energy produced by combustion can be converted to work, i.e. an engine cycle which converts heat to work cannot be 100 per cent efficient. This leads to the *Second Law of Thermodynamics*, which states that 'it is impossible to construct a system which will operate in a cycle, extract heat from a source, and do an equivalent amount of work on the surroundings'.

In order to receive heat, the system must be in contact with a thermal reservoir at a temperature higher than that of the fluid at some point during the cycle. For heat to be rejected from the system, the fluid must, at some point of the cycle, be in contact with a thermal sink at lower temperature. Thus, if a system is to undergo a cycle and produce work, it must operate between at least two reservoirs at different temperatures (Fig. 4.5).

For an expansion process to be *reversible*, the path traced during expansion must be exactly retraceable during compression. If heat is transferred between the expanding fluid and the environment during the expansion process, a reverse compression process cannot bring the fluid back to its original state. A process can only be completely reversible if it is *fully resisted*, and is either: (a) *adiabatic*—i.e. no heat transfer between the system and its surroundings takes place; or (b) *isothermal*—i.e. the system temperature remains constant and so the heat supplied to the system is converted totally into work. On the return path, the work supplied is transferred totally into heat.

Conditions which prevent a process being fully reversible are:

- The presence of friction
- Heat transfer to the environment

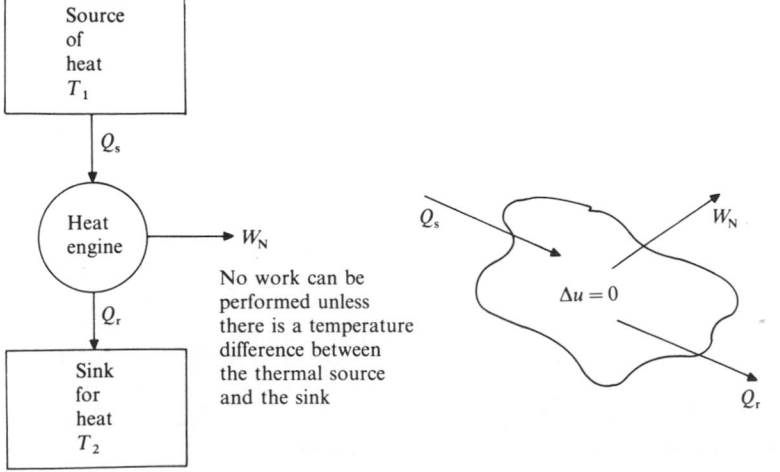

Figure 4.5 Schematic ideal Carnot cycle.

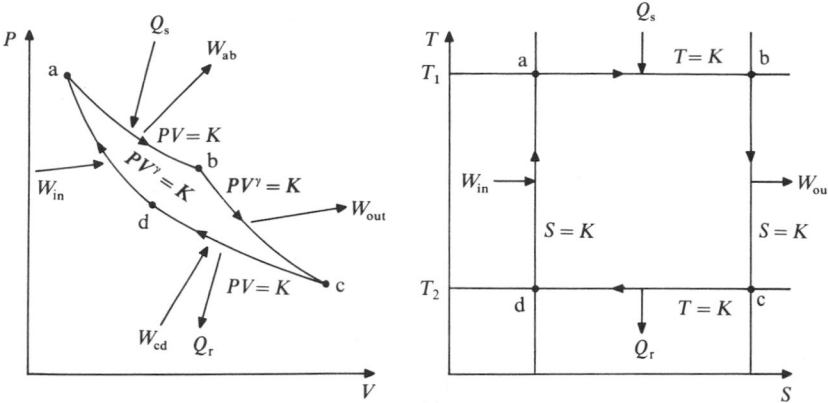

Figure 4.6 Thermodynamics of the ideal Carnot cycle.

- Unresisted expansion
- Paddle work

The *Carnot Cycle* (Fig. 4.6) is an imaginary ideal cycle constructed from reversible adiabatic (*isentropic*, or constant entropy) and isothermal expansions and compressions. Because all the net heat supplied is converted to work, this cycle gives the maximum work output possible for a system operating between two reservoirs at temperatures T_1 and T_2 (K). Its efficiency, η, is given by

$$\eta = \text{net work output/net heat supplied} = \frac{T_1 - T_2}{T_1} \tag{4.8}$$

Heat Exergy, X (J or W when the process is continuously operating between an infinite source and an infinite sink), is that part of a quantity of heat, Q (J or W), at a temperature, T (K), which would be converted to work, were the substance containing the heat reduced to a state in equilibrium with the environmental datum at T_0 (K).

Then, from the Carnot efficiency

$$\text{heat exergy, } X = \text{maximum work possible} = \left(\frac{T - T_0}{T}\right) Q \tag{4.9}$$

The term $(T - T_0/T)$ is known as the *exergetic potential* of the heat source Q and is given the symbol γ.[1]

Figure 4.7 shows that, as the temperature of the substance reduces, less work can be extracted from it.

The coefficient of performance for a *refrigeration* cycle, operating between two temperatures, T_1 and T_2 (K) is given by

$$COP = \text{cooling supplied/work input} = \frac{T_2}{(T_1 - T_2)} \tag{4.10}$$

where T_2 is the lower temperature—that of the evaporator.

The coefficient of performance for a *heat pump* cycle, operating between two temperatures,

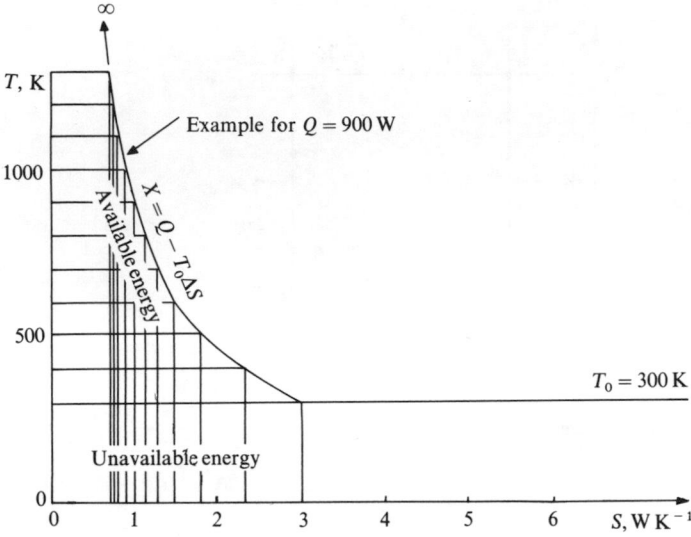

Figure 4.7 Available energy.

T_1 and T_2 (K) is given by

$$COP = \text{heat supplied/work input} = \frac{T_1}{(T_1 - T_2)} \qquad (4.11)$$

where T_1 is the higher temperature—that of the condenser.

If the work from an ideal heat engine cycle were supplied to an ideal heat pump cycle, Fig. 4.8 shows that heat exergy, $X = \gamma Q$, is conserved. Thus, it is possible to obtain *more* heat at a lower temperature than is supplied to the system at a higher temperature. Conversely,

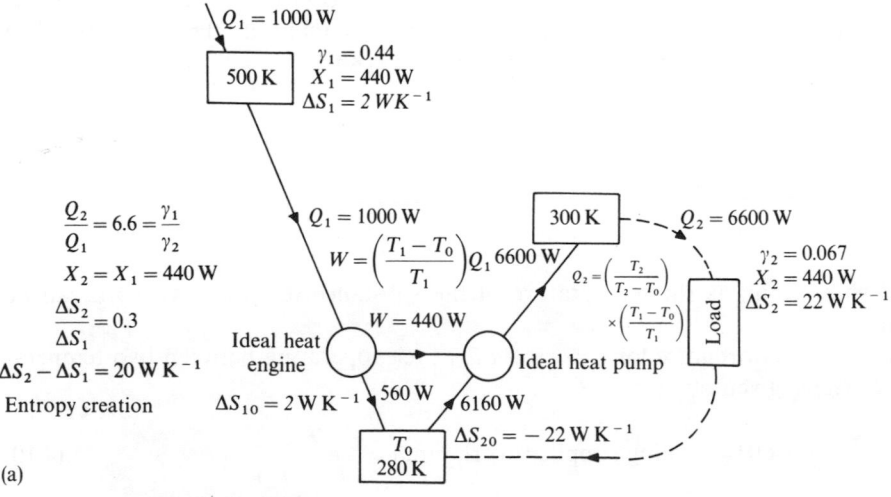

Figure 4.8 The energy transducer: (a) As an 'energy creator'.

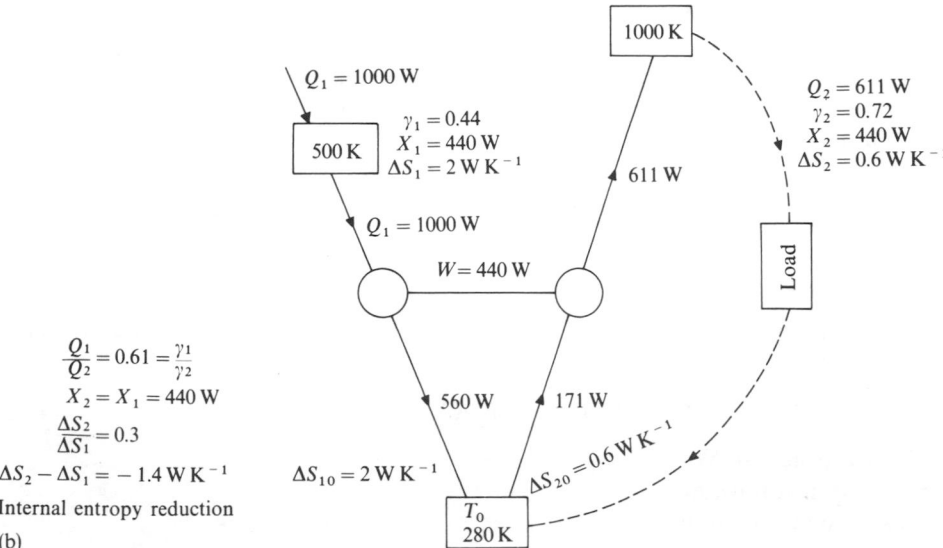

Figure 4.8b As an entropy reducer.

low-grade or low-temperature heat can be upgraded to a higher temperature for re-introduction to a thermal process.

The *value* of an energy source increases the higher its exergetic potential. Electricity has $\gamma = 1$, whilst reject heat at 350 K in an environment at 300 K has $\gamma = 0.167$.

Energy managers should attempt to rate energy flows in terms of their *qualities*, or exergetic potentials, and so produce *exergy* audits. Economic possibilities for *energy cascading* may then be identified. The cost of heat exergy is its monetary value divided by its exergetic potential. Thus, if electricity costs 6p per kWh, its exergy cost is 6p per kWh, whereas if the 'waste heat' at 350 K costs 2p per kWh, its exergy cost is 12p per kWh.

Exergy economics must be employed when assessing the potentials of heat recovery and combined heat and power schemes, or others involving low-grade energy, such as solar energy utilization.

4.3 CONDUCTIVE HEAT TRANSFER

Conductive heat flow is a process by which thermal energy is transmitted by direct molecular communication. It is the only mechanism by which heat flows in an opaque solid. Conduction in a translucent solid is accompanied by radiation, whilst heat transfer through quiescent gases and liquids takes place by conduction with some radiation. Convection enhances the thermal equilibrium process in moving fluids.

In general, conduction occurs as a result of collisions between atoms or molecules—atoms in the hotter region of the material have higher kinetic energies and communicate this energy to neighbouring atoms. Conduction of heat by this mechanism occurs in most liquids, gases and non-metallic solids. A dielectric material is a non-conductor of electricity. These substances are usually poor conductors of heat. A *phonon* is a quantum of lattice vibration in a solid. Where pure phonon transfer occurs, e.g. in pure crystals of diamond, conductive heat transfer is very good. Phonon mismatches in non-pure solids result in low conductivities. In metals, the mechanism is different. The energy is carried by electrons, moving through the lattice and colliding with atoms, imparting kinetic energy. Thus good electrical conductors are also good conductors of heat. The thermal conductivity, $k\,(\mathrm{W\,m^{-1}\,K^{-1}})$, characterizes the material. Values of k range from 0.024 for air to $600\,\mathrm{W\,m^{-1}\,K^{-1}}$ for copper.

One-directional heat flow

When a temperature difference is applied across a solid, the rate of heat which flows through it is calculated from

$$Q = kA\Delta T/\Delta L + mc\Delta T/\Delta t + qV \tag{4.12}$$

where Q = heat transferred, W
$\quad\quad A$ = area to heat flow, $\mathrm{m^2}$
$\quad\quad L$ = distance for heat flow, m
$\quad\quad T$ = temperature, K
$\quad\quad t$ = time, s
$\quad\quad m$ = mass of the solid, kg
$\quad\quad c$ = specific heat of the solid, $\mathrm{J\,kg^{-1}\,K^{-1}}$
$\quad\quad q$ = rate of internal heat generation per unit volume, $\mathrm{W\,m^{-3}}$
$\quad\quad V$ = the volume of the solid, $\mathrm{m^3}$.

In the steady-state (invariant with time), and with no internal heat generation, the equation becomes

$$Q = kA\,\Delta T/\Delta L = \Delta T/R = C\Delta T \tag{4.13}$$

where $R = \Delta T/kA$, the *conductive thermal resistance* $(\mathrm{K\,W^{-1}})$. The *conductive thermal conductance* $C\,(\mathrm{W\,K^{-1}})$ is the reciprocal of R.

Composite walls

If n layers of different materials in series comprise a solid wall, the total resistance to heat flow is obtained by simply summing the resistances of each section.

If n layers of different materials in parallel comprise a solid wall, the total conductance for heat flow is obtained by summing the conductances of each section.

Directional conductances

Consider the block of copper and insulant of depth 10 units (into the page) shown in Fig. 4.9. The thermal conductance, CX, in the X direction is calculated as follows:

$$CX = CI + CC = 0.04 \times 10/10 + 600 \times 90/10 = 5400$$

The effective thermal conductivity in the X direction, $k_{\mathrm{eff}}X$, is then calculated from

$$CX = k_{\mathrm{eff}}X \times 100/10 = 5400$$

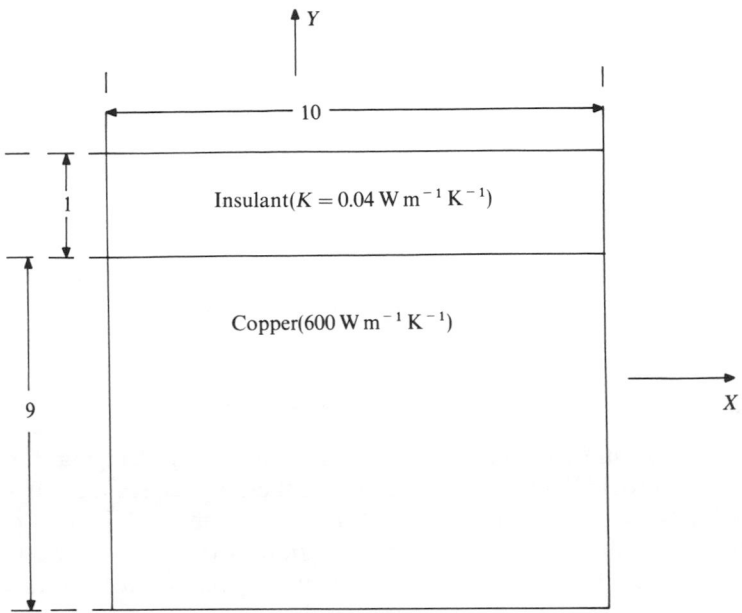

Figure 4.9 Directional thermal conductivity.

Hence

$$k_{eff} X = 540 \, \text{W m}^{-1} \, \text{K}^{-1}$$

The thermal resistance, RY, in the Y direction is calculated as follows:

$$RY = RI + RC = 1/(0.04 \times 100) + 9/(600 \times 100) = 0.25 + 0.000\,15 = 0.250\,15$$

The effective thermal conductivity in the Y direction, $k_{eff} Y$, is then calculated from

$$RY = 10/(k_{eff} Y \times 100) = 0.250\,15$$

Hence

$$k_{eff} Y = 0.4 \, \text{W m}^{-1} \, \text{K}^{-1}$$

The thermal conductivity in the X direction is 1350 times that in the Y direction. This illustrates the effects of thin striations or air in heat transfer paths. If a system is to be cooled, very small air gaps can cause substantial overheating.

Radial heat conduction

Because the area, through which the heat passes, increases with radius, a logarithmic mean thermal resistance, obtained from integration, is employed in radial heat flow situations.

For a hollow tube of length L(m), the thermal resistance is calculated from

$$R = \log_e(r_o/r_i)/2\pi kL \tag{4.14}$$

where r_o and r_i are external and internal radii respectively. Then $Q = \Delta T/R$ as usual. Resistances are simply added to obtain the overall thermal resistance for composite pipes.

Spherical shells

The rate of heat conduction is given by[2]

$$Q = k(A_o A_i)^{0.5} \Delta T/(r_o - r_i) \qquad (4.15)$$

when the material is homogeneous, where A_o and A_i and r_o and r_i are outside and inside surface areas (m²) and radii (m) respectively. Resistances may again be added for composite shells.

Transient heat conduction

The conditions at the boundaries of most thermal structures are continually changing. For this reason, steady-state analyses provide only approximations of thermal behaviour.

In general, the heat flowing between a system and its surroundings alters the amount of heat stored within the system boundary. An approximate indication of the time-dependent response to changes in boundary conditions may be deduced by assuming that the temperature of the system is uniform at any instance of time. If this temperature changes by amount ΔT during a time interval Δt, then the change in internal energy is equal to the net rate of heat flow across the boundary during the same time interval; i.e. heat entering in time ΔT = heat stored in time Δt; i.e.

$$hA(T - T_o)\Delta t = c\rho V \Delta T \qquad (4.16)$$

where $h\,(\mathrm{W\,m^{-2}\,k^{-1}})$ is the heat transfer coefficient at the boundary (discussed later) and T and T_o are the system and surrounding temperatures (K) respectively.

By separating variables and integrating, the temperature response of the system is given by

$$\frac{T - T_o}{T_{t=0} - T_o} = \exp(-hAt/c\rho V) \qquad (4.17)$$

where $c\rho V/hA$ is the *time constant* of the system, or the ratio of internal to external resistances.

Systems having large time constants are slow to respond to changes in external boundary conditions. Systems having small time constants respond rapidly to changes in external boundary conditions.

4.4 FLUID FLOW

By definition, a fluid can offer no permanent resistance to shearing stresses. The application of a force causes a fluid to flow. Fluids are divided into liquids, which are considered to be incompressible, and gases, which are easily compressed. A liquid presents a free surface at the boundary with a gas, whereas a gas expands to fill any space within which is confined. A fluid in motion consists of a very large number of submicroscopic particles moving in the general direction of the flow. The velocity of any particle is a vector quantity which varies from moment to moment. The path followed by a particle is called a *streamline*.

Laminar (or streamline) flow can be thought of as having parallel layers, each gliding over its neighbours. The resistance to flow caused by viscous forces is a tangential force opposing the motion of a layer with respect to its neighbour. Laminar flow occurs when sufficient viscous drag is present in the flow to damp down transflow disturbances. The fluid streamlines do not cross one another. The phenomenon of laminar flow is governed by *Newtons Law of viscosity* when momentum and viscous drag are exchanged from layer to layer. The law states that the viscous force per unit area opposing the motion of the fluid is proportional to the velocity gradient between the layers of fluid; i.e. shear force, τ, is given by

$$\tau = \mu \Delta u / \Delta y \tag{4.18}$$

where u is the velocity ($\mathrm{m\,s^{-1}}$) in the x-direction and y is the y direction. Fluids that exhibit this behaviour are known as *Newtonian fluids*.

The *kinematic viscosity*, or *momentum diffusivity*, $v\,(\mathrm{m^2\,s^{-1}})$ of a fluid is a derived transport property, namely

$$v = \mu / \rho \tag{4.19}$$

Turbulent flow takes place when inertia forces are sufficient to overcome viscous drag and so disturbances are able to grow. The fluid particles then move in irregular paths, resulting in confused streamlines. Momentum is transferred in eddies as the particles move from one plane to another. Rates of shear stress and momentum transfer in a direction perpendicular to the general direction of flow are much greater than those encountered in laminar flow.

The *eddy viscosity*, ε_{m}, $(\mathrm{m^2\,s^{-1}})$, is defined by

$$\tau = \rho(v + \varepsilon_{\mathrm{m}})\Delta u / \Delta y \tag{4.20}$$

where τ is the effective shear stress ($\mathrm{N\,m^{-2}}$) produced in turbulent flow.

The *Reynolds number*, *Re*, is a dimensionless quantity which characterizes the nature of the flow, being defined as the ratio of inertia to viscous forces present; i.e.

$$\begin{aligned} Re &= \text{inertia force/viscous force} \\ &= \rho u L / \mu \end{aligned} \tag{4.21}$$

where u is the fluid velocity ($\mathrm{m\,s^{-1}}$) and L is a characteristic length defining the system (i.e. the diameter of a pipe or the length along a plate). Regions in fluid flow where transition to turbulence occurs depends upon the magnitude of the local *Re*.

Boundary layers

A fluid flowing over a solid is retarded by viscous forces as the fluid 'sticks' to the surface. Adjacent layers are slowed down less and less at layers successively further from the solid wall. The viscous retarding force is given by

$$\text{force} = \tau A \tag{4.22}$$

where $A\,(\mathrm{m^2})$ is the area over which the shear force acts.

The region in the vicinity of the surface where the velocities are less than 99 per cent of the free-stream velocity, u_0, is termed the *hydrodynamic boundary layer* which has a thickness δ at any point x (Fig. 4.10).

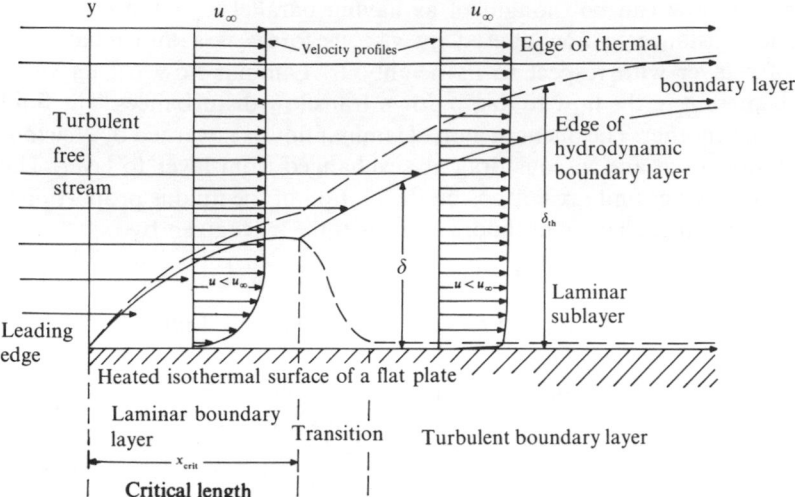

Figure 4.10 Growth of the hydrodynamic and thermal boundary layers for uniform fluid flow over a heated isothermal flat plate.

There are two forces acting on this boundary layer—inertia forces (acting from left to right in the figure) and viscous forces (acting from right to left in the figure). As the thickness δ increases, the ratio of the viscous forces to the inertia forces ($= 1/Re$) decreases. When the viscous forces are so relatively small that they no longer sufficiently damp down the turbulent tendencies in the free-stream flow, the boundary layer itself becomes unstable and turns turbulent. In the turbulent flow region, fluid particles adjacent to the plate are still arrested by the powerful local viscous forces at the surface of the plate, and so a *laminar sub-layer* persists very close to the surface.

The velocity in the laminar boundary layer varies parabolically from zero at the wall to 99 per cent of the free-stream velocity at the edge of the boundary layer.[3] Thus

$$u/u_0 = 1.5(y/\delta) - 0.5(y/\delta)^3 \qquad (4.23)$$

For the turbulent section, most of the momentum of the free stream is destroyed in the laminar sub-layer resulting in a characteristic velocity distribution according to[3]

$$u/u_0 = (y/\delta)^{1/7} \qquad (4.24)$$

The local Reynolds number at a position x from the leading edge of the plate is calculated from

$$Re_x = \rho u_0 L/\mu \qquad (4.21)$$

When $Re_x < 80\,000$, laminar flow prevails. When $Re_x > 5 \times 10^5$, turbulent flow dominates. The region between these boundaries is known as the *transition region*, where unstable flow occurs, hopping from laminar to turbulent and vice versa.

The thickness, δ, of the hydrodynamic boundary layer at any point, x, may be calculated from[2,3,4]

$$\delta/x = 4.64\,Re^{-0.5} \qquad \text{for laminar flow}$$
$$\delta/x = 0.376\,Re^{-0.2} \qquad \text{for turbulent flow}$$

Inside tubes and ducts

The boundary layer has zero thickness at the entrance to a tube but builds up through laminar and turbulent modes until the retarded annular region meets at the axis of the tube and the flow is said to be *fully developed*. If the annular boundary layer fills to the axis before transition to turbulence, the resulting boundary layer in the fully developed flow is laminar. If the annular boundary layer fills to the axis after transition to turbulence, the resulting boundary layer in the fully developed flow is turbulent (Fig. 4.11).

The velocity profile for fully developed laminar flow is parabolic, being given by[3]

$$u = 2U(1 - (r/r_{\mathrm{w}})^2) \tag{4.25}$$

where U is the average velocity in the tube, and r_{w} is the radius to the wall.

The velocity distribution for fully developed turbulent flow undergoes its major change in the laminar sub-layer.

For non-circular ducts, a *hydraulic diameter*, D_{h}, for use in the Reynolds number, is defined as

$$D_{\mathrm{h}} = 4 \times \frac{\text{flow cross-sectional area}}{\text{the 'wetted' perimeter of the duct}} \tag{4.26}$$

When $Re_x < 2100$, laminar flow prevails. When $Re_x > 10^5$, turbulent flow dominates.

Energy equations

Energy (W) may be contained in a flow system as

potential energy	mgh
kinetic energy	$mu^2/2$
pressure energy	mP/ρ
heat energy	$mc_{\mathrm{v}}T$

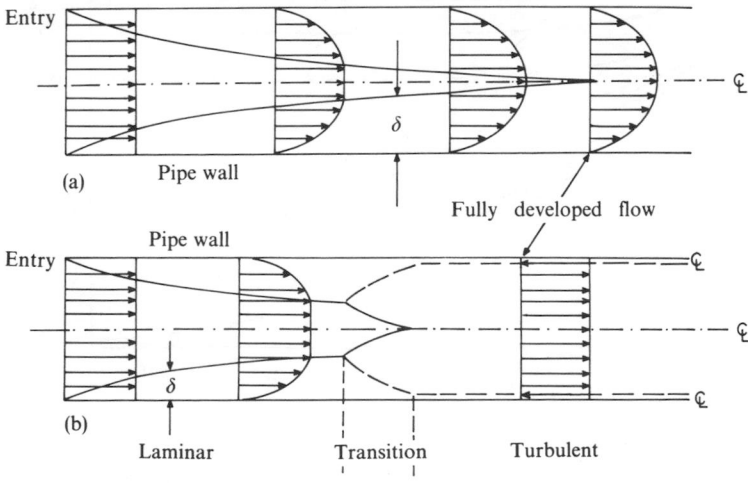

Figure 4.11 Boundary layer growth at the entrance of a pipe: (a) with a laminar, fully developed flow profile; (b) with a turbulent fully developed flow profile.

where m = mass flow, $\mathrm{kg\,s^{-1}}$

g = acceleration due to gravity ($= 9.81\,\mathrm{m\,s^{-2}}$)

u = mean velocity, $\mathrm{m\,s^{-1}}$

P = absolute pressure, $\mathrm{N\,m^{-2}}$

c_v = specific heat at constant volume, $\mathrm{J\,kg^{-1}\,K^{-1}}$

T = absolute temperature, K.

Note that 1 Newton is the SI unit of force which gives a mass of 1 kg an acceleration of $1\,\mathrm{m\,s^{-2}}$. It thus has the units of $\mathrm{kg\,m\,s^{-2}}$.

$1\,J = 1\,\mathrm{Nm}$ and so has the alternative units $\mathrm{kg\,m^2\,s^{-2}}$.

$1\,W = 1\,\mathrm{Nm\,s^{-1}}$ and so has the alternative units $\mathrm{kg\,m^2\,s^{-3}}$.

For flow in pipes, these forms of energy are expressed in terms of head (m) of fluid; i.e.

potential head $\qquad h$

kinetic head $\qquad u^2/2g$

pressure energy $\quad P/\rho g$

heat energy $\qquad c_v T/g$

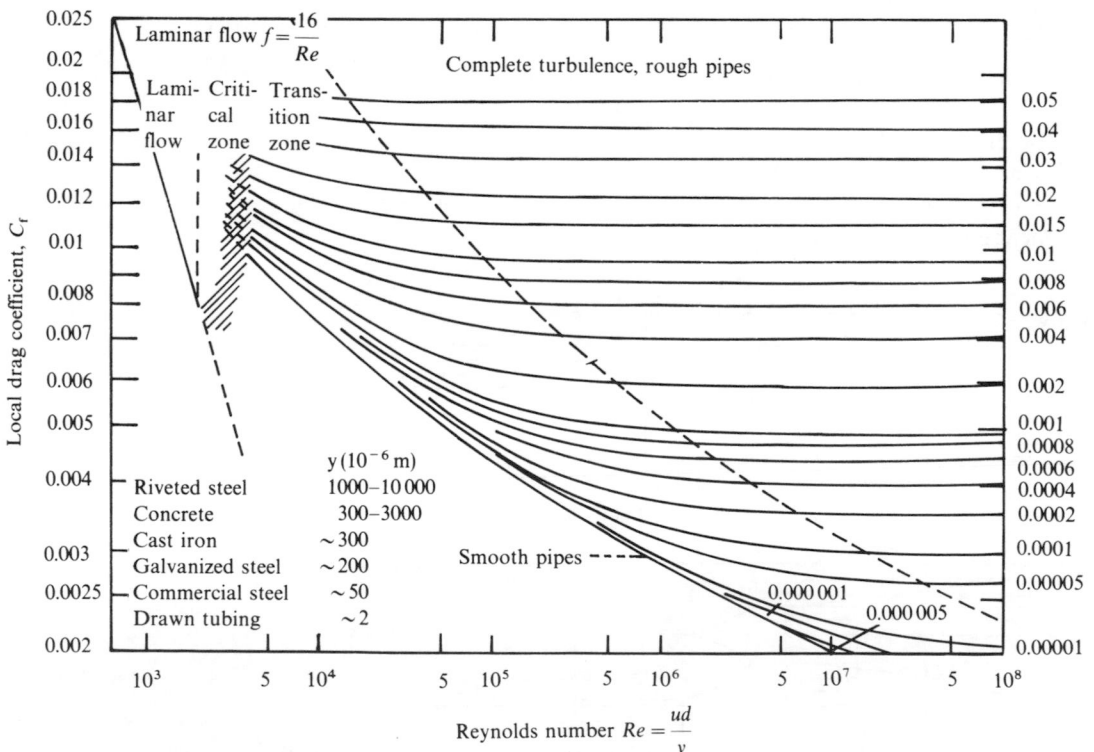

Figure 4.12 Friction factors versus Reynolds number for fluid flowing inside pipes (reproduced with the permission of the American Society of Mechanical Engineers, L. S. Moody, *ASME Trans.*, **66**, 1944).

The *skin friction* or *drag* coefficient, c_f, is defined as the ratio of the viscous shear stress at a surface to the kinetic energy per unit volume, $\rho u^2/2$.

The pressure drop through a pipe of length, L, and hydraulic diameter, D, is calculated from

$$\Delta P = f \frac{L}{D} \rho \frac{U^2}{2} \tag{4.27}$$

where f is the Fanning friction coefficient. By equating pressure drop and viscous drag offered by the pipe surface, it may be proved that $f = 4c_f$.

Figure 4.12 shows the *Moody Diagram*, which relates local drag coefficient, c_f to the Reynolds number for flow inside pipes. Note that the drag coefficient is not affected by pipe internal roughness in the laminar flow region, because the thickness of the laminar boundary layer far exceeds the height of the roughness asperities. In the turbulent flow region, because surface asperities can protrude through the thin laminar sub-layer, surface roughness can substantially affect the pressure drop to be overcome.

For fully developed laminar flow, the friction factor is given by

$$f = 64/Re \tag{4.28}$$

For fully developed turbulent flow (smooth pipe)

$$f = 0.184\, Re^{-0.2} \tag{4.29}$$

4.5 CONVECTIVE HEAT TRANSFER

Convection is a process of energy transport by the combined modes of heat conduction, energy storage and mixing motion in a fluid. Rates of convective heat transfer depend upon the fluid transport properties and the speed and nature of the flow.

If an external force causes the flow (a pump or blower, or the wind), the nature of convection is termed *forced*.

If the fluid moves as a result of internal *buoyancy* forces, arising from temperature and hence density differences, then *natural* or *free* convection is said to occur.

The rate of heat transfer by convection between surface and a fluid is calculated from

$$Q = hA\,\Delta T \tag{4.30}$$

where h is the *convective heat transfer coefficient* $(\mathrm{W\,m^{-2}\,K^{-1}})$ (Table 4.6) which depends upon the following parameters:

- Geometry of the system
- Flow velocity
- Mode of flow (laminar or turbulent)
- Fluid transport properties
- Temperature difference in free convection

While heat transfer coefficients are usually measured,[5,6] analytical derivations which lead to equations for local and overall mean heat transfer coefficients for flows over simple surfaces, such as flat plates and inside tubes are possible. These use the analogy between momentum

Table 4.6 Orders of magnitude of heat transfer coefficients, $W\,m^{-2}\,K^{-1}$

Air in free convection	5–25
Air in forced convection	25–250
Oil in forced convection	50–150
Water in forced convection	250–10 000

transfer in fluid flow and convective heat transfer.[3] They thus generally start from a study of the dynamics of the fluid flow. When fluid is in contact with a surface at a different temperature, heat is transferred from the surface by molecular conduction on a submicroscopic scale. In laminar flow, adjacent layers of fluid continue to transfer this heat by conduction as the layers slide over one another. In turbulent flow, the conduction mechanism is considerably enhanced by eddy motion, which carry lumps of fluid and their heat contents across the streamlines to mix and exchange heat with colder lumps in direct contact.

The Nusselt number, Nu

Heat transfer coefficients are usually expressed in dimensionless form, using the Nusselt number, defined as

$$Nu = hL/k \tag{4.31}$$

where $k\,(W\,m^{-1}\,K^{-1})$ is the thermal conductivity of the fluid and L is the significant dimension (m) as used in the Reynolds number.

The *thermal diffusivity*, $\alpha\,(m^2\,s^{-1})$, of a fluid is a combination of transport properties defined by

$$\alpha = k/\rho c_p \tag{4.32}$$

where $c_p\,(J\,kg^{-1}\,K^{-1})$ is the specific heat of the fluid at constant pressure.

The thermal diffusivity, which dictates how heat spreads, is analogous to the kinematic viscosity which dictates how momentum forces are retarded.

The Prandtl number, Pr

This is the ratio of the kinematic viscosity and the thermal diffusivity.

$$Pr = v/\alpha = \mu c_p/k \tag{4.33}$$

The value of the Prandtl number determines the relative rates of momentum and heat transfer.

Thick treacle-like substances (e.g. heavy fuel oils) have good viscous retardation (i.e. momentum destruction) properties in combination with bad heat transfer characteristics. They thus have high values of Prandtl numbers. Thin slippery substances (e.g. liquid metals) have bad viscous retardation (i.e. momentum destruction) properties in combination with good heat transfer characteristics. They thus have low values of Prandtl numbers.

The region in the vicinity of a heated or cooled surface where the temperatures differ by greater than 1 per cent of the free-stream temperature is known as the *thermal boundary layer* (see Fig. 4.10).

For fluids having $Pr \gg 1$, the extent of the hydrodynamic boundary layer exceeds that of the thermal boundary layer. For fluids having $Pr \ll 1$, the reverse occurs.

If the Prandtl number of a fluid is unity, the hydrodynamic and thermal boundary layers coincide and there is then a direct analogy between momentum and heat transport. Values for the Nusselt number, and hence the convective heat transfer coefficient, may be deduced[3] from local and overall mean drag coefficients—better heat transfer, more drag.

The combination $Re\,Pr$ is known as the *Peclet number, Pe*, and the ratio Nu/Pe is the *Stanton number, St*.

Forced convection

Figure 4.13 shows the flow over a flat plate. The heat transfer coefficient is largest at the leading edge, where the laminar boundary layer has zero thickness. As the laminar boundary layer thickens, the heat transfer coefficient decreases in value until the transition region is reached. As the flow turns turbulent, the value of the heat transfer coefficient jumps. Then as the turbulent boundary layer thickens, there is a small decrease in the heat transfer coefficient. As most of the resistance to heat flow occurs in the laminar sub-layer, the turbulent heat transfer coefficient does not vary much.

The Stanton number is given by

$$St = c_{\mathrm{f}}/2$$

and since $f = 0.184\,Re^{-0.2}$ for turbulent flow inside a pipe

$$St = f/8 = 0.023\,Re^{-0.2} \tag{4.34}$$

or, since $Pr = 1$,

$$Nu = 0.023\,Re^{0.8} \tag{4.35}$$

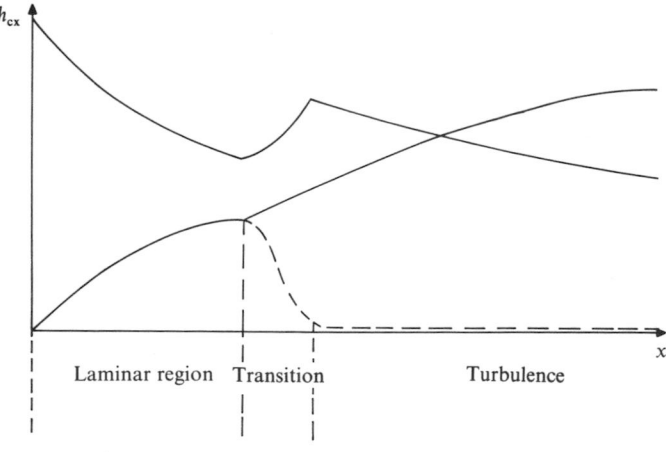

Figure 4.13 Variation of the magnitude of the local surface heat transfer coefficient for fluid flowing in forced convection over a flat surface.

A further dimensionless group is the Colburn j-factor[2] defined by

$$j = St\,Pr^{0.66} \tag{4.36}$$

Thus most data for heat transfer coefficients obtained analytically, empirically, or as empirically-corrected analytical derivations, appear as

Nu vs function (Re, Pr)
St vs function (Re, Pr), or
j vs function (Re, Pr)

The *eddy diffusivity*, ε_h is analogous to the eddy viscosity, ε_m, being defined by

$$Q/A = -\rho c_p (\alpha + \varepsilon_h)\,dT/dy \tag{4.37}$$

A 'turbulent' Prandtl, Pr_t number can be defined as

$$Pr_t = \frac{v + \varepsilon_m}{\alpha + \varepsilon_h} \tag{4.38}$$

Because $\varepsilon_m \gg v$ and $\varepsilon_h \gg \alpha$, $Pr_t = \varepsilon_m / \varepsilon_h$ approximately, and since an individual eddy carries both heat and momentum $Pr_t = 1$ approximately, making analogies possible to apply in the turbulent regions for a greater range of fluids.

Figure 4.14 shows the build up of laminar, transitional and turbulent boundary layers at the entrance section of a pipe. It is assumed that the flow entering the pipe is turbulent. At the leading lip of the section, the fluid near the wall is slowed by viscous shear to form the laminar boundary layer. This thickens as progressive layers of fluid are retarded by adjacent layers. The heat transfer reduces until transition to turbulence as for the flat plate. In the fully developed flow regime, the value of the heat transfer coefficient remains constant throughout the length of the pipe. If fully developed flow is established whilst the boundary layer is still laminar, the boundary layer will remain laminar through all the pipe, resulting in bad heat transfer.

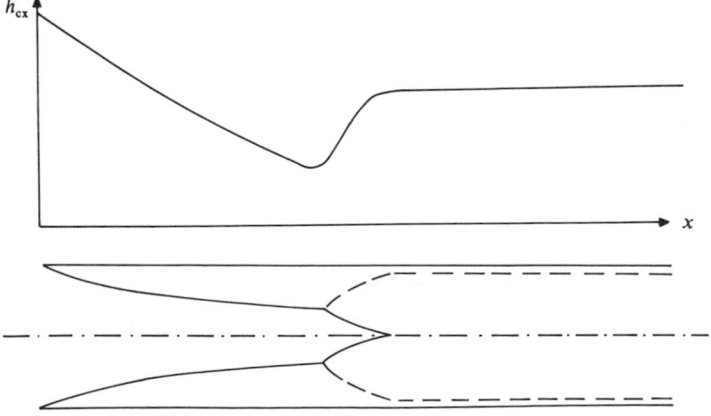

Figure 4.14 Variation of the magnitude of the local surface heat transfer coefficient at the entrance section of a pipe.

Table 4.7 Summary of useful Nusselt number relationships for forced convective flows

Configuration	Flow mode	Relationship and conditions
Flat plate	Laminar	$Nu_x = 0.331\, Re_x^{0.5}\, Pr^{0.33}$ $Nu_L = 0.662\, Re_x^{0.5}\, Pr^{0.33}$ } (†) Constant wall temperature $Re < 3 \times 10^5$ $Pr = 1$, approx. Heating starts at leading edge $T_f = 0.58(T_w - T_o) + T_o$
Flat plate	Turbulent	$Nu_x = 0.0292\, Re_x^{0.8}$ $Nu_L = 0.0366\, Re_x^{0.8}$ (‡) Constant wall temperature $Re > 5 \times 10^5$ Heating starts at leading edge $T_f = (T_w + T_o)/2$
Inside a pipe	Laminar	Developing profile $Nu_x = 1.077(Re\,Pr)_x^{0.33}(D/x)^{0.33}$ Constant wall temperature $100 < Re_x\,Pr\,D/x < 5000$ Heating starts at entrance Fully developed profile $Nu_D = 1.86(Re_D\,Pr\,D/L)^{0.33}(\mu_b/\mu_w)^{0.14}$ Constant wall temperature $Re_D < 2100$ Heating starts at entrance liquids only
Inside a pipe	Turbulent	Fully developed profile $Nu_D = 0.023\, Re_D^{0.8}\, Pr^{0.33}$ Constant wall temperature $Re > 10\,000$ Heating starts at leading section $0.5 < Pr < 100$ Entrance effects neglected (§) $T_f = (T_w + T_o)/2$

(†) The subscript x denotes local values at x, whilst the subscript L indicates the average value over the length L.

(‡) This equation assumes that the boundary layer is turbulent from the leading edge. A correction for the presence of the laminar leading portion yields

$$Nu_L = 0.036(Re_x^{0.8} - 23\,200)Pr^{0.33} \tag{4.39}$$

(§) For gases and liquids flowing in short circular tubes, $(2 < L/D < 60)$, the heat transfer coefficients obtained for fully-developed turbulent flow should be modified by factors according to

$$1 + (D/L)^{0.7} \text{ for } 2 < L/D < 20$$

or

$$1 + 6D/L \text{ for } L/D > 20$$

Table 4.7 provides a summary of Nusselt number relationships for convective heat transfers in common system configurations.

Forced convection in flow over cylindrical surfaces

Figures 4.15 shows the complex boundary layer formed when a turbulent (in the approach stream) fluid flows over a cylindrical surface. The heat transfer coefficient is large at the leading edge, or stagnation point, where the boundary layer is of zero thickness. As the laminar boundary layer thickness of the transition point A, the heat transfer coefficient decreases in value until transition to turbulence takes place. A further transition occurs at the *separation* point B, where the boundary layer separates from the surface, viscous forces being overcome by inertia forces as the fluid flow round the surface. A highly turbulent *vortex street* is formed on the leeward side in a wake, which results in high local heat transfer coefficients associated

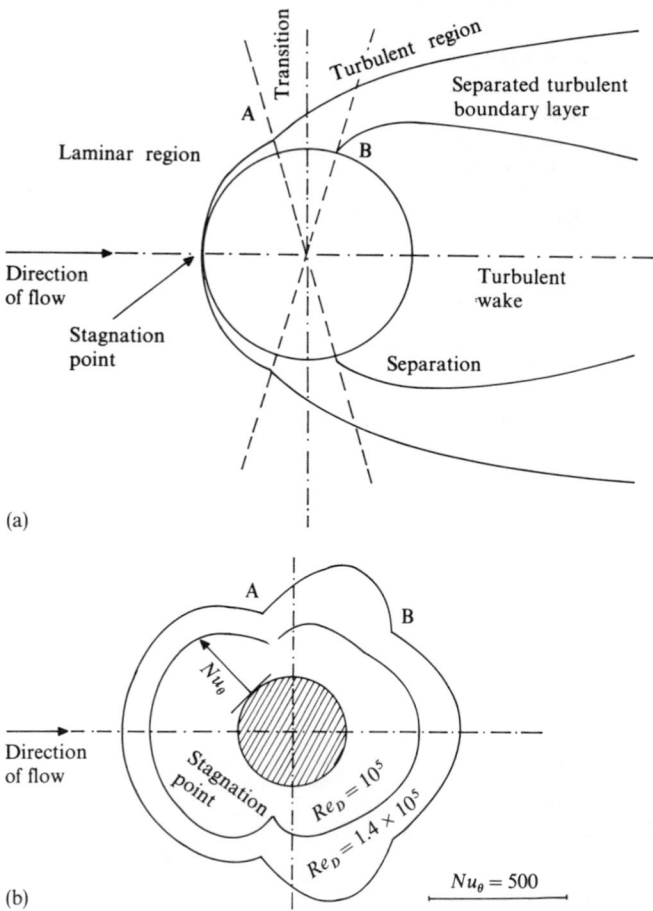

(a)

(b)

Figure 4.15 (a) Boundary layer growth; transition and separation for fluid flowing over a cylinder; (b) Approximate variation of the local convective Nusselt number for fluid flowing over a tube.

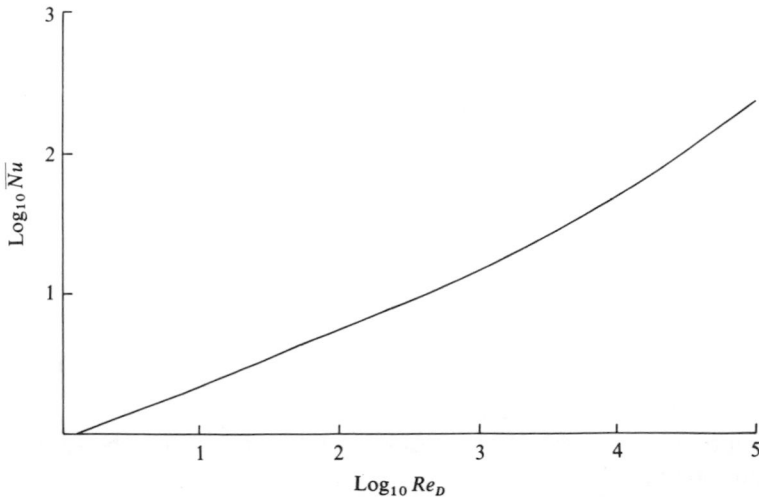

Figure 4.16 Average values for the Nusselt numbers for transverse fluid flow over cylinders.

with vortex shedding. The locations of transition and separation points and the local magnitudes of the heat transfer coefficients depend upon the entry Reynolds number based upon the velocity of the entry stream and the diameter of the tube.

In many instances, such as for flow over heat exchanger tubes, it is not necessary to know the variation of the surface heat transfer coefficient, only the overall mean value.

For air, empirical data can be correlated by

$$Nu = K(Re)^n$$

as in Fig. 4.16.[2] Values for K, n at a range of Reynolds numbers are shown at Table 4.8.

For liquids

$$Nu = (0.35 + 0.56\, Re_D^{0.5})Pr^{0.31} \qquad (4.40)$$

Table 4.8 Values of the constant, K, and the index n in the equation $Nu = K(Re)^n$ for heat transfer involved in air flowing over cylindrical surfaces

Re_D	K	n
0.4–4	0.891	0.330
4–40	0.821	0.385
40–4000	0.615	0.466
4000–40000	0.174	0.618
40000–400000	0.024	0.805

Tube bundles in cross-flow

Laminar flow Because local velocities vary, correlating equations are written in the form

$$Nu = 0.33K(G_{max}D/\mu)^n Pr^{0.33} \tag{4.41}$$

for $Re < 200$, where Nu is the average Nusselt number for a bank of 10 or more transverse rows, G_{max} is the mass flow rate per unit minimum free flow area, i.e. where the maximum local velocity occurs, and K and n are empirical constants depending upon the arrangement. The Reynolds number is also based upon this maximum velocity.

Turbulent flow For $Re > 6000$

$$Nu = 0.33K(G_{max}D/\mu)^{0.6} Pr^{0.33} \tag{4.42}$$

for staggered or in-line rows and for 10 or more transverse rows.

In many cases K is approximately unity. Corrections for entry effects must be applied when less than ten rows are present.[2,6]

Free convection

When an initially stationary fluid is in contact with a hotter (or colder) surface, fluid particles in contact with the wall are heated by molecular conduction. These particles heat adjacent

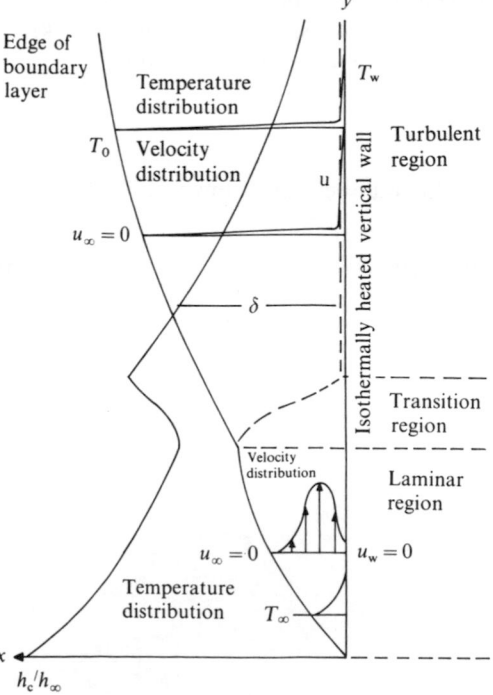

Figure 4.17 Velocity and temperature distributions and the variation of the natural convective heat transfer coefficient for fluid in contact with a vertical isothermally-heated wall.

Table 4.9 Summary of useful Nusselt number relationships for free convective flows

Configuration	Flow mode	Relationship and conditions
Vertical flat plate (†)	Laminar	$Nu_x = 0.51(0.95 + Pr)^{-0.25}Pr^{0.5}Gr_x^{0.25}$ (‡) $Nu_L = 0.68(0.95 + Pr)^{-0.25}Pr^{0.5}Gr_x^{0.25}$ (‡) Constant wall temperature $Gr\,Pr < 10^9$ $0.01 < Pr < 1000$ Heating starts at boundary layer start $T_f = (T_w + T_o)/2$
or alternatively		$Nu_x = 0.411(Gr\,Pr)^{0.25}$ $Nu_L = 0.548(Gr\,Pr)^{0.25}$
Horizontal plate	Laminar	Heated plate facing upwards Cooled plate facing downwards $Nu_L = 0.54(Gr\,Pr)^{0.25}$ $10^5 < Gr\,Pr < 10^7$ L is the dimension of a side $3 \times 10^5 < Gr < 3 \times 10^{10}$ Heated plate facing downwards Cooled plate facing upwards $Nu_L = 0.27(Gr\,Pr)^{0.25}$ $10^5 < Gr\,Pr < 2 \times 10^7$ L is the dimension of a side $3 \times 10^5 < Gr < 3 \times 10^{10}$
Vertical enclosed air spaces	Laminar	$Nu_\delta = 0.2(Gr\,Pr)^{0.25}/(L/\delta)^{0.11}$ $2 \times 10^4 < Gr\,Pr < 2 \times 10^3$ δ is the width of the airspace Below $Gr = 2000$, natural convection is suppressed and conduction through the fluid controls, then $h = k/\delta$
Horizontal enclosed air spaces	Laminar	$Nu_\delta = 0.21(Gr\,Pr)^{0.25}/(L/\delta)^{0.11}$ $10^4 < Gr\,Pr < 3.2 \times 10^5$ δ is the width of the airspace Below $Gr = 2000$, natural convection is suppressed and conduction through the fluid controls, then $h = k/\delta$
Horizontal cylinders	Laminar	$Nu_D = 0.53(Gr\,Pr)^{0.25}$
Vertical flat plate	Turbulent	$Nu_x = 0.0295Gr_x^{0.4}Pr^{0.47}(1 + 0.494Pr^{0.66})^{-0.4}$ $Nu_L = 0.0246Gr_L^{0.4}Pr^{0.47}(1 + 0.494Pr^{0.66})^{-0.4}$ Constant wall temperature $Gr\,Pr > 10^9$ Heating starts at boundary layer start Boundary layer assumed all turbulent $T_f = (T_w + T_o)/2$
Horizontal plate	Turbulent	Heating plate facing upwards Cooled plate facing downwards $Nu_L = 0.14(Gr\,Pr)^{0.33}$ $Gr > 3 \times 10^{10}$ L is the dimension of a side

Table 4.9 (*continued*)

Configuration	Flow mode	Relationship and conditions
		Heated plate facing downwards
		Cooled plate facing upwards
		Does not occur
Vertical enclosed air spaces	Turbulent	$Nu_\delta = 0.071(Gr\,Pr)^{0.33}/(L/\delta)^{0.11}$
		$2.1 \times 10^5 < Gr\,Pr < 1.1 \times 10^7$
		δ is the width of the airspace
Horizontal enclosed air spaces	Turbulent	$Nu\delta = 0.075(Gr\,Pr)^{0.33}$
		$3.2 \times 10^5 < Gr < 10^7$
		δ is the width of the airspace

[†] If the plate is inclined at an angle to the vertical, simply multiply the Grashof number of the cosine of the angle.
[‡] The subscript x denotes local values at x, whilst the subscript L indicates the average value over the length L.

particles. As the particles heat up, they become less dense, relative to the bulk of the fluid. Then buoyancy forces cause the particle to rise (or fall). In the steady-state, a boundary layer is established (Fig. 4.17), which exhibits laminar, transitional and turbulent regions (with laminar sub-layer) as for forced convective boundary layers. The fluid is always stationary at the wall, being held by viscous forces. It is also stationary in the quiescent bulk of the fluid outside the free convective boundary layer. The velocity distributions in both the laminar and turbulent regions thus exhibit maxima, where the ratio of buoyancy force to viscous force is largest. The temperature of the fluid, however, changes continuously from that at the wall to that of fluid far from the wall. The ratio of buoyancy forces to viscous forces is embodied in the *Grashof number*.

$$\text{Grashof number } Gr = \rho^2 g\beta\Delta TL^3/\mu^2 \qquad (4.43)$$

where g = the acceleration due to gravity ($= 9.81\,\mathrm{m\,s^{-2}}$), $\Delta T\,(K)$ is the temperature difference between that of the wall and that of the bulk of the fluid, and $L\,(\mathrm{m})$ is the distance along the wall. The Grashof number is used instead of the Reynolds number in free, or natural, convection.

Transition to turbulence within the boundary layer takes place when $Gr > 10^9$.

In cases of combined forced and free convection, for example, when the wind blows across a heated surface (such as the roof of a factory), the ratio, Gr/Re^2, indicates which mechanism dominates. When $Gr/Re^2 > 1$, then natural convection dominates. When $Gr/Re^2 < 1$, then forced convection dominates.

Energy management calculations should be made assuming free convection to ascertain worst cases of overheating, and assuming forced convection to estimate extreme heating loads. In general, for natural convective flows,

$$Nu = \text{function}(Gr, Pr) \qquad (4.44)$$

The nature of the function being dependent upon the configuration and correlated from analytical and/or empirical equations.

Table 4.9 lists some relationships between Nusselt number and Grashof number for natural convective heat transfer in commonly encountered configurations.

4.6 THE U-VALUE

In the majority of situations, heat is transferred from one fluid to another by a three-step process (Fig. 4.18):

- From the warmer fluid to the solid wall through the hot side boundary layer by convection

$$Q = h_{\mathrm{h}} A(T_1 - T_2) \tag{4.45}$$

- Through the wall by conduction

$$Q = (kA/\delta)(T_2 - T_3) \tag{4.46}$$

- From the solid wall to the colder fluid through the cold side boundary layer by convection

$$Q = h_{\mathrm{c}} A(T_3 - T_4) \tag{4.47}$$

This process is known as *recuperative heat transfer*. It is customary to employ an overall heat transfer coefficient, U, based upon the overall temperature difference between the two fluids, ΔT. The heat transferred between the fluids is given by

$$Q = U A \Delta T \tag{4.48}$$

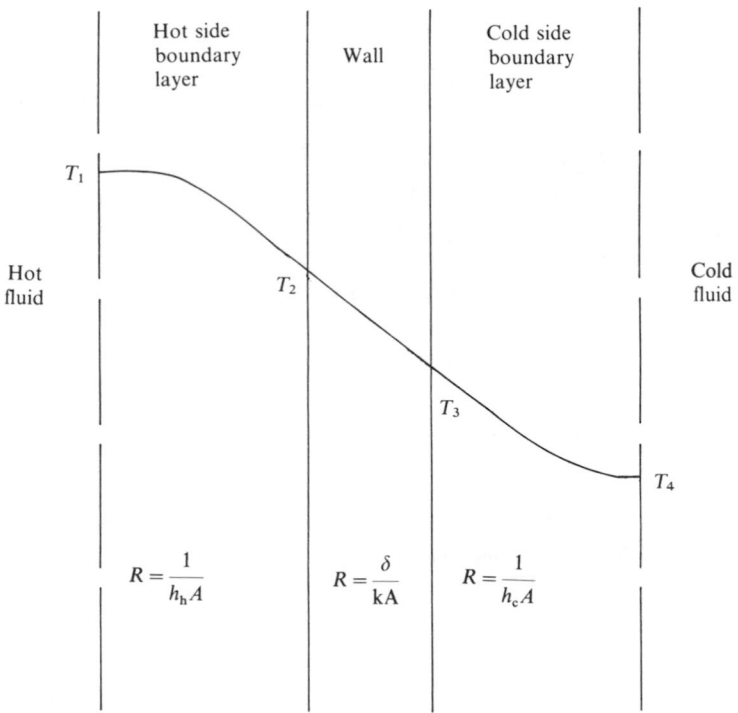

Figure 4.18 Components of the U-value—the overall heat transfer coefficient for recuperative heat transfer.

From the above equations

$$(T_1 - T_2) = Q/(h_h A)$$
$$(T_2 - T_3) = Q/(kA/\delta)$$
$$(T_3 - T_4) = Q/(h_c A)$$

Adding the three equations,

$$(T_1 - T_4) = Q/(h_h A) + Q/(kA/\delta) + Q/(h_c A)$$
$$= Q/UA$$

Therefore

$$1/UA = 1/(h_h A) + \delta/(kA) + 1/(h_c A)$$

or

$$1/U = 1/h_h + \delta/k + 1/h_c \tag{4.49}$$

4.7 HEAT TRANSFER ACROSS THE WALLS OF BUILDINGS

In the case of heat flows across the walls of buildings, the partial heat transfer coefficients either side of the wall are due to free convection. The process is dominated by the resistance of the wall itself.

For the 3 m-high brick wall detailed in Fig. 4.19,

Inside air temperature	T_i	25	°C
Density	ρ	1.185	$\mathrm{kg\,m^{-3}}$
Dynamic viscosity	μ	1.825×10^{-5}	$\mathrm{kg\,m^{-1}\,s^{-1}}$
Grashof number, Gr	$\dfrac{\rho^2 g\beta\Delta T L^3}{\mu^2}$	$\dfrac{1.185^2 9.81(1/298)\Delta T L^3}{(1.825 \times 10^{-5})^2}$	
		$0.139 \times 10^9\,\Delta T L^3$	
Assumed inside surface temperature	T_{is}	20	°C
Then Gr becomes		$0.695 \times 10^9\,L^3$	
At the transition point	$Gr =$	10^9	
and	$L =$	1.13	m

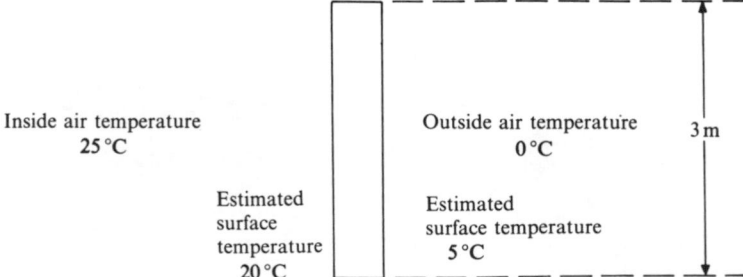

Inside air temperature
25 °C

Estimated surface temperature 20 °C

Outside air temperature
0 °C

Estimated surface temperature 5 °C

3 m

Figure 4.19 Section of a wall.

Thus the boundary layer will be laminar for the upper ≈ 1 m of the inner wall and turbulent below this.

Outside air temperature	T_o	0	°C
Density	ρ	1.286	kg m^{-3}
Dynamic viscosity	μ	1.7×10^{-5}	kg m^{-1} s^{-1}

Grashof number, Gr
$$\frac{\rho^2 g \beta \Delta T L^3}{\mu^2} \quad \frac{1.286^2 9.81(1/273)\Delta T L^3}{(1.7 \times 10^{-5})^2}$$
$$= 0.205 \times 10^9 \, \Delta T L^3$$

Assumed outside surface temperature	T_{os}	0	°C
Then Gr becomes		$1 \times 10^9 \, L^3$	
At the transition point	$Gr =$	10^9	
and	$L =$	1.0	m

Thus the boundary layer will be laminar for the lower 1 m (approx.) of the outer wall and turbulent above this.

The appropriate boundary layer equations are given in Fig. 4.20(a).

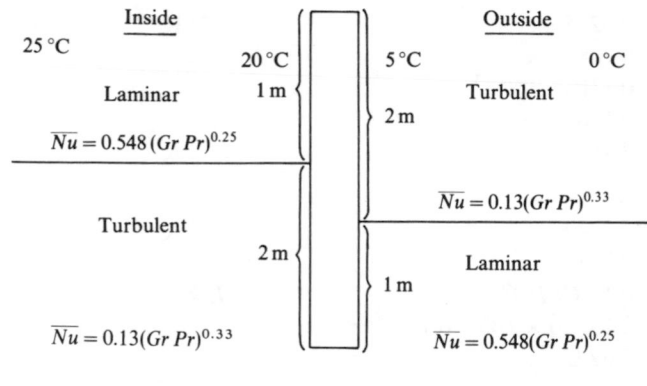

(a)

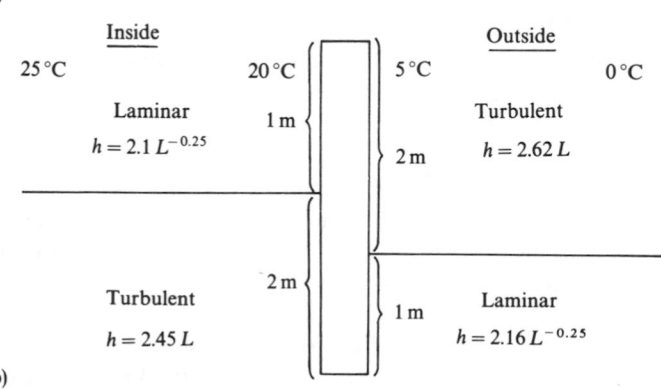

(b)

Figure 4.20 (a) Laminar and turbulent regions either side of the wall; (b) Variation of the mean surface heat transfer coefficient over the length, L, at the inside surface of the vertical wall.

The equation for the laminar natural convective boundary layers can be reduced as follows:

For the inside surface

Thermal conductivity	k	0.0255	$\text{W m}^{-1}\text{K}^{-1}$
Prandtl number	Pr	0.7085	
Nusselt number	$Nu =$	$0.548(Gr\,Pr)^{0.25}$	$= hL/k$
		$0.548(0.695 \times 10^9\,L^3 0.7085)^{0.25}$	
heat transfer coefficient	$h =$	$2.1\,L^{-0.25}$	$\text{W m}^{-2}\text{K}^{-1}$

For the outside surface

Thermal conductivity	k	0.0242	$\text{W m}^{-1}\text{K}^{-1}$
Prandtl number	Pr	0.713	
Nusselt number	$Nu =$	$0.548(Gr\,Pr)^{0.25}$	$= hL/k$
		$0.548(1.0 \times 10^9\,L^3 0.713)^{0.25}$	
heat transfer coefficient	$h =$	$2.16\,L^{-0.25}$	$\text{W m}^{-2}\text{K}^{-1}$

The equation for the turbulent natural convective boundary layers can be reduced as follows:

For the inside surface

Thermal conductivity	k	0.0255	$\text{W m}^{-1}\text{K}^{-1}$
Prandtl number	Pr	0.7085	
Nusselt number	$Nu =$	$0.13(Gr\,Pr)^{0.33}$	$= hL/k$
		$0.13(0.695 \times 10^9\,L^3 0.7085)^{0.33}$	
heat transfer coefficient	$h =$	$2.45\,L$	$\text{W m}^{-2}\text{K}^{-1}$

For the outside surface

Thermal conductivity	k	0.0242	$\text{W m}^{-1}\text{K}^{-1}$
Prandtl number	Pr	0.713	
Nusselt number	$Nu =$	$0.13(Gr\,Pr)^{0.33}$	$= hL/k$
		$0.13(1.0 \times 10^9\,L^3 0.713)^{0.33}$	
heat transfer coefficient	$h =$	$2.62\,L$	$\text{W m}^{-2}\text{K}^{-1}$

These simplified relationships (Fig. 4.20(b)) are plotted in Fig. 4.21(a).

The mean heat transfer coefficient for the laminar section of the *inside* surface is $3.03\,\text{W m}^{-2}\text{K}^{-1}$ over one metre at the top of the wall. The mean heat transfer coefficient for the turbulent section of the *inside* surface is $4.61\,\text{W m}^{-2}\text{K}^{-1}$ over the lower two metres of the wall.

The average *inside* surface heat transfer coefficient is therefore $4.08\,\text{W m}^{-2}\text{K}^{-1}$.

The mean heat transfer coefficient for the laminar section of the *outside* surface is $2.67\,\text{W m}^{-2}\text{K}^{-1}$ over 1 metre at the bottom of the wall. The mean heat transfer coefficient for the turbulent section of the *outside* surface is $5.37\,\text{W m}^{-2}\text{K}^{-1}$ over the upper 2 metres of the wall.

The average *outside* surface heat transfer coefficient is therefore $4.47\,\text{W m}^{-2}\text{K}^{-1}$.

The thermal conductivity for dry brick is $0.45\,\text{W m}^{-2}\text{K}^{-1}$, and so a thickness of 10 cm would present a thermal resistance of $(0.1/0.45) = 0.222\,\text{m}^2\,\text{K W}^{-1}$. The overall U-value

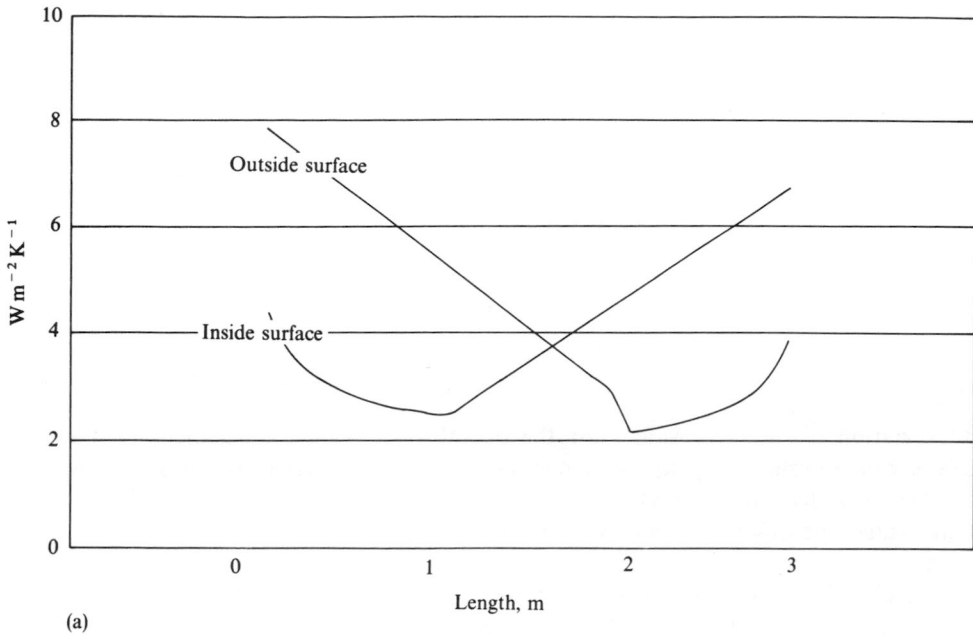

(a)

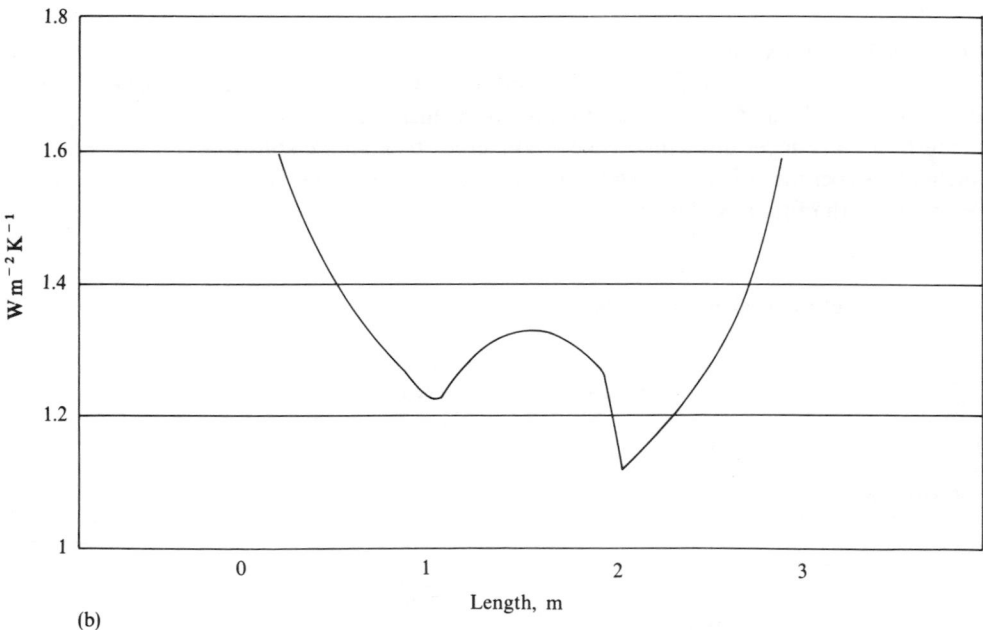

(b)

Figure 4.21 (a) Heat transfer coefficients at the inside and outside surfaces of a wall; (b) U-value variation along a wall.

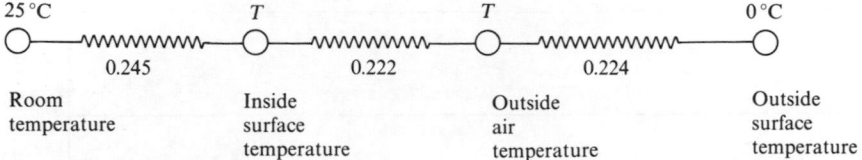

Figure 4.22 Thermal network across the wall.

$(\mathrm{W\,m^{-2}\,K^{-1}})$ is thus calculated from

$$1/U = 1/4.08 + 0.222 + 1/4.47$$
$$= 0.69 \text{ or } U = 1.45\,\mathrm{W\,m^{-2}\,K^{-1}}$$

The variation of the local values for the overall heat transfer coefficient is shown in Fig. 4.21(b), which perhaps explains why condensation often starts to occur at the top and bottom edges of walls and windows.

At this stage, the original estimates of the inside and outside surface temperatures must be verified.

The calculated thermal resistances of the inside air layer, the wall and the outside air layer are 0.245, 0.222 and 0.224 $\mathrm{m^2\,K\,W^{-1}}$ respectively and the overall thermal resistance is 0.69 $\mathrm{m^2\,K\,W^{-1}}$. Applying potentiometric ratios (Fig. 4.22),

$$\frac{25 - T_{is}}{25 - 0} = 0.235/0.69 = 0.34$$

Thus $T_{is} = 16.5\,°C$ and similarly $T_{os} = 8.15\,°C$.

The exact solution is an iterative process and must now be repeated (see Table 4.10) using the new values for T_{is} and T_{os} and correcting the values for the mean transport properties within the boundary layer until the process converges to a stable solution.

The heat loss per metre length of wall is calculated from $Q = UA(T_i - T_o) = 1.45 \times 3 \times 25 = 109.0\,\mathrm{W\,m^{-1}}$ for the first calculation.

Table 4.10 The iterative calculation process

	Summary data from the first calculation	Summary data from the first iteration
T_{is} (°C)	20	17
T_{os} (°C)	5	9
Extent of inner laminar region (m)	1.13	0.96
Extent of outer laminar region	1.0	0.83
U-value ($\mathrm{W\,m^{-2}\,K^{-1}}$)	1.45	1.66
New T_{is} (°C)	16.5	17.0
New T_{os} (°C)	8.15	7.5
etc.		

Effects of moisture

The dry brick wall considered had a thermal conductivity of $0.45\,\mathrm{W\,m^{-2}\,K^{-1}}$, resulting in a thermal resistance of $0.222\,\mathrm{m^2\,K\,W^{-1}}$ and an overall U-value of $1.45\,\mathrm{W\,m^{-2}\,K^{-1}}$. A 9 per cent moisture content increases the thermal conductivity to $0.8\,\mathrm{W\,m^{-2}\,K^{-1}}$, changing the thermal resistance to $0.125\,\mathrm{m^2\,K\,W^{-1}}$ and the overall U-value to $1.68\,\mathrm{W\,m^{-2}\,K^{-1}}$.

Effect of external wind speeds

A high wind speed (greater than $15\,\mathrm{m\,s^{-1}}$) can eliminate the external resistance. For the first calculation above, the U-value then becomes $2.14\,\mathrm{W\,m^{-2}\,K^{-1}}$, increasing the heat loss to $160.5\,\mathrm{W\,m^{-1}}$ ($+46$ per cent).

If the brickwork contains 9 per cent moisture, the U-value further increases to $2.7\,\mathrm{W\,m^{-2}\,K^{-1}}$. Because internal air movements are typically of the order of $0.3\,\mathrm{m\,s^{-1}}$, the internal mean heat transfer coefficient will not diverge appreciably from 4–$5\,\mathrm{W\,m^{-2}\,K^{-1}}$.

Table 4.11 lists measured U-values for some commonly-encountered structures.[7]

Effects of thermal insulants in cavity walls

The iterative calculation is even more complex for a cavity wall. Then the inner and outer surface temperatures within the cavity must also be estimated. Table 4.12 summarizes the results of one such computation and shows the effects of introducing various cavity wall insulations.

Table 4.11 Commun U-values, $\mathrm{W\,m^{-2}\,K^{-1}}$

Component	U-value
105 mm solid brickwork	3.3
220 mm solid brickwork	2.3
335 mm solid brickwork	1.7
260 mm cavity brickwork	0.8 – 1.5
Corrugated sheeting	5.3[†]
150 mm solid concrete	3.4
6 mm single glazing	5.6[†]
2 mm airspace double glazing	2.9
Uninsulated flat roof	≈ 3.0
Uninsulated pitched roof (35° slope)	≈ 1.5
Solid floor	≈ 0.3[‡]

[†] These values are heavily dependent upon the values of film-side heat transfer coefficients adopted. The solid U-values (i.e. neglecting boundary resistances) of 5 mm metal sheeting and 6 mm glazing are respectively $\approx 9000\,\mathrm{W\,m^{-2}\,K^{-1}}$ and $\approx 130\,\mathrm{W\,m^{-2}\,K^{-1}}$.

[‡] Referred to inside minus outside temperature difference.

Table 4.12 Effects of thermal insulants in cavity walls

	Surface resistances, $m^2 K W^{-1}$	Solid resistances, $m^2 K W^{-1}$	Cavity resistances, $m^2 K W^{-1}$	U-value, $W m^{-2} K^{-1}$
100 mm air-filled cavity with two layers of dry brickwork	0.20	0.44	0.20	1.19
As above but with convection in the air-gap completely suppressed	0.20	0.44	0.96	0.96
As above and with radiation shielding	0.20	0.44	4.00	0.22
100 mm glass-fibre-filled cavity with two layers of dry brickwork	0.20	0.44	2.70	0.30
100 mm expanded polystyrene-filled cavity with two layers of dry brickwork	0.20	0.44	2.80	0.29
100 mm mineral-wool-filled cavity with two layers of dry brickwork	0.20	0.44	3.00	0.27

Effects of intercavity and surface radiation heat transfers

The radiative heat transfers are of the order of the convective heat transfers (section 4.11) provided that the surfaces can *see* each other. Thus the resistance of an air-filled cavity reduces from 0.2 to 0.1 $m^2 K W^{-1}$ and the surface resistances might also be halved.

4.8 HEAT EXCHANGERS

In heat exchangers, the partial heat transfer coefficients in the U-value should be forced convective to ensure good heat transfer. This is achieved by increasing velocities, and hence pumping power required. It is usual to compromise on overall U-values and pumping power by arranging that Re is in the region of 50 000.

The values given in Table 4.13 are indicative of the order of magnitude of overall heat transfer coefficients to be expected for clean tubes. The values are commonly found in practice and may be predicted using partial heat transfer coefficients predicted from

$$Nu = 0.023 \, Re^{0.8} Pr^{0.33}$$

for turbulent flow. The thermal resistance of the solid tube material is insignificant in this case. The values apply for parallel or counterflow concentric tube heat exchangers with a tube wall thickness 3.2 mm and thermal conductivity (steel) 45.5 $W m^{-1} K^{-1}$. The tube resistance is $0.0032/45.5 = 0.000\,07 \, K W^{-1}$.

The nature of the reciprocal equation is such that the *overall heat transfer coefficient, U,* is dominated by the lowest heat transfer coefficient in the series chain. The overall heat transfer coefficient is always less than the smaller partial heat transfer coefficient.

Table 4.13 Heat transfer coefficients

Hot side to cold side	h_h $\mathrm{W\,m^{-2}K^{-1}}$	v_h $\mathrm{m\,s^{-1}}$	h_c $\mathrm{W\,m^{-2}K^{-1}}$	v_c $\mathrm{m\,s^{-1}}$	U $\mathrm{W\,m^{-2}K^{-1}}$
Air to air	28.4	4.58	28.4	4.58	14.2
Air to air	56.8	12.2	56.8	12.2	28.4
Air to water	28.4	4.58	4090	1.52	28.1
Air to water	56.8	12.2	4090	1.52	56.8
Water to water	4090	1.52	4090	1.52	1760
Oil to oil	510	1.52	510	1.52	256
Condensing steam to boiling water	11400	—	5678	—	2980
Condensing steam to water	11400	—	1136	—	960
Condensing steam to oil	11400	—	510	1.52	477
Oil to water	510	1.52	4090	1.52	440
Oil to air	510	1.52	28.4	4.58	25.8

Secondary surfaces

When one partial heat transfer coefficient is appreciably smaller than the other, as in heat exchange between air and water, the overall heat transfer coefficient is controlled by the partial, film-side heat transfer coefficient on the air side. It is thus little influenced by altering the value of the water-side coefficient by speeding the water flow. In such circumstances, it often becomes economical to fit more surface area (fins, or *gills*) to the surface in contact with the fluid (air) having the lower partial heat transfer coefficient. The resistance on the air side can then be increased by the ratio of the area of the extended surface to that of the bare surface.

The equation for the U-value then becomes

$$1/UA_c = 1/(h_h A_h) + \delta/(kA) + 1/(h_c A_c) \tag{4.50}$$

or

$$1/U = A_c/(h_h A_h) + \delta A_c/(kA) + 1/h_c. \tag{4.51}$$

Additional resistances due to layers of dirt should be added to this chain. The U-value is referred to the surface in contact with the smaller surface area, in this case the water side, by convention. This equation assumes that the temperature of the finning is equal to that of the surface to which the finning is attached. To take into account temperature drop between the bare surface and the root of the finning and through the fins to the fluid, a *fin effectiveness factor*, F (0–1), the value of which depends upon the arrangement, must be applied to the equation; i.e.

$$1/U = A_c/(h_h F A_h) + \delta A_c/(kA) + 1/h_c \tag{4.52}$$

Mean temperature difference

If the temperatures of the fluids, either side of the wall, were invariant, as in condensing steam to boiling water, the rate of heat transfer is then calculated from

$$Q = UA\Delta T \tag{4.48}$$

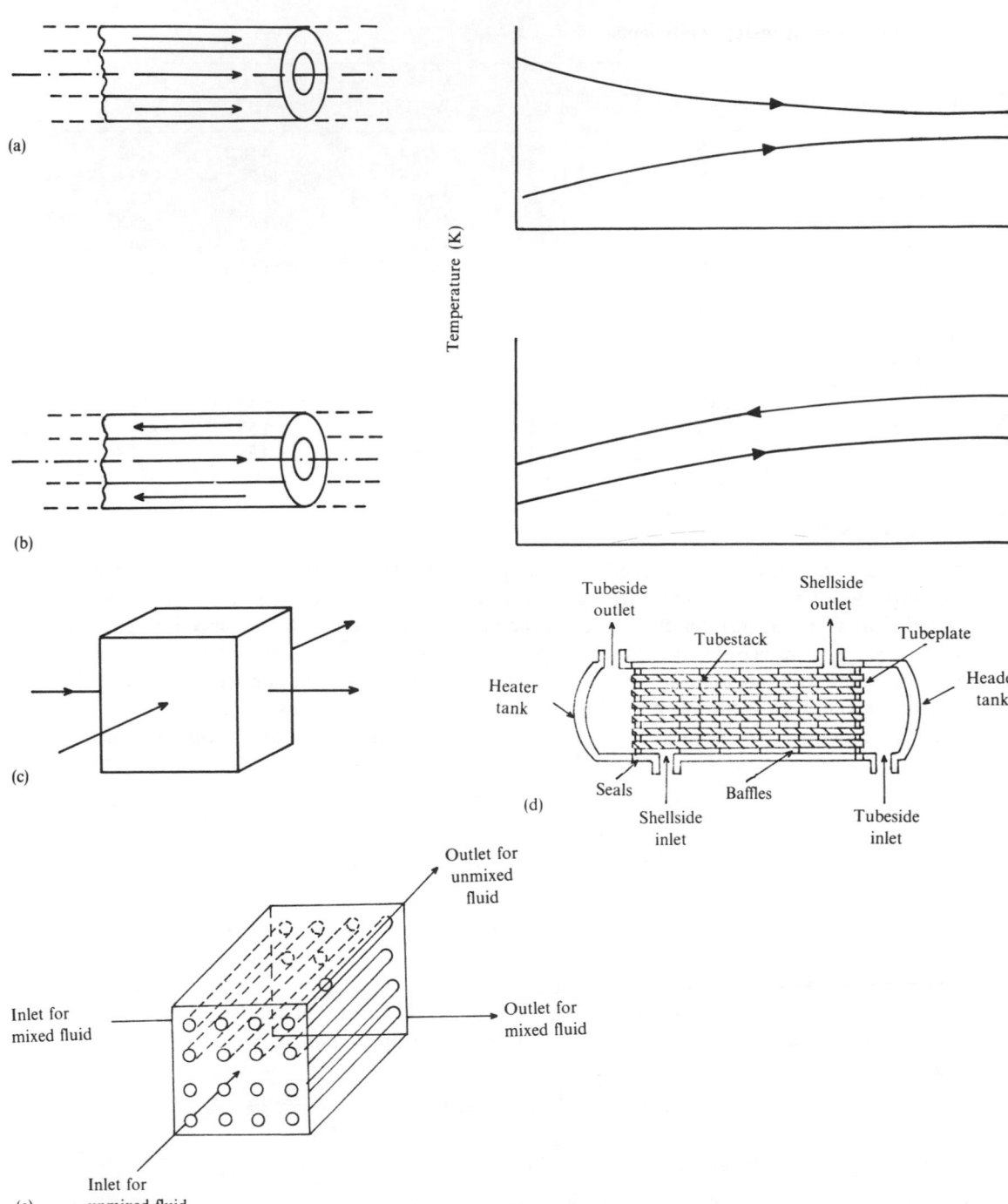

Figure 4.23 Recuperative heat exchangers: (a) in parallel flow; (b) in counterflow; (c) in crossflow; (d) as shell-in-tube; (e) in crossflow with one fluid mixed, the other unmixed.

Because the temperatures of the fluids vary throughout heat exchangers from sensibly heating to sensibly cooling fluids, a *logarithmic mean temperature difference* must be used. The value for this depends upon the operational type of the heat exchanger.

Parallel flow operation (Fig. 4.23(a)), in which the two fluids flow in the same direction along parallel paths separated by a wall.

$$\Delta T = \frac{(T_{hi} - T_{ci}) - (T_{ho} - T_{co})}{\log_e((T_{hi} - T_{ci})/(T_{ho} - T_{co}))} \qquad (4.53)$$

Contraflow, or counterflow operation (Fig. 4.23(b)), in which the two fluids flow in opposite directions along parallel paths separated by a wall. Contraflow operation is obviously more effective than parallel flow operation, where the exit hot side temperature can be less than the cold side inlet temperature, thus allowing a greater proportion of the heat in the hot side to be transferred to the cold side.

$$\Delta T = \frac{(T_{hi} - T_{co}) - (T_{ho} - T_{ci})}{\log_e((T_{hi} - T_{co})/(T_{ho} - T_{ci}))} \qquad (4.54)$$

where the subscripts denote: hi, hot side entry; ho, hot side exit; ci, cold side entry; co, cold side exit.

In both parallel and contraflow operation,

$$\Delta T = \frac{LTTD - STTD}{\log_e(LTTD/STTD)} \qquad (4.55)$$

where $LTTD$ is the largest terminal temperature difference (K), and $STTD$ is the smallest terminal temperature difference (K).

Practical recuperators are far more complex then simple tube-in-tubes. Figure 4.23(c) to (e) show schematic diagrams of shell-in-tube and crossflow arrangements. Fortunately, the mean temperature differences to be used in designing such heat exchangers are obtained empirically and correction factors, $F'(0-1)$ to be applied to the *contraflow* ΔT are available.[2]

When sizing a heat exchanger, all the terminal temperatures are usually known, so the heat transfer area required may be determined, from

heat lost by the hot side = heat transferred = heat gained by the cold side
$$(mc)_h \Delta T_h = U A \Delta T = (mc)_c \Delta T_c \qquad (4.56)$$

The overall heat transfer coefficient, U, is usually evaluated at a mean section halfway between the ends of the heat exchanger. Mass flow rates and areas to flow usually depend upon the permissible pressure drops compatible with structural limitations, pumping arrangements and economic factors.

Heat exchanger effectiveness

The logarithmic mean temperature difference approach can be adopted only when all four terminal temperatures are known. In off-design operation, the temperatures of the fluids leaving the heat exchanger are not known. It is known that the heat lost by the hot side equals that gained by the cold side, eliminating one unknown temperature. To obtain the other using the logarithmic mean temperature difference method involves a *trial and error* procedure.

The concept of *heat exchanger effectiveness*, ω avoids this complication. It is defined as

$$\omega = \frac{\text{actual heat transfer accomplished}}{\text{maximum heat transfer possible}} \qquad (4.57)$$

Figures 4.24 show that the maximum heat transfer possible would occur when the outlet hot-side temperature reached the cold-side outlet temperature in parallel flow operation, or the cold-side inlet temperature in contraflow operation. Both these equilibrations would need heat exchangers having infinite area. Then the actual heat transfer accomplished is given by

$$Q = (mc)_{hs}(T_{hsi} - T_{hso})$$
$$= (mc)_{cs}(T_{cso} - T_{csi})$$
$$= \omega(mc)_{min}(T_{hsi} - T_{csi}) \qquad (4.58)$$

Values for ω for various types of heat exchanger, usually empirically derived, are presented[2] in the form of graphs as functions of the *number of transfer units*, NTU of a heat exchanger,

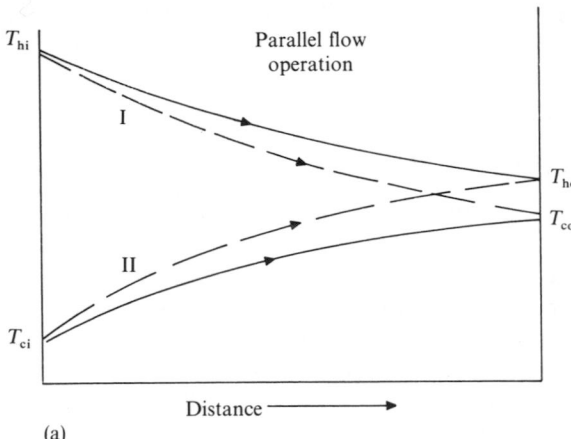

(a)

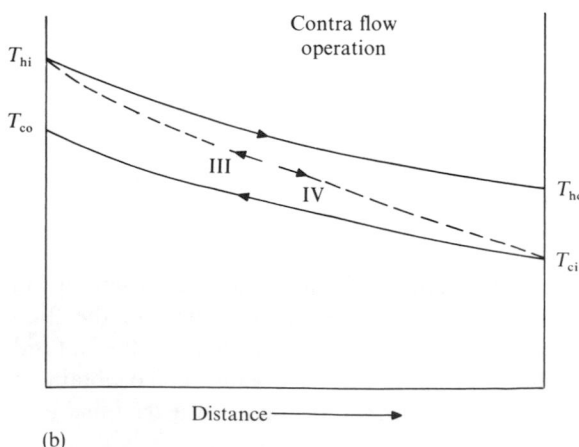

(b)

Figure 4.24 Temperature variations along: (a) parallel flow; (b) contraflow operations.

where

$$NTU = UA/(mc)_{\min} \qquad (4.59)$$

Heat exchanger selection

In order to demonstrate the heat exchanger selection process, a simplified building measuring $30\,\text{m} \times 40\,\text{m} \times 15\,\text{m}$ high, requiring one fresh air change per hour will be considered, detailed as follows:

Inside environmental temperature	T_i	25	°C
Outside environmental temperature	T_o	0	°C
Internal volume	V	18 000	m³
Density of air	ρ	1.24	kg m^{-3}
Specific heat of air	c	1 004	J kg^{-1}
Thermal conductivity of air	k	0.025	$\text{W m}^{-1}\,\text{K}^{-1}$
Dynamic viscosity of air	μ	1.76×10^{-5}	$\text{kg m}^{-1}\,\text{K}^{-1}$
Prandtl number of air	Pr	0.71	
Volume flow of air	vf	5	$\text{m}^3\,\text{s}^{-1}$
Mass flow of air	m	6.2	kg s^{-1}
$M_{\max} = M_{\min}$	mc	6 224.8	W K^{-1}
Rate of heat rejection	$mc\Delta T$	155	kW

Thus the incoming fresh air must be heated at the rate of 155 kW. Some of the heat rejected in the used exhaust air may be used to pre-heat the fresh air.

It is assumed that an air-to-air crossflow recuperator having the parameters detailed in Fig. 4.25 is available.

Details of recuperator

Volume	vr	0.125	m³
Number of air channels per side	n	25	
Area to flow per side	A	0.125	m²
Air velocities each side	$u = vf/A$	40	m s^{-1}

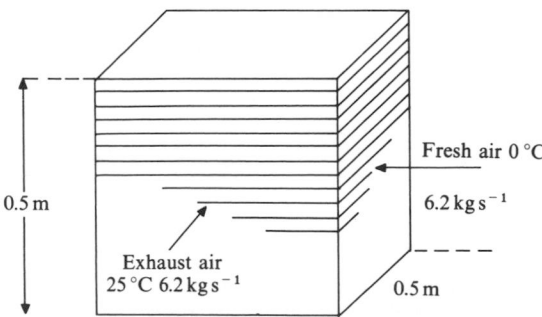

Figure 4.25 Details of a crossflow recuperator.

Wetted perimeters per side	L	25	m
Extended surface factor (corrugated surfaces)	F	5	
Heat transfer area	$ah = (n-1) \times 2 \times 0.25 \times F$	60	m^2
Hydraulic diameter per side	$D = 4A/L$	0.02	m

Overall U-value

Reynolds number	Re	$\rho u D/\mu$	56 686 (turbulent)	
Nusselt number	Nu	$hD/k = 0.023\, Re^{0.8} Pr^{0.33}$	130	
Partial heat transfer coefficient	h		163	W m^{-2} K^{-1}
Overall heat transfer coefficient	U	$1/(2/h)$	81	W m^{-2} K^{-1}
Number of transfer units	NTU	$ahU/M_{\min}$	0.78	

Referring to the effectiveness chart (Fig. 4.26) for a crossflow heat exchanger yields, for

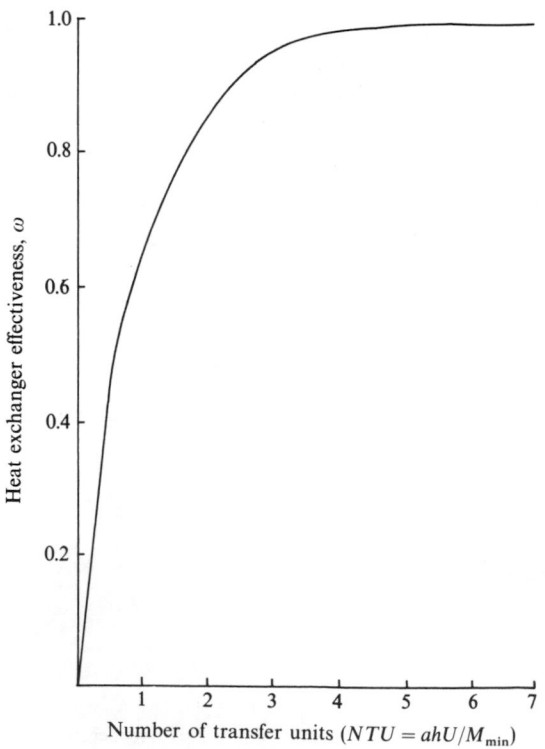

Number of transfer units ($NTU = ahU/M_{\min}$)

Figure 4.26 Heat exchanger effectiveness versus number of transfer units for the crossflow recuperator.

$NTU = 0.78$ and $M_{max}/M_{min} = 1$, a heat exchanger effectiveness, ω, of 0.57. Thus the heat transfer accomplished is

$$Q = \omega M_{min}(T_i - T_o) = 88\,kW$$

Power requirements

$$W = \Delta p A u = \frac{fL\rho u^2 Au}{2D} \tag{4.60}$$

From Fig. 4.12, for smooth surfaces, $f = 0.0185$ and so the pumping requirement, $W = 2.3\,kW$ per side.

Thus the incoming air will be pre-heated by 14.25 °C, using this heat exchanger. An optimally designed heat exchanger could reclaim more reject heat.

Regenerative heat exchangers

In a *regenerative heat exchanger*, hot and cold fluids pass alternatively over the same heat transfer surface in rotary or valved packed bed systems containing solid pellets, for heat recovery from flue gases, or metal matrices, sometimes plastic coated, for low temperature applications.

During one half of the cycle, heat from the hot fluid is transferred by direct contact to the solid bed. This heat is given up to the cold fluid when it comes into contact with the heated solids.

The design theory for regenerators is complex, involving the thermal mass flows in the fluids and the thermal mass of the intermediate heat carrier. Fortunately, empirically-derived values for heat exchanger effectivenesses are tabulated[8] and plotted as functions of matrix capacity rate, M_{rot}, the thermal capacities (*mc*) of the fluids and a modified number of transfer units, NTU, where, $M_{rot} = $ (mass rate) × (specific heat of the solid) (= r.p.m. × matrix mass × *c*) for a rotary system.

$$NTU = (1/(hA)_{cs} + 1/(hA)_{hs})^{-1}/M_{min} \tag{4.61}$$

Once the effectiveness has been established, the design procedure is exactly as for recuperators.

An energy management checklist for reject heat recovery is provided in Appendix 2.

4.9 ANALOGOUS EQUILIBRIUM PROCESSES

All energy forms 'flow' from a high-grade state to a lower-grade state. Thus the flows of electricity, heat, liquids, gases and solids are all essentially manifestations of the same equilibrium process. Tables 4.14 and 4.15 summarize some of these equilibrium processes.

The particular analogous equilibrium processes relating heat, mass and momentum transport are summarized in Table 4.16.

With simultaneous diffusion of heat and momentum (convection), Prandtl number, $Pr = v/\alpha = $ viscous diffusion/thermal diffusion.

Table 4.14 Analogous equilibrium processes

Form of energy	Equilibrium	Process	Stored energy
Electricity	$I = \Delta V/R$	Ohms law	Capacitance C
Heat	$Q = \Delta T/R$	Fourier's law	$mc_p\Delta T$
Pipe flow	$Q = \Delta P/R$	Fluid flow equations	$mgh, mP/\rho$
Mass flow	$G = \Delta c/R$	Fick's law	mP_v/ρ
Fluid stress	$\tau = \Delta u/R$	Newton's law of viscosity	mP/ρ
Solid stress	$\tau = \Delta u/R$	Stress equations	$\tau/2$
Chemical energy	Change of state	= Chemical potential resistance to reaction	Fuels, etc.
Nuclear energy	Radioactive decay	= Radioactive potential resistance to reaction	Radioactive materials

Table 4.15 Equilibrium processes

Equilibrium process	Useful energy	→ Entropy	Stores
Flow	Potential Difference	Resistance (energy dissipater)	
Electric current	Voltage	Resistor	Capacitance
Heat flow	Temperature	Thermal resistance	Thermal capacity
Fluid flow	Pressure	Flow resistance	Reservoir
Mass flow	Concentration Pressure Height Partial pressure Relative humidity	Mass flow resistance	Concentration
Fluid stress	Velocity	Resistance to shear	
Solid stress	Movement	Resistance to deformation	Springs
Chemical energy	Chemical potential	Resistance to dissipation	Fuels
Nuclear energy	Nuclear potential	Resistance to radiation decay	Fissionable materials

With simultaneous diffusion of heat and mass (e.g. psychrometry), Lewis number, $Le = \alpha/k^* =$ thermal diffusion/mass diffusion.

With simultaneous diffusion of momentum and mass without heat transfer (e.g. two-phase flow at the same temperature), Schmidt number, $Sc = v/k^* =$ momentum diffusion/mass diffusion.

Analogies apply when Pr or Le or $Sc =$ unity. Then if $Nu = 0.023\,Re^{0.8}\,Pr^{0.33}$ and $Sc \approx 1.0$, $Sh = 0.023\,Re^{0.8}\,Sc^{0.33}$, etc.

4.10 PSYCHROMETRY

Atmospheric air always contains water vapour, usually up to 0.03 kg of water vapour for each kg of dry air. A typical, thermally comfortable, room, 10 m × 10 m × 2 m, contains 280 kg

Table 4.16 Heat, mass and momentum transport

	Heat transfer	Mass transfer	Momentum transfer
Forcing function	Temperature gradient $\Delta T(\mathrm{K})$	Concentration gradient $\Delta c\,(\mathrm{kg\,m^{-3}})$	Pressure gradient $\Delta P\,(\mathrm{N\,m^{-2}})$
Equilibrium process	Conduction Convection Radiation	Molecular diffusion Convection	Momentum transfer by viscous forces
What moves	Heat, $Q\,(\mathrm{W})$	Mass, $m\,(\mathrm{kg\,s^{-1}})$	Momentum, strains $(\mathrm{N\,m^{-2}})$
Analogies	Fourier equation (1-D form of Laplace equation) $Q = kA\,\mathrm{d}T/\mathrm{d}x$ $Q = \dfrac{k}{\rho c_{\mathrm{p}}}\mathrm{d}(\rho c_{\mathrm{p}}T)/\mathrm{d}x$ $Q = \alpha\,\mathrm{d}(\rho c_{\mathrm{p}}T)/\mathrm{d}y$ $\alpha = $ thermal diffusivity Rate of flow	Fick's law $m = k^*A\,\mathrm{d}c/\mathrm{d}x$ $k^* = $ mass diffusivity (permeability) $= $ diffusivity $\times$ gradient of forcing function	Newtons law $\tau = \mu\,\mathrm{d}u/\mathrm{d}y$ $\tau = \dfrac{\mu}{\rho}\mathrm{d}(\rho u)/\mathrm{d}y$ $\tau = v\,\mathrm{d}(\rho u)/\mathrm{d}y$ $v = $ momentum diffusivity (kinematic viscosity)
2-D differential equations	$u\dfrac{\mathrm{d}T}{\mathrm{d}x} + v\dfrac{\mathrm{d}T}{\mathrm{d}y} = \alpha\dfrac{\mathrm{d}^2 T}{\mathrm{d}x^2}$ Energy equation	$u\dfrac{\mathrm{d}c}{\mathrm{d}x} + v\dfrac{\mathrm{d}c}{\mathrm{d}y} = \alpha\dfrac{\mathrm{d}^2 c}{\mathrm{d}x^2}$ Diffusion equation	$u\dfrac{\mathrm{d}u}{\mathrm{d}x} + v\dfrac{\mathrm{d}u}{\mathrm{d}y} = \alpha\dfrac{\mathrm{d}^2 u}{\mathrm{d}x^2}$ Momentum equation
Dimensionless numbers	Nusselt $Nu = hL/k$ $Nu = f(Re, Pr)$ $h = $ heat transfer Prandtl number, $Pr = v/\alpha$	Sherwood $Sh = h^*L/k^*$ $Sh = f(Re, Sc)$ $h^* = $ mass transfer Schmidt number, $Sc = \mu/\rho k^*$	Reynolds $Re = \rho uL/\mu$

of dry air and about 2 kg of water, which will condense on cold surfaces. This is a small fraction of the total mass and, because the transport properties of water vapour are of the same order as those for dry air, little error results in heat transfer calculations if the presence of water is neglected. If, however, a change in phase of the water occurs during a cooling or heating process, the concomitant latent heat transfers can contribute substantially to the heat exchange process. If, for example, 1 kg of *saturated* air at 40 °C, containing 0.03 kg of water vapour is cooled to 10 °C, 0.022 kg of water would condense, releasing 55 kJ of latent heat, compared with 30 kJ lost in sensibly cooling the dry air simultaneously.

Psychrometry applies the ideal gas laws separately to each of the two substances (water and air) in an atmospheric mixture to produce tables and charts of humidities, enthalpies and densities to aid air conditioning design calculations. The ideal gas laws are:

Dalton's Law of Partial Pressures, which states that, if a mixture of two gases occupy the same volume at a given temperature, the total pressure exerted by the mixture is equal to the sum of the pressures exerted by each constituent gas.

Table 4.17 Properties at room temperatures and pressures

Property	Dry air	Water vapour
Molecular weight, M, kg	28.97	18.02
Gas constant, R, $J\,kg^{-1}\,K^{-1}$	287	462
Specific heat, $J\,kg^{-1}\,K^{-1}$		
at constant pressure, c_p	1005	4210
at constant volume, c_v	718	1810
Adiabatic index, γ	1.4	2.33
Density, ρ, $kg\,m^{-3}$	1.22	0.013

The Characteristic Gas Law, which states

$$PV = mRT$$

Thus the pressure ($N\,m^{-2}$) exerted by a gas having characteristic gas constant, R ($J\,kg^{-1}\,K^{-1}$) at temperature T (K) and in a volume, V (m³), is given by

$$P = mRT/V$$

Some properties of dry air and water vapour at $T = 288.5\,K$, $P = 10^5\,N\,m^{-2}$ are given in Table 4.17. Properties at other conditions are given in the Databank provided in Appendix 1.

The density of ice is $913\,kg\,m^{-3}$, and the latent heat of melting/solidification, $333\,500\,J\,kg^{-1}$. The density of liquid water is $1000\,kg\,m^{-3}$ and the latent heat of vapourization/condensation, $250\,0000\,J\,kg^{-1}$.

Absolute values for enthalpy, h ($J\,kg^{-1}$) are related to a datum at 273.15 K.

The volume of 1 mol of the perfect gases (at $101\,325\,N\,m^{-2}$, the standard atmospheric pressure, and 0 °C) is $22.4136\,m^3$.

Pure liquid water boils at 100 °C under an atmospheric pressure of $101\,325\,N\,m^{-2}$. If the total pressure is reduced, boiling occurs at a lower temperature. The relationship between pressure and temperature at which the change of phase occurs is known as the *saturation curve*. According to the National Engineering Laboratory Steam Tables, 1964, the following equation may be used for the vapour pressure ($N\,m^{-2}$) of steam over water up to 100 °C:

$$\log(P) = 28.590\,51 - 8.2\log(T) + 0.002\,480\,4T - 3142.31/T \tag{4.62}$$

where T is degrees Kelvin. This function is plotted in Fig. 4.27.

Liquid water at a state-point located on the saturation curve will evaporate with the application of heat. Thus a psychrometric mixture of water and air at a given temperature can hold no more water vapour than the corresponding saturation pressure will permit; i.e.

$$\text{if } T = 30\,°C, \quad P_{sat} = 4240\,N\,m^{-2}$$

The maximum mass of water vapour that the air can hold is known as the *specific humidity*, or *moisture content*, γ:

$$\gamma = \text{mass of water/mass of dry air}$$

$$= m_w/m_a$$

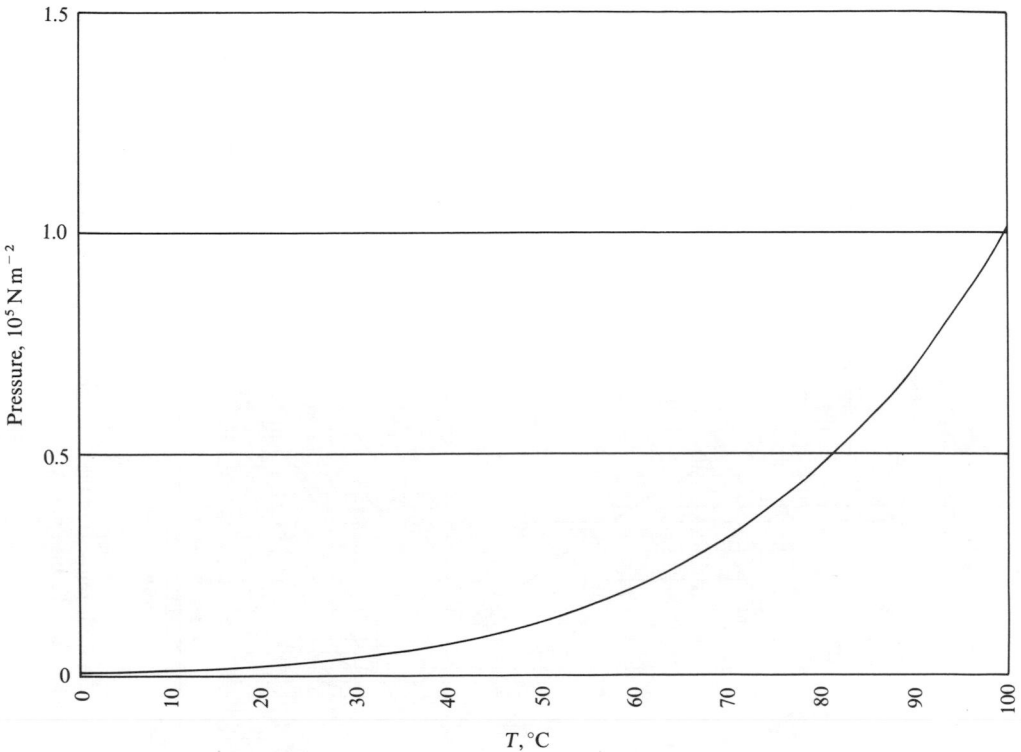

Figure 4.27 Saturation curve for steam.

From the characteristic gas law, $PV = mRT$ and so

$$m_w/m_a = (PV/RT)_w/(PV/RT)_a$$

If the total pressure of the mixture is $101\,325\,\mathrm{N\,m^{-2}}$, then the partial pressure of the air is $101\,325 - 4240 = 97\,985\,\mathrm{N\,m^{-2}}$.

Cancelling the common temperature and volume occupied by each constituent (Dalton's Law):

$$\gamma = m_w/m_a = 4340R_a/97985R_w$$
$$\gamma = (4340 \times 287)/(97985 \times 462) = 0.0274\,\text{kg of moisture per kg of dry air}$$

If more than this amount of water is present, the excess will not evaporate. If however, less than this amount of water is present, all will evaporate providing that enough heat is available to fuel the process. The resulting mixture will then have a *relative humidity* (i.e. relative to saturated conditions), ϕ, defined by

$$\phi = P_v/P_{sat} \tag{4.63}$$

where P_v $(\mathrm{N\,m^{-2}})$ is the vapour pressure of the water.

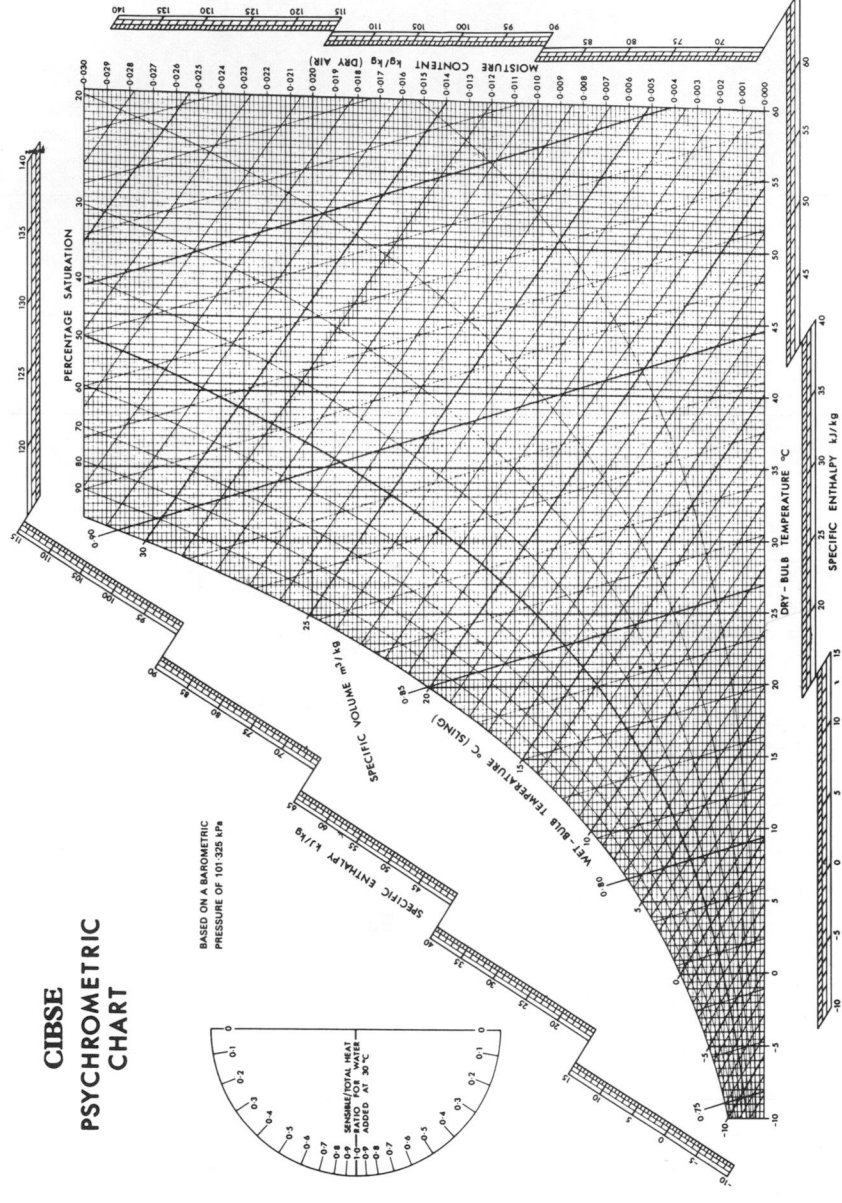

Figure 4.28 The CIBSE psychrometric chart (reproduced with the permission of the Chartered Institution of Building Services. Pads of charts size A3 suitable for permanent records are available from the Institution).

From the characteristic gas equation,

$$\phi = P_\text{v}/P_\text{sat} = (mR)_\text{v}/(mR)_\text{v, sat} \qquad (4.64)$$

i.e. ϕ depends on the mass of water present.

When a psychrometric mixture is cooled at constant pressure and specific humidity, its relative humidity rises until it reaches 100 per cent at the saturation curve. Further cooling causes droplets to condense from the air/water mixture. The temperature at which this occurs is known as the *dew-point* temperature—the saturation temperature corresponding to the mixture's specific humidity, i.e. when

$$\gamma = m_\text{w}/m_\text{a} = P_\text{sat} R_\text{a}/P_\text{a} R_\text{v} \qquad (4.65)$$

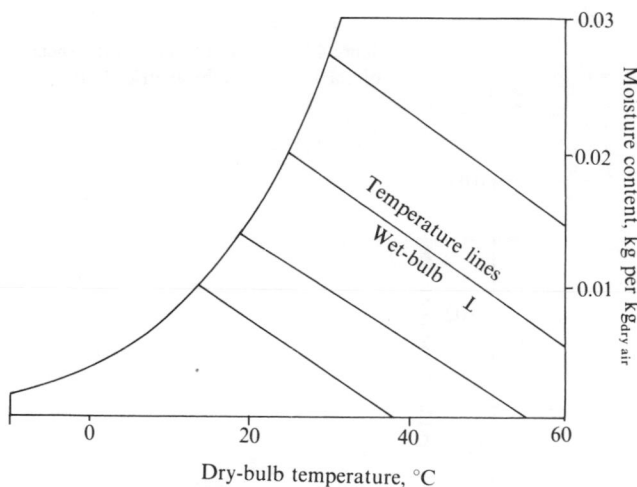

Figure 4.29 Wet-bulb temperature lines on the psychrometric chart.

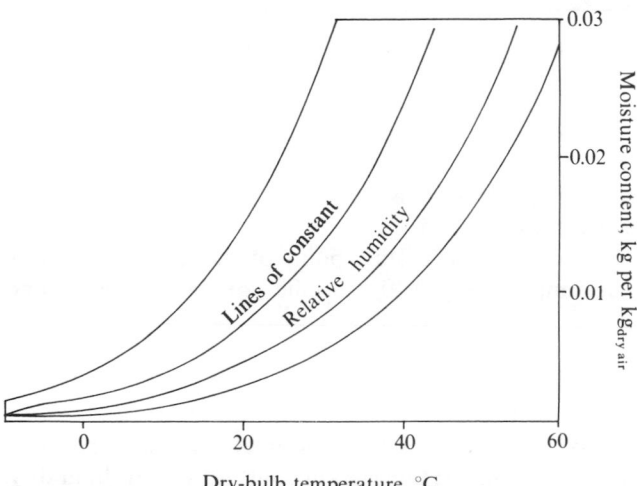

Figure 4.30 Lines of constant relative humidity on the psychrometric chart.

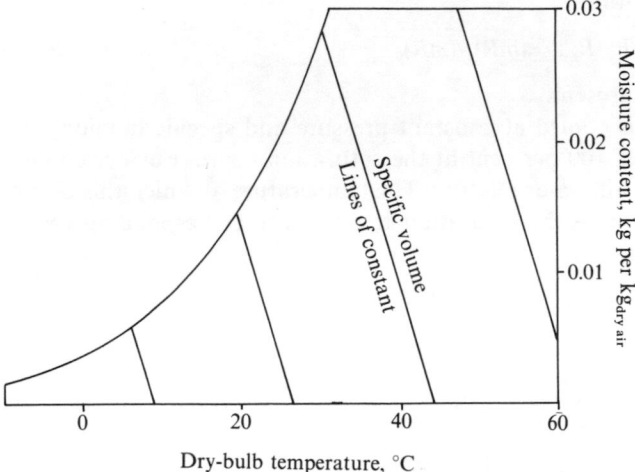

Figure 4.31 Lines of constant specific volume on the psychrometric chart.

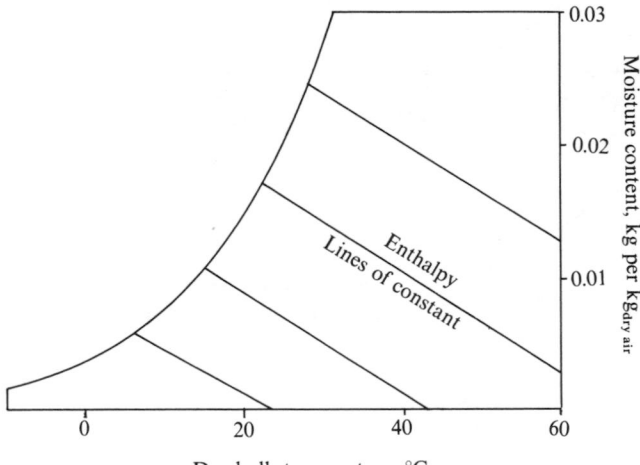

Figure 4.32 Lines of constant enthalpy on the psychrometric chart.

The *psychrometric chart* (Fig. 4.28) has *dry-bulb temperature*, T_{db} and specific humidity, γ, as coordinates. The *wet-bulb temperature*, T_{wb} (Fig. 4.29) is the value indicated by an ordinary thermometer, having a wettened wick wrapped around the bulb. The evaporation of water extracts sensible heat from the bulb and cools it. The state-point of the air may thus be located on the chart, and the relative humidity (Fig. 4.30), specific volume (Fig. 4.31) and enthalpy (Fig. 4.32) may be read off.

Psychrometric processes

Sensible heating or cooling When a psychrometric mixture, originally at state-point A (Fig. 4.33) is heated to state-point B, its dry bulb temperature rises at constant specific humidity, its relative humidity decreases and the air becomes 'drier', i.e. the partial pressure of water

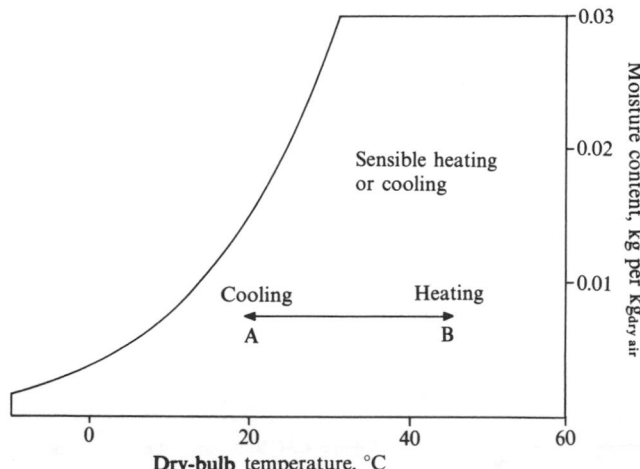

Figure 4.33 Sensible heating and cooling illustrated on the psychrometric chart.

vapour in the air decreases, allowing a greater rate of evaporation of liquid water from wet surfaces, including human skin, the rate of evaporation, G_v ($kg\,m^{-2}\,s^{-1}$), being given by

$$G_v = h^* \Delta P / R_v T \tag{4.66}$$

where ΔP is the difference between the saturation pressure corresponding to the liquid water temperature and the vapour pressure of the psychrometric mixture, which is calculated from

$$P_v = \phi P_{sat}$$

When a psychrometric mixture, originally at state-point B (Fig. 4.33) is cooled to state point A, which has a dry-bulb temperature above the dew-point of the mixture, its dry-bulb temperature falls at constant specific humidity, its relative humidity increases and the air becomes 'wetter', i.e. the partial pressure of water vapour in the air increases, allowing a lesser rate of evaporation of liquid water from wet surfaces.

Dehumidification

When a psychrometric mixture, originally at state-point C (Fig. 4.34) is cooled to state-point O, which has a dry-bulb temperature below the dew-point of the mixture, the process follows the route CI until the dew-point is reached. Further reductions in temperature cause water to condense from the mixture, which follows the saturation curve, IO. Its moisture content decreases, but its relative humidity increases. In practice, the mixture is passed through a cooler coil, the surface of which is maintained at state-point O. The path during dehumidification is said to follow CD. The effectiveness of dehumidification, is quantified by the *contact factor* of the cooler coil, defined by A/(A + B) in the figure. In summer, warm dry air must be cooled and dehumidified. This is accomplished by following a dehumidification process with sensible reheat, as shown in the figure.

Humidification

When a psychrometric mixture, originally at state-point C (Fig. 4.35) passes through a spray chamber, the spraying water being maintained at the wet-bulb temperature of the mixture,

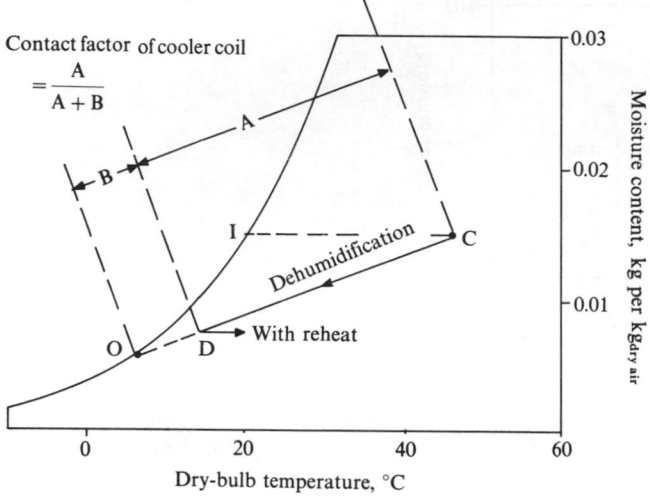

Figure 4.34 Summer operation—dehumidification with reheat illustrated on the psychrometric chart.

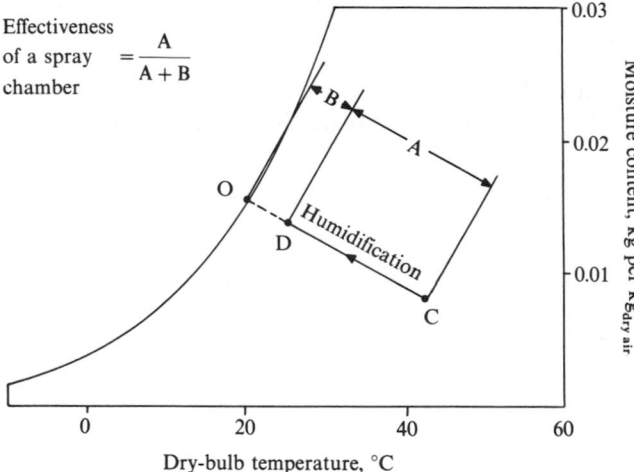

Figure 4.35 The process of humidification illustrated on the psychrometric chart.

O, water evaporates into the mixture, raising its moisture content and its relative humidity. The path during dehumidification follows CD. The effectiveness of humidification, is quantified by the *effectiveness of a spray chamber*, defined by A/(A + B) in the figure. In winter, cold wet air must be heated and humidified. This is accomplished by sensible heating, followed by humidification with sensible reheat, as shown in Fig. 4.36.

4.11 VAPOUR MIGRATION AND CONDENSATION

The *vapour permeability*, k^* ($\mathrm{kg\,m\,N^{-1}\,s^{-1}}$) of a material is analogous to thermal conductivity. The rate of vapour flow through a material, G_v ($\mathrm{kg\,m^{-2}}$), is given by

$$G_v = \Delta p_v k^*/\delta \qquad (4.67)$$

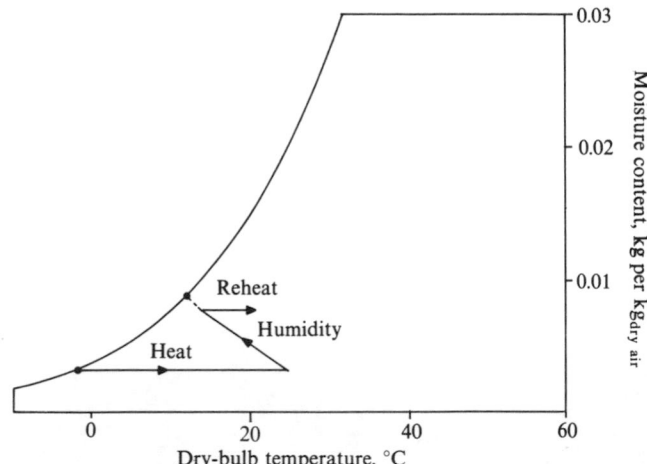

Figure 4.36 Winter operation—preheating, humidification with reheating illustrated on the psychrometric chart.

where Δp_v (N m^{-2}) is the difference in vapour pressure across the material section, given by the relative humidities × saturation pressures at the respective dry bulb temperatures, and δ is the thickness (m) of the section of material. The process is analogous to conduction. The parameter k^*/δ is the permeance, C^* (kg N^{-1} s^{-2}), which is analogous to conductance.

The permeability of a material is often expressed as a fraction of the permeability of air ($k_a^* = 20 \times 10^{-11}$ kg m N^{-1} s^{-1}), i.e.

$$k^* = k_a^*/\kappa \tag{4.68}$$

where κ is the diffusion resistance factor.

Just to confuse things a little further, the permeability of water vapour in air is given by

$$k_a^* = D^*/R_v T \tag{4.69}$$

Table 4.18 Diffusion resistance factors

Material	Diffusion resistance factor, κ
Brick	10
Brickwork	35
Cement, mortar	45
Clinker block	420
Concrete	40
Cork slab	10
Foam glass	zero
Hardboard	150
Insulating board	5
Mineral wool	1
Plaster	10
Plasterboard	8
Plywood	200–700
Woods	20–300

Table 4.19 Permeances

Material	Permeance, C^* $10^{-11}\,\mathrm{kg\,N^{-1}\,s^{-1}}$
Aluminium foil	0–0.6
Building paper	1.7–45.0
Painted insulating board	5.7–290
Kraft paper	170–460
Painted plaster	5.7–17.0
Polythene	0.6
Roofing felt	1.0–23.0
Painted wood	1.7–11.0
Aluminium painted wood	25.0–54.0

where D^* is the diffusion coefficient for water vapour in air ($= 2.8 \times 10^{-5}\,\mathrm{m^2\,s^{-1}}$).

Thus rates of water diffusion through different materials can be calculated using values for diffusion resistance factors, κ given in Table 4.18.

Vapour pressure gradients may be calculated exactly as the temperature gradients through structures. Condensation will occur wherever the local vapour pressure is equal to the local saturation vapour pressure corresponding to the local temperature. A thin solid coating or foil (e.g. a vapour barrier) has a permeance C^* ($\mathrm{kg\,N^{-1}\,s^{-1}}$) (Table 4.19), analogous to conductance.

For water vapour diffusion through air, a surface coefficient for heat transfer, h^* ($\mathrm{m\,s^{-1}}$),

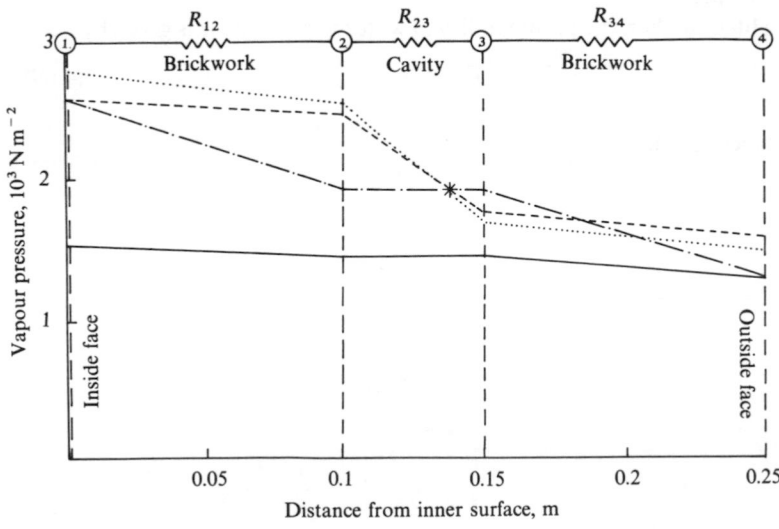

Figure 4.37 Water vapour transmission and condensation within a wall.

is related via analogy to the heat transfer coefficient using the Lewis relationship

$$h^* = h_c/\rho c_p \tag{4.70}$$

The amount of mass transfer, G_v $(\mathrm{kg\,m^{-2}})$ is then given by

$$G_v = h^* \Delta p / R_v T \tag{4.71}$$

Condensation or evaporation is accompanied by the release or absorption of latent heat h $(\mathrm{J\,kg^{-1}})$. For water at $20\,^\circ\mathrm{C}$, $h = 2.5 \times 10^6\,\mathrm{J\,kg^{-1}}$, and so the total rate of heat dissipation is calculated from

$$Q = G_v h\,(\mathrm{W\,m^{-2}}) \tag{4.72}$$

Examples of application

(a) Uninsulated air cavity Two 100 mm layers of brickwork are separated by a 50 mm thick air layer. Inside and outside air conditions are respectively $T_i = 25\,^\circ\mathrm{C}$, $\phi = 80$ per cent and $T_o = 11\,^\circ\mathrm{C}$, $\phi = 100$ per cent. Using the nodal numbering sequence adopted in Fig. 4.37, the results as shown in Table 4.20, are obtained.

The total thermal resistance, $R_{tot} = 2.64\,\mathrm{m^2\,K\,W^{-1}}$, and the total vapour resistance, $R_v^* = 353 \times 10^8\,\mathrm{m\,s\,kg^{-1}}$.

(b) Mineral-wool-insulated cavity The results are shown at Table 4.21.

The total thermal resistance, $R_{tot} = 3.12\,\mathrm{m^2\,K\,W^{-1}}$, and the total vapour resistance, $R_v^* = 365.5 \times 10^8\,\mathrm{m\,s\,kg^{-1}}$.

The introduction of mineral wool increases the temperatures, and hence the saturation vapour pressures, on the warmer side of the wall, whilst depressing temperatures and saturation vapour pressures on the colder side. Because the vapour resistance of the mineral wool contributes little to the overall vapour resistance, the vapour pressure distribution is not significantly affected. Condensation within the insulant will occur and the risk of condensation (and freezing at colder outside air temperatures) in the outer layer is increased. Venting-off the moisture under cyclic conditions is somewhat inhibited by the presence of the insulant.

Figure 4.37 shows the condensation will occur at the inner face and inner section of the outer layer of brickwork. This may be avoided by venting the cavity or by introducing a vapour barrier on the warmer side of the structure.

(c) Insulated cavity with a vapour barrier The introduction of a sheet of foil-backed building paper, having permeance, $C^* = 0.6 \times 10^{-11}\,\mathrm{kg\,N^{-1}\,s^{-1}}$, at the inner surface of the inner wall does not affect the saturated vapour pressure distribution, but does lower local vapour pressures (see Table 4.22).

The total thermal resistance, $R_{tot} = 3.12\,\mathrm{m^2\,K\,W^{-1}}$, the vapour resistance of the vapour barrier $= 1.66 \times 10^{11}\,\mathrm{N\,s\,kg^{-1}}$ and the total vapour resistance, $R_v^* = 2022 \times 10^8\,\mathrm{m\,s\,kg^{-1}}$.

It may be seen that the vapour pressure is lower than the saturation vapour pressure at all positions in the structure. Thus internal condensation will be avoided. Alternative means must be provided however to remove water vapour from the inside space.

Table 4.20 Results: uninsulated air cavity

Node	i	1	2	3	4	o	Units
Temperature, T	25	22.5	20.7	15.3	13.5	11	°C
Thermal resistance, R	0.48	0.33	1.02	0.33	0.48		$\mathrm{m^2\,K\,W^{-1}}$
Saturation vapour pressure, P_{sat}	3.17	2.60	2.45	1.75	1.57	1.33	$1000\,\mathrm{N\,m^{-2}}$
Relative humidity, ϕ	80					100	%
Vapour resistance, R_v^*	0.77	175		175	0.77		$10^8\,\mathrm{N\,s\,kg^{-1}}$
Vapour pressure, P_v	2.54	2.54	1.94	1.93	1.65	1.33	$1000\,\mathrm{N\,m^{-2}}$

Table 4.21 Results: mineral-wool-insulated cavity

Node	i	1	2	3	4	o	Units
Temperature, T	25	22.8	21.4	14.6	12.2	11	°C
Thermal resistance, R	0.48	0.33	1.50	0.33	0.48		$\mathrm{m^2\,K\,W^{-1}}$
Saturation vapour pressure, P_{sat}	3.17	2.75	2.52	1.67	1.47	1.33	$1000\,\mathrm{N\,m^{-2}}$
Relative humidity, ϕ	80					100	%
Vapour resistance, R_v^*	0.77	175	5.00	175	0.77		$10^8\,\mathrm{N\,s\,kg^{-1}}$
Vapour pressure, P_v	2.54	2.54	1.94	1.93	1.33	1.33	$1000\,\mathrm{N\,m^{-2}}$

Table 4.22 Results: insulated cavity with vapour barrier

Node	i	1	2	3	4	o	Units
Temperature, T	25	22.8	21.4	14.6	12.2	11	°C
Thermal resistance, R	0.48	0.33	1.50	0.33	0.48		$\mathrm{m^2\,K\,W^{-1}}$
Saturation vapour pressure, P_{sat}	3.17	2.75	2.52	1.67	1.47	1.33	$1000\,\mathrm{N\,m^{-2}}$
Relative humidity, ϕ	80					100	%
Vapour resistance, R_v^*	1666	175	5.00	175	0.77		$10^8\,\mathrm{N\,s\,kg^{-1}}$
Vapour pressure, P_v	2.54	1.54	1.43	1.43	1.33	1.33	$1000\,\mathrm{N\,m^{-2}}$

4.12 RADIATIVE HEAT TRANSFER

A body emits radiation when part of its internal energy is converted into electromagnetic waves. These waves travel in free space at the speed of light ($c = 2.997\,925 \times 10^8\,\mathrm{m\,s^{-1}}$) and transmit energy without the need for an intervening transport medium. When they encounter another body, part of the electromagnetic energy is absorbed and converted into internal energy in that body, raising its temperature. Electromagnetic waves are divided into classes according to their wavelengths, λ (m), and frequencies ω (cycles/s or Hertz). The wavelength of the radiation is defined as the ratio of the propagation velocity to the frequency via

$$\omega = \frac{c}{\lambda} \tag{4.73}$$

Figure 4.38 shows the electromagnetic wave spectrum. All bodies emit radiation, to which we are exposed at all times. Humans can sense radiation in the band 0.1 to 100 μm, within which wavelengths in the range 0.38 to 0.76 μm, are detected as visible light.

Figure 4.39 shows the visible spectrum of colours, the ultraviolet and the infra-red regions.

Absorption, reflection and transmission

When radiation encounters a surface, some of it is reflected, some absorbed and some transmitted according to

$$\alpha + \rho + \tau = 1 \tag{4.74}$$

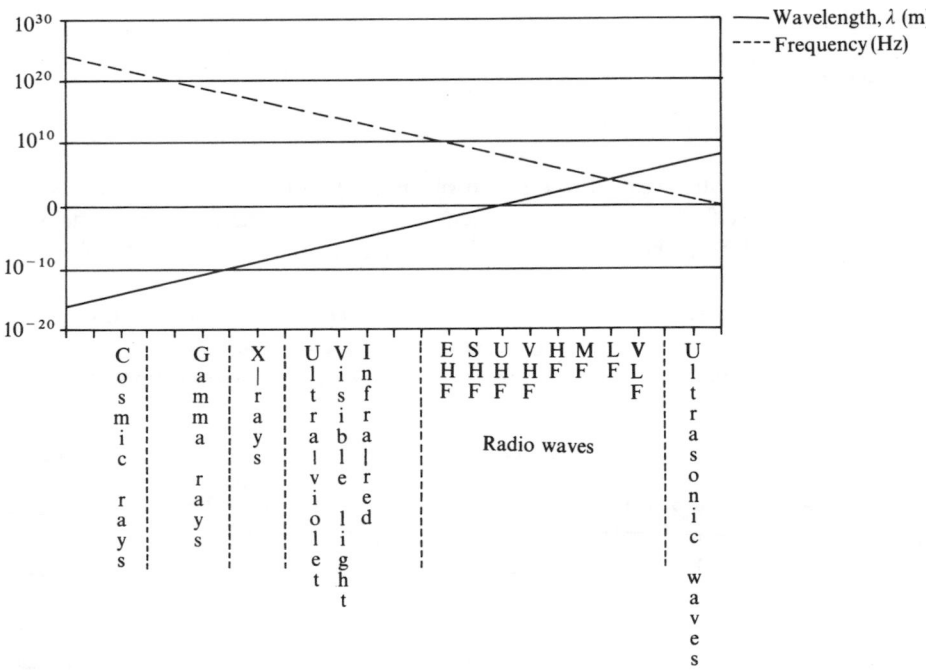

Figure 4.38 The electromagnetic wave spectrum.

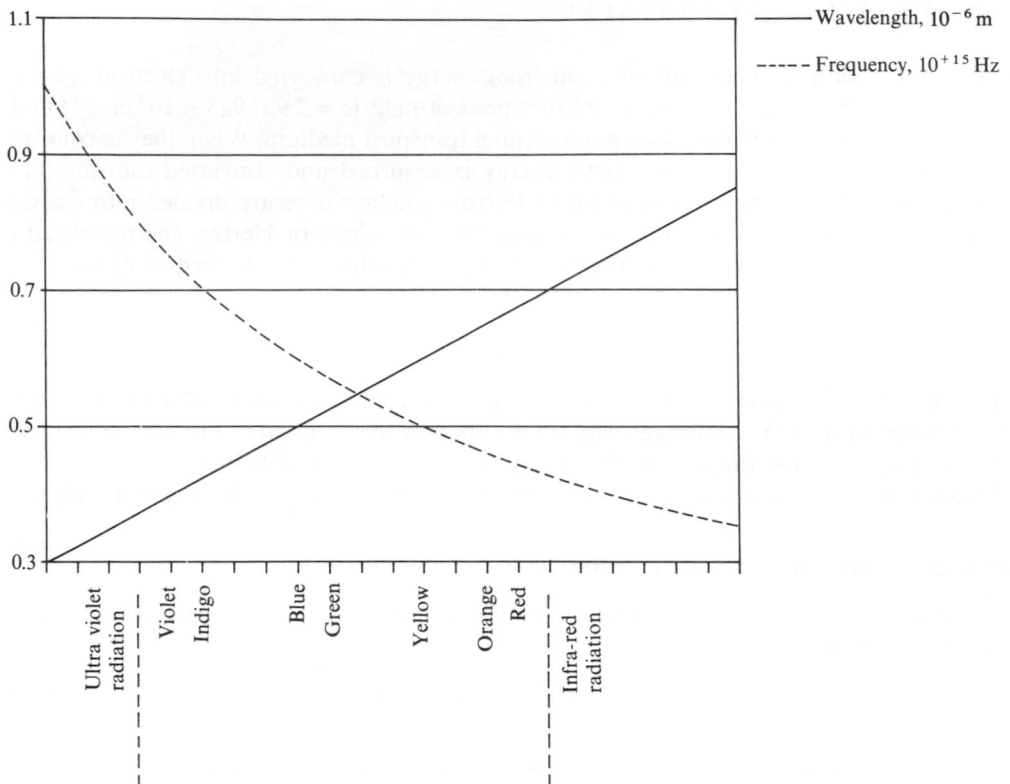

Figure 4.39 The visible spectrum.

where α is the surface absorptivity, i.e. the fraction of incident radiation which is absorbed by the body, raising its specific heat; ρ is the surface reflectivity, i.e. the fraction of incident radiation which is reflected from the body; and τ is the surface transmissivity, i.e. the fraction of incident radiation which passes through the body.

Most solid materials absorb radiation in a very thin surface layer (less than 1 mm deep). The exceptions are the translucent materials, i.e. glass, for which some transmission occurs at certain wavelengths. Many liquids and all gases are transparent. Opaque bodies do not transmit radiation, then

$$\alpha + \rho = 1 \tag{4.75}$$

The relative magnitudes of α, ρ and τ depend upon the material, its thickness and also the wavelength of the radiation. Table 4.23 lists the emissivities for the surfaces of various materials, at various wavelengths and temperatures. Further values are provided in reference 2 and the various databooks and guidebooks.

The reflection of radiation may be *specular* (or *regular*), when a coherent incident beam of radiation is reflected at an angle of reflection equal to the angle of incidence or *diffuse*, when the reflected rays spread out in all directions (such as the reflection from white paper).

Table 4.23 Surface emissivities

Wavelength (μm)	9.30	5.40	3.60	1.80	0.60
Temperature (K)	311.83	537.04	805.56	1611.11	Solar
Metals:					
Polished aluminium	0.04	0.05	0.08	0.19	0.30
Oxidized aluminium	0.11	0.12	0.18		
Polished brass	0.10	0.10			
Oxidized brass	0.61				
Polished iron	0.06	0.08	0.13	0.25	0.45
Oxidized iron	0.63	0.66	0.76		
Polished silver	0.01	0.02	0.03		0.11
Polished stainless steel	0.15	0.18	0.22		
Weathered stainless steel	0.85	0.85	0.85		
Building and insulating materials:					
Asbestos paper	0.93	0.93			
Asphalt	0.93		0.9		0.93
Red brick	0.93				0.70
Fire clay	0.90	0.70	0.75		
White paper	0.95		0.82	0.25	0.28
Plaster	0.91				
Paints:					
Aluminized lacquer	0.65	0.65			
Cream	0.95	0.88	0.70	0.42	0.35
Lampblack	0.96	0.97		0.97	0.97
Red	0.96				0.74
Yellow	0.95		0.50		0.30
White	0.95		0.91		0.18

Black bodies

The *black body* is a theoretical ideal radiator, which absorbs all the radiation incident upon it, and also emits at a given temperature the maximum possible amount of thermal radiation at all wavelengths.

Kirchhoff's Law

Suppose that two bodies, having surface areas A_1 and A_2 are placed inside a large evacuated enclosure which is perfectly insulated from its surroundings. Radiation is exchanged between the bodies and the walls of the enclosure until equilibrium is attained and both bodies and the walls are at equal temperatures. At this steady-state condition, the rate at which each body emits radiation = the rate at which each body receives radiation.

If α_1 and α_2 are the absorptivities of the surfaces, E_1 and E_2 are their total (i.e. over all wavelengths) emissive powers per square metre and G is the rate at which radiant energy from the walls of the enclosure falls upon the bodies,

$$A_1 E_1 = \alpha_1 A_1 G$$

and

$$A_2 E_2 = \alpha_2 A_2 G$$

from which

$$G = \frac{E_1}{\alpha_1} = \frac{E_2}{\alpha_2} \qquad (4.76)$$

i.e. the ratio of the emissive power of a surface to its absorptivity is the same for all bodies. This is known as Kirchhoff's Law. Since α is limited to an upper limit of unity, this places an upper limit on the emissive power. The imaginary concept of the black body has an absorptivity of unity and it is also an ideal radiator having an emissive power of unity.

The emissive power of a body other than a black body is less than unity, the ratio of its emissive power to that of a black body at the same wavelength is called the emissivity of the surface, ε, and, from Kirchhoff's Law, this also equals the absorptivity, α at the same wavelength.

A black body, or ideal radiator, may be defined therefore as either a body which absorbs radiation incident upon it and reflects or transmits none, or as a radiator which emits at any specified temperature the maximum possible amount of thermal radiation at all wavelengths. A diffusely reflecting surface reflects with the same spatial energy distribution as a black body.

The distribution of radiant energy over the electromagnetic wave spectrum for a black body radiator is given by Plank's law; i.e.

$$E(\mathrm{W\,m^{-2}\,\mu m^{-1}}) = \frac{K_1 \lambda^{-5}}{\exp\left(\dfrac{K_2}{\lambda T}\right) - 1} \qquad (4.77)$$

where $K_1 = 3.74 \times 10^8 \, \mathrm{W\,m^{-2}\,\mu m^{-4}}$, $K_2 = 1.44 \times 10^4 \, \mu\mathrm{m\,K}$ and E, the radiant energy per unit wavelength, is expressed in $\mathrm{W\,m^{-2}\,\mu m^{-1}}$ when T is in degrees K.

Figure 4.40 shows the distributions for various absolute temperatures. The top curve in the figure corresponds to the sun as a black body radiator. When viewed from the earth through the atmosphere, its effective radiating temperature is 6000 K—its actual temperature is, of course, far in excess of this ($\sim 10\,000\,000$ K).

The bottom curve corresponds to radiating surfaces at terrestrial temperatures. Note that the bulk of the radiating energy shifts from the ultraviolet to the infra-red end of the spectrum as the temperature of the radiating surface decreases. The '*greenhouse effect*' is a thermal radiation rectifier, which allows short-wave (or short-wavelength radiation), emanating from a high temperature source, such as the sun, to pass through a radiation barrier (the glass of a horticultural greenhouse or the blanket of greenhouse gases in the troposphere) but inhibits the passage of long-wave (or long-wavelength) radiation emitted by low temperature surfaces.

The wavelength at which maximum emission occurs thus shifts with decreasing temperatures to longer wavelengths. The relationship between this maximum value for E is described by Wien's displacement law, viz.

$$\lambda_{max} T = 2900 \, \mu\mathrm{m\,K} \qquad (4.78)$$

The maximum monochromatic emissive power at any temperature is then given by

$$E_{max} = \frac{K_1 \lambda_{max}^{-5}}{\exp\left(\dfrac{K_2}{\lambda_{max} T}\right) - 1} \qquad (4.79)$$

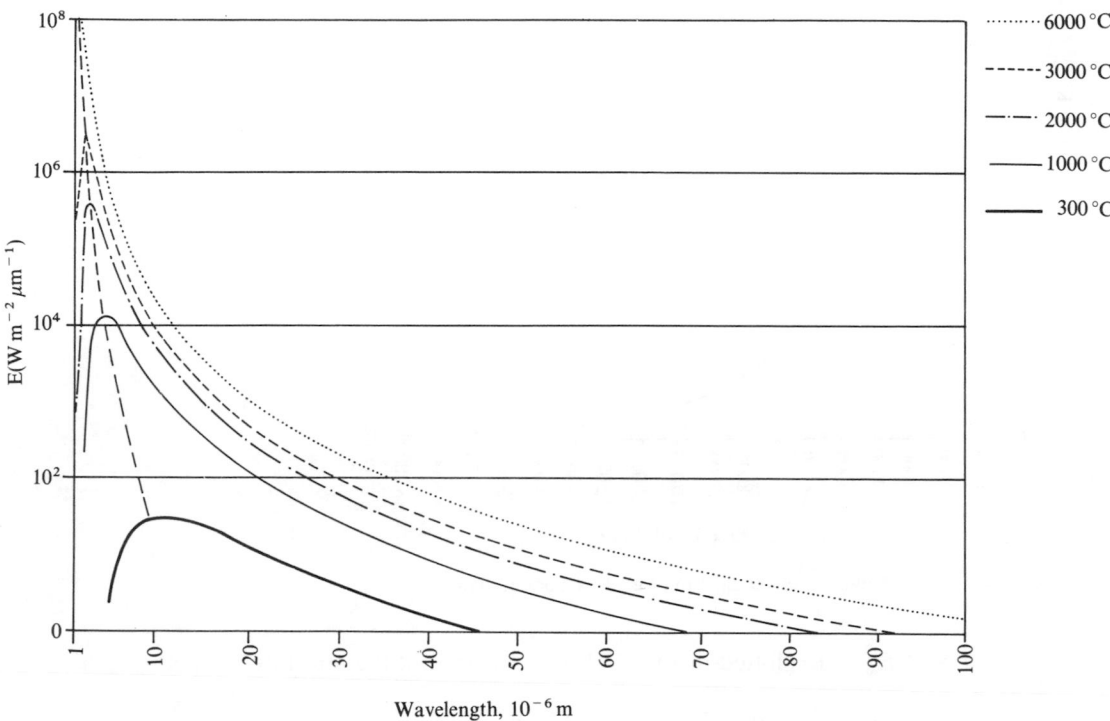

Figure 4.40 Spectroradiometric curves illustrating Plank's law and Wien's displacement law.

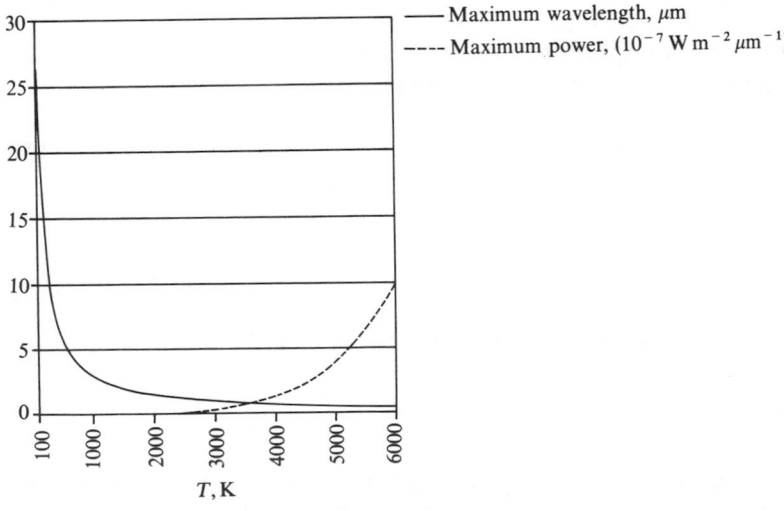

Figure 4.41 Maximum radiant emission—variation with temperature.

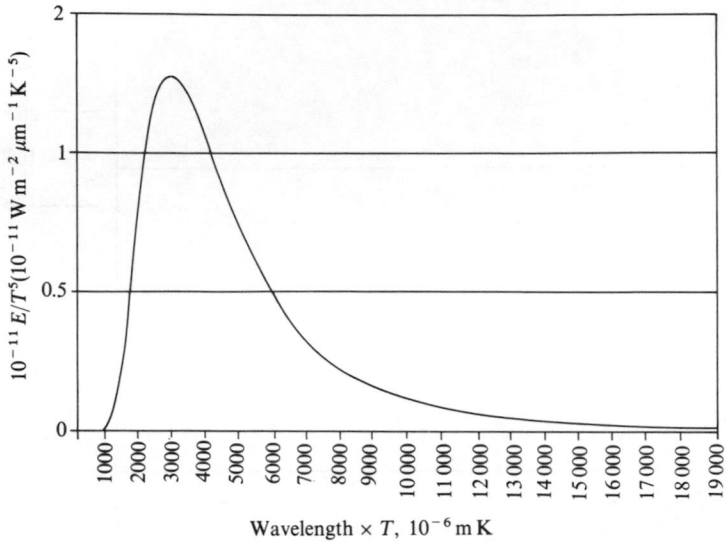

Figure 4.42 Monochromatic radiant emission for black body radiators.

and $\lambda_{max} T$ and E_{max} are plotted in Fig. 4.41 as a function of the absolute temperature of the radiating surface.

Dividing both sides of the equation by T^5, Plank's law may be expressed as

$$\frac{E}{T^5} = \frac{K_1}{(\lambda T)^5 \left(\exp\left(\frac{K_2}{\lambda T}\right) - 1 \right)} \qquad (4.80)$$

When this is plotted, the curves of Fig. 4.40 condense to a single curve, Fig. 4.42. Peak radiant emission corresponding to $(E/T^5)_{max} = 1.28 \times 10^{-11}\,\mathrm{W\,m^{-2}\,\mu m^{-1}\,K^{-5}}$ occurs at $\lambda T = 2900\,\mu m\,K$.

The total area enclosed by any spectroradiometric curve (Fig. 4.41) represents the total energy radiated at the temperature considered. Thus, from Plank's law,

$$\text{total energy(W\,m}^{-2}) = \left(\frac{K_1 \lambda^{-5}}{\exp\left(\frac{K_2}{\lambda T}\right) - 1} \right) d\lambda \qquad (4.81)$$

Integrating this,[2]

$$E_{tot} = \sigma T^4 \qquad (4.82)$$

where σ is the Stefan–Boltzmann constant ($= 5.67 \times 10^{-8}\,\mathrm{W\,m^{-2}\,K^{-4}}$).

Radiation intensity and total emissive power

The radiation intensity, I, is defined[2] as the radiant energy propagated in a particular direction per unit solid angle per unit area, as projected on a plane perpendicular to the direction of propagation. By integrating I for all angles over a hemisphere, Kreith shows that the total emissive power, E, of a diffuse surface is given by

$$E = \pi I \tag{4.83}$$

Grey surfaces

Radiation from real surfaces is modified by the value of the surface emissivity, ε:

$$E_{\text{tot}} = \varepsilon \sigma T^4 \tag{4.84}$$

For any surface, receiving or emitting radiation over a band of wavelengths, its emissivity, σ, and absorptivity, α, are dependent upon the wavelength of the radiation. At a single wavelength, $\sigma = \alpha$, but integrating over the range of wavelengths the resultant effective mean values of σ and α are not equal.

From Table 4.23, emissivity at a single wavelength is equal to absorptivity. Thus when designing to inhibit solar gains, surfaces should be chosen to have a low absorptivity to short-wave radiation whilst also having a high emissivity at low temperatures, minimizing the absorption of solar radiation whilst maximizing the loss of heat by radiation. Table 4.24 lists ratios of absorptivity for short-wave radiation to emissivity for long-wave radiation. The lower the value of this index, the better the surface is for inhibiting solar gains.

In general, emissivity increases with increasing wavelength for electrical conductors and decreases with increasing wavelength for electrical non-conductors. Figure 4.43 shows typical characteristics. It is seen that the best opaque surface in Table 4.23 for remaining cool in sunshine is a white painted, or white stone surface, as adopted in Mexican haciendas. The 'mirrored' polished metal surfaces have high absorptivities for solar radiation coupled with

Table 4.24 Ratios of absorptivity for short-wave radiation to emissivity for long-wave radiation

Material	$\dfrac{\alpha_{\text{sort}}}{\varepsilon_{\text{long}}}$
Electrical conductors:	
Polished zinc	23.00
Polished aluminium	7.50
Polished iron	7.50
Polished silver	11.00
Electrical non-conductors:	
Asphalt	1.00
Red brick	0.75
Cream paint	0.37
Lampblack	0.98
Red paint	0.32
White paint	0.19
Glass	0.01

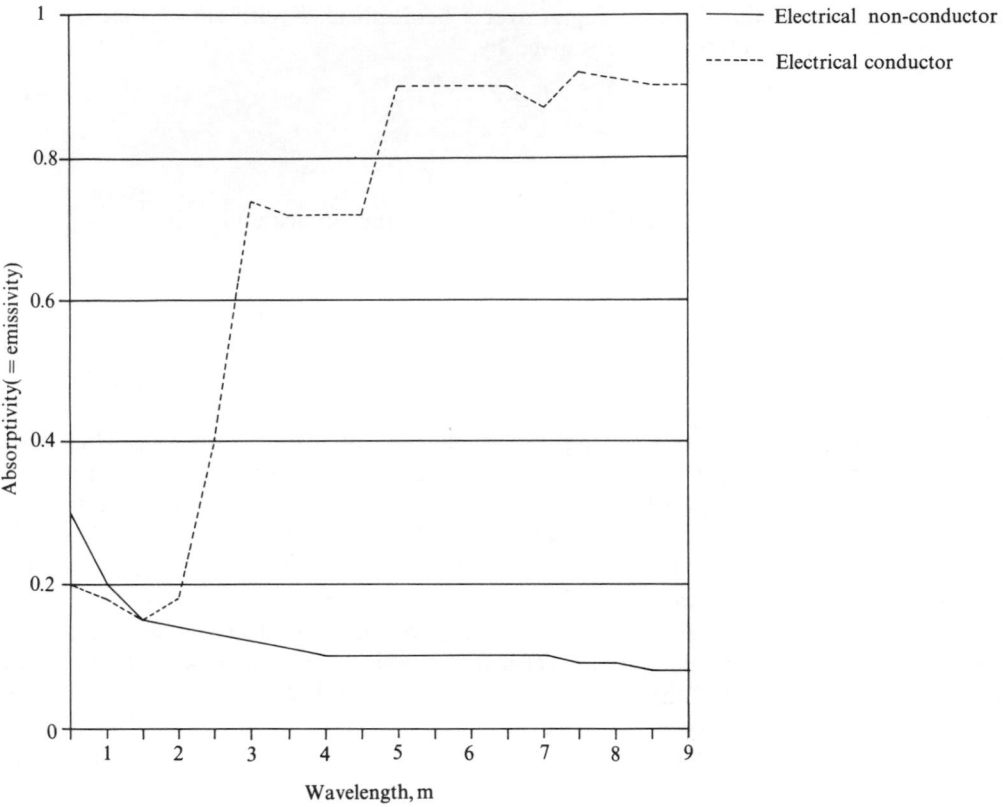

Figure 4.43 Typical variations of monochromatic absorptivities for electrical conductors and non-conductors.

low emissivities for longwave radiation and so would reach higher equilibrium temperatures than the white painted surface.

Polished metal surfaces are best for solar collecting surfaces. Selective absorbers absorb short-wave radiation well but emit longer waves poorly. In solar collectors they consist of polished metal surfaces, having low emissivities (≈ 0.1) for low temperature long-wave radiation, coated with thin deposits of black salts, such as nickel oxide or copper oxide. These films have high absorptivities (≈ 0.9) for short-wave solar radiation. If the film is thin enough, it is transparent to radiation longer than the coating thickness. Thus, for the coated surface, its emissivity is close to that of the metal beneath, whilst its absorptivity is that of the surface coating.

Radiation exchange between two surfaces, 1 and 2, can be estimated from

$$Q = \mathscr{F} A \sigma (T_1^4 - T_2^4) \tag{4.85}$$

where $\mathscr{F}$ is a configuration factor, the value of which (0–1) depends upon the emissivities of the surfaces and the relative view (or 'shape') factor, F, between the surfaces.

Figure 4.44 shows two surfaces of black bodies, between which radiation is being exchanged, in three-dimensional space. To calculate the fractional amount of radiant energy

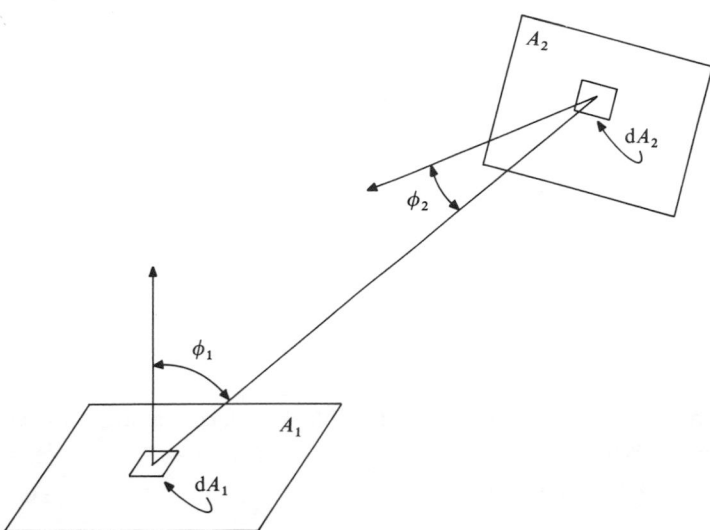

Figure 4.44 Two surfaces exchanging radiation.

leaving surface A_1 that strikes surface A_2, we first consider the elemental surfaces dA_1 and dA_2, separated by the distance r. The rate at which radiation from dA_1, dq_{12}, is received by dA_2 is given by

$$qA_{12} = I_1 \cos \phi_1 dA_1 d\omega_{12} \tag{4.86}$$

where I_1 is the intensity of radiation from dA_1; $dA_1 \cos \phi_1$ is the projection of area element dA_2 as seen from dA_2 and $d\omega_{12}$ is the solid angle subtended by the receiving area dA_2 with respect to the centre point of dA_1. This angle is given by the *projected* area, $dA_2 \cos \phi$, of the receiving surface in the direction of the incident radiation divided by the square of the distance, r (i.e. from the definition of a solid angle).

Substituting for $d\omega_{12}$, and repeating the analysis for the rate at which radiation from dA_2, dq_{21}, is received by dA_1, the net rate of radiant heat transfer between dA_1 and dA_2 becomes

$$q_{12} = (E_{b1} - E_{b2}) \frac{\cos \phi_1 \cos \phi_2 dA_1 dA_2}{\pi r^2} \tag{4.87}$$

The net radiation between the entire surfaces is obtained by integrating this equation over both surfaces:

$$q_{12} = (E_{b1} - E_{b2}) \int \int \frac{\cos \phi_1 \cos \phi_2 dA_1 dA_2}{\pi r^2} \tag{4.88}$$

The double integral above is usually written as $A_1 F_{12}$ or $A_2 F_{21}$ and is termed the *shape factor* for radiation between the surfaces.

The reciprocity theorem applies in that

$$A_1 F_{12} = A_2 F_{21}$$

Thus, for radiation exchange between two surfaces,

$$q_{12} = (E_{b1} - E_{b2})A_1 F_{12} = (E_{b1} - E_{b2})A_2 F_{21} \qquad (4.89)$$

where F_{12} is the fraction of the total radiation leaving A_1 which is intercepted by A_2.

Analytical determinations of shape factors are impossible for all but the simplest of shapes, although computational techniques of numerical integration are available. Catalogues of shape factors[9,10] are available for use by designers and analysts. The shape factor between infinite parallel planes of black bodies (or for black body completely surrounded by another) is unity.

Radiation networks

Black bodies The net rate of radiation interchange between two black surfaces having emissive powers E_{b1} and E_{b2} (Fig. 4.45) is given by

$$Q = A_1 F_{12}(E_{b1} - E_{b2}) \qquad (4.90)$$

The net rates of radiation interchange among three black surfaces having emissive powers E_{b1}, E_{b2} and E_{b3} (Fig. 4.46) is given by solving the simultaneous equations

$$Q_{12} = A_1 F_{12}(E_{b1} - E_{b2})$$
$$Q_{13} = A_1 F_{13}(E_{b1} - E_{b3}) \qquad (4.91)$$
$$Q_{23} = A_2 F_{23}(E_{b2} - E_{b3})$$

Thermal networks can be built up in this way for complex configurations. The result is a set of simultaneous linear equations for computer solution.

Grey bodies If the radiating surfaces are not black, they can be considered as being grey. A grey surface follows Lambert's cosine law and reflects diffusely. The *radiosity*, J of a grey

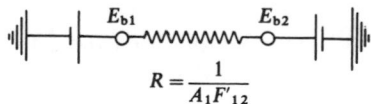

$$R = \frac{1}{A_1 F'_{12}}$$

Figure 4.45 Thermal network for radiation exchange between two black body surfaces.

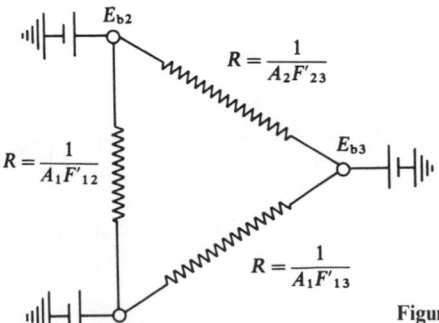

$$R = \frac{1}{A_1 F'_{12}}$$

$$R = \frac{1}{A_2 F'_{23}}$$

$$R = \frac{1}{A_1 F'_{13}}$$

Figure 4.46 Thermal network for radiation exchange among three black body surfaces.

surface is defined as the rate at which radiation leaves a given surface per unit area. It is the sum of the radiation emitted, reflected and transmitted.

For an opaque body

$$J = \rho G + \varepsilon E_b \qquad (4.92)$$

where J = radiosity, $W\,m^{-2}$

G = irradiation incident on the surface, $W\,m^{-2}$

E_b = black-body emissive power, $W\,m^{-2}$

ρ = reflectivity

ε = emissivity.

The net rate at which radiation leaves a grey surface is then given by

$$dQ/dA = J - G \qquad (4.93)$$

(i.e. the leaving radiation minus the incident radiation). Substituting for the irradiation G,

$$\frac{dQ}{dA} = J - J/\rho - E_b J \varepsilon/\rho \qquad (4.94)$$

and because for a grey surface, ρ is constant over all wavelengths and $\rho + \varepsilon = 1$,

$$\frac{dQ}{dA} = \frac{\varepsilon}{\rho} E_b - \left(\frac{1-\rho}{\rho}\right) J = \frac{\varepsilon}{\rho}(E_b - J) \qquad (4.95)$$

Integrating over the surface,

$$Q = \frac{\varepsilon}{\rho} A(E_b - J) \qquad (4.96)$$

This is equivalent to an equilibrium equation which can be interpreted as a flow of radiation, Q, across a potential difference, $(E_b - J)$, connected by a resistance, $\rho/\varepsilon A$ (Fig. 4.47).

Thermal circuits for grey bodies therefore can be constructed by treating the potential differences, $(E_b - J)$, as thermal *springs* at the nodes. These may be included in the analyses of radiation interchanges among grey bodies using equivalent conductances

$$A_1 \mathscr{F}_{12} = \frac{1}{\rho_1 A_1 \varepsilon_1 + \dfrac{1}{A_1} + \dfrac{\rho_2}{A_2 \varepsilon_2}} \qquad (4.97)$$

in the equation

$$Q = (E_{b1} - E_{b2}) A_1 \mathscr{F} E_{12} \qquad (4.98)$$

where A_1 is the area of the smaller surface, and $\mathscr{F}$ is the *grey body shape factor* for two surfaces.

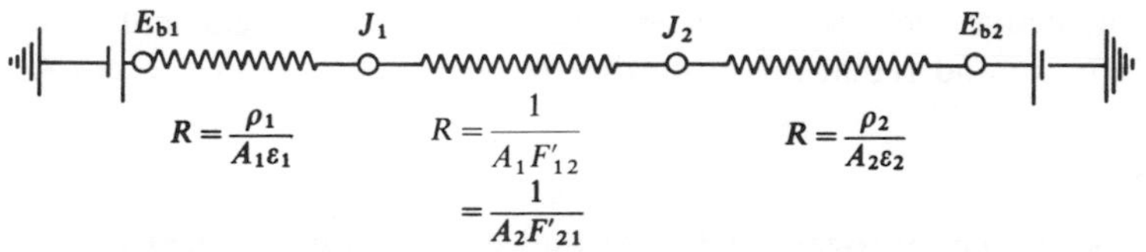

Figure 4.47 Thermal network for radiation exchange between two grey surfaces.

Substituting for $E_b = \sigma T^4$, the general equation for radiative heat transfer between two grey surfaces becomes

$$Q = \rho A_1 \mathscr{F}(T_1^4 - T_2^4) \tag{4.99}$$

where

$$A_1 \mathscr{F}_{12} = \left[\frac{1}{A_1}\left(\frac{1}{\varepsilon_1} - 1\right) + \frac{1}{A_2}\left(\frac{1}{\varepsilon_2} - 1\right) + \frac{1}{A_1 F_{12}} \right]^{-1} \tag{4.100}$$

For radiation between two parallel flat plates, *grey body shape factor* reduces to

$$\mathscr{F}_{12} = \left[\frac{1}{\varepsilon_1} + \frac{1}{\varepsilon_2} - 1 \right]^{-1} \tag{4.101}$$

For a small grey body in black surroundings

$$A_1 \mathscr{F}_{12} = A_1 \varepsilon_1 \quad \text{or} \quad \mathscr{F}_{12} = \varepsilon_1 \tag{4.102}$$

Linearized equation

Many problems may be simplified by linearizing the equation for radiative heat transfer over the range of interest, as follows:

$$Q = \mathscr{F} A \sigma(T_1^4 - T_2^4) \equiv hA(T_1 - T_2) \tag{4.103}$$

where h is a linear heat transfer coefficient $(W\,m^{-2}\,K^{-1})$

$$h = \mathscr{F}\left(\frac{\sigma(T_1^4 - T_2^4)}{T_1 - T_2} \right) \tag{4.104}$$

Now

$$\left(\frac{(T_1^4 - T_2^4)}{(T_1 - T_2)} \right) = (T_1^2 + T_2^2)(T_1^2 - T_2^2)/(T_1 - T_2)$$

$$= (T_1^2 + T_2^2)(T_1 + T_2) = \approx 4T_{\text{average}}^3$$

Therefore

$$Q = \mathscr{F} A \sigma(T_1^4 - T_2^4) \approx \mathscr{F} A 4 T_{\text{average}}^3 (T_1 - T_2) \tag{4.105}$$

Within the range 270–320 K, for example,

$$Q = 5.8 \mathscr{F} A(T_1 - T_2) \tag{4.106}$$

giving an equivalent radiative heat transfer coefficient of

$$h = 5.8 \mathscr{F} \, (W\,m^{-2}\,K^{-1}) \tag{4.107}$$

This can be added to the convective coefficient to produce a heat transfer coefficient which combines the effects of convection and radiation. Note that temperatures are in degrees K.

Radiation involving gases

Many common gases, such as oxygen, nitrogen, hydrogen and dry air, are practically transparent to thermal radiation. Carbon dioxide, water vapour, sulphur dioxide, carbon monoxide,

Table 4.25 Emission and absorptance bands for common gases and vapours

Substance	Wavelength band λ, μm
Carbon dioxide	2.36–3.02
	4.01–4.80
	12.5–16.5
Water vapour	2.24–3.27
	4.80–8.50
	12.00–25.00

ammonia, hydrocarbon and alcohol vapours emit and absorb radiation only between narrow bands (see Table 4.25).

The greenhouse gases, carbon dioxide, CO_2, nitrous oxide, N_2O, the CFCs, methane, CH_4, tropospheric ozone, O_3 and water vapour, H_2O, all impede the transmission of long-wavelength radiation from the earth.

Stratospheric ozone inhibits the transmission of short-wavelength ultraviolet radiation from the sun.

Solar radiation

The sun is a fusion reactor in which hydrogen is being converted to helium, releasing nuclear energy. Internal sun temperatures are of the order of $10^7\,°C$, with an effective surface temperature, when viewed from the earth, of 6000 °C. The solar energy flux at the sun's surface (1.383×10^6 km diameter) is $70\,MW\,m^{-2}$. The earth (11 700 km diameter) intercepts solar energy at the rate of $1.362\,kW\,m^{-2}$ at the boundary of the upper atmosphere (10^{21} MW in total).

The solar radiation incident at the perimeter of the earth's atmosphere contains 5 per cent ultraviolet, 53 per cent visible and 43 per cent infra-red radiation. When the sun is directly overhead in a cloudless sky, $1.025\,kW\,m^{-2}$ of specular radiation can reach the earth's surface. The remaining $0.337\,kW\,m^{-2}$ is reflected by gas molecules which filter out the shorter wavelengths and selectively transmit light in the blue end of the visible spectrum. The modified radiation contains 1 per cent ultraviolet, 39 per cent visible and 60 per cent infra-red radiation.

Solar radiation which has been scattered, or absorbed and reradiated, by a combination of gas molecules, water vapour and dust particles, becomes white and diffuse. Layers of clouds reflect direct and diffuse 'sky' radiation away from the earth but also insulate against direct radiation losses from the earth's surface. At night, in the absence of cloud cover, up to $6000\,W\,m^{-2}$ can be lost from the surface of the earth by direct radiation to space. In the steady-state, all solar radiation received by the earth is finally received by the vault of deep space. The current rate of fossil fuel consumption corresponds to about 0.001 per cent of the rate of solar energy received by the earth.[1]

Insolation on land masses represents about 29 per cent of the total reaching the surface of the earth. Solar radiation falling upon vegetation is possibly 0.01 per cent of this. As plants are approximately 1 per cent efficient in converting solar radiation to fuel,[11] if it is assumed that an average of $300\,W\,m^{-2}$ of solar radiation occurs during the growing period, then it can be calculated that the earth's fossil fuel store has taken 100 000 years of sunlight to accumulate. The United Kingdom, with a land area of 245 000 km² consumes 9×10^{12} MJ each year and receives solar radiation at an average of 200 W m⁻². Thus 0.5 per cent of total UK solar radiation influx would have to be harnessed for self-sufficiency.[1]

Solar calculations

As the sun is far from the earth, its rays may be considered as being straight and parallel to each other at the earth's surface. Because the axis of the earth's rotation is not parallel to its axis of rotation about the sun, the sun is vertically above the earth's equator on two occasions during the year at the vernal and autumnal equinoxes (Fig. 4.48). The summer and winter solstices are the times when the sun is furthest from the earth's equator.

The angular displacement of the sun from the plane of the earth's equator is termed the *declination* of the sun, d'. This angle varies between $+23.5$ and -23.5 degrees as the earth

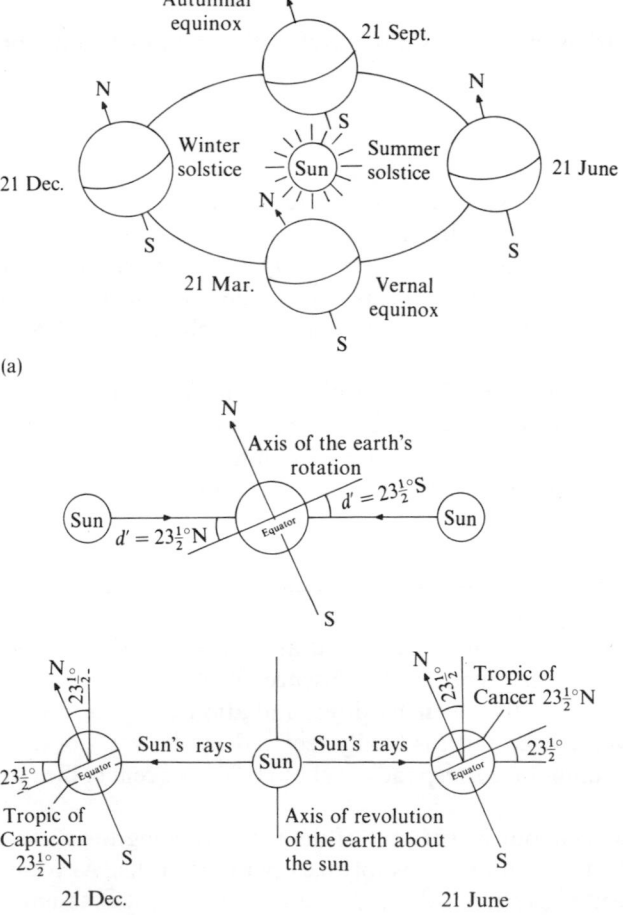

(a)

(b)

Figure 4.48 (a) Position of the earth with respect to the sun at solstices and equinoxes; (b) Definition of declination and the tropics.

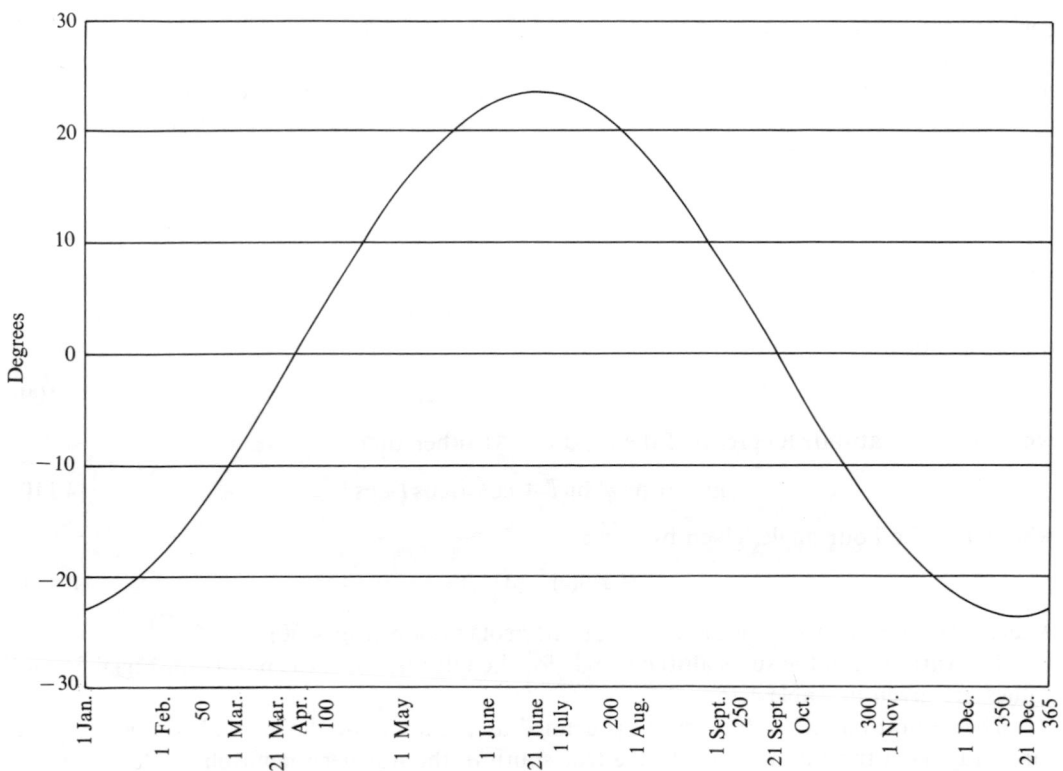

Figure 4.49 Annual variation of the declination of the sun from the earth.

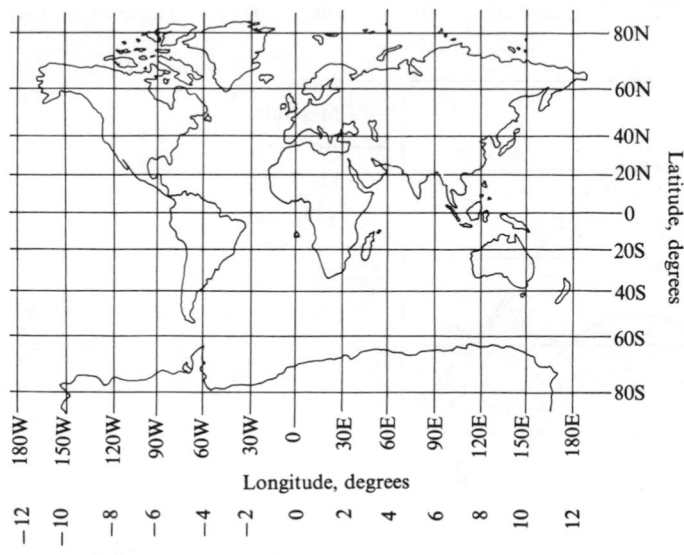

Figure 4.50 Difference between local sun-time and GMT for all parts of the world.

performs its annual circumnavigation about the sun. Thus the latitudes $+23.5$ degrees and -23.5 degrees are respectively known as the Tropic of Cancer, where the sun is vertically overhead at the summer solstice (21 June—the longest day in the northern hemisphere), and the Tropic of Capricorn, where the sun is vertically overhead at the winter solstice (21 December—the shortest day in the northern hemisphere).

For calculation purposes, the declination can be estimated from

$$d' = 23.5 \sin 2\pi(t - 80)/365 \qquad (4.108)$$

where the day number, $t = 1$ on 1 January. This relationship is plotted in Fig. 4.49.

The altitude of the sun, α', is the angle a direct ray makes with the horizontal at a particular location. At noon

$$\alpha' = 90 - (l' - d')\text{degrees} \qquad (4.109)$$

where l' is the latitude (degrees) of the location. At other times

$$\sin \alpha' = \sin d' \sin l' + \cos d' \cos l' \cos h' \qquad (4.110)$$

where h' is the hour angle, given by

$$h' = 360t'/24 = 15t' \qquad (4.111)$$

where t' is the *sun-time* in hours before or after solar noon (Fig. 4.50).

The variation of the sun's altitude angle for the latitude 51.7 N is plotted in Fig. 4.51 and values are given in Table 4.26.

The azimuth angle of the sun, z', is defined as the angle the horizontal component of a direct ray from the sun makes with the true south in the northern hemisphere. At noon, the sun lies directly south and so the azimuth angle is zero. At other times during the day

$$\tan z' = \sin h'/(\sin l' \cos h' - \cos l' \tan d') \qquad (4.112)$$

This function is plotted in Fig. 4.52 and values are given in Table 4.26. Note that, when the

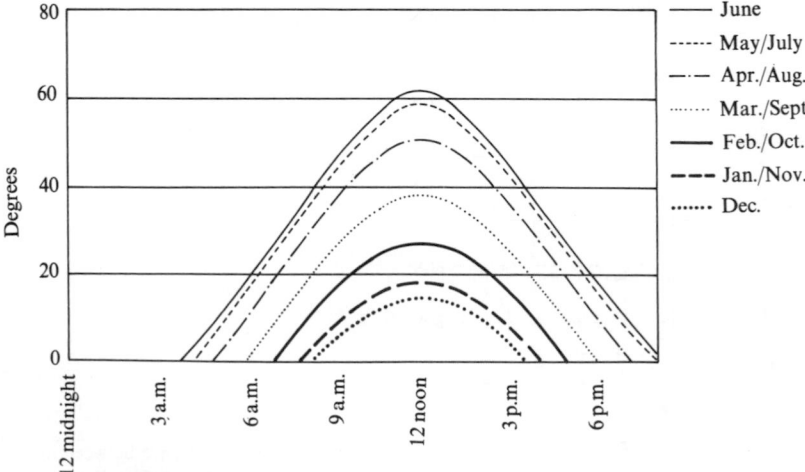

Figure 4.51 Variation of the altitude angle of the sun with time of day for each month of the year for the latitude 51.7 °N.

Table 4.26 Sun angles for latitude 51.7 °N

Sun-time (= GMT)	Solar altitude angle, α'						
	22 Dec.	21 Jan. and 21 Nov.	20 Feb. and 23 Oct.	22 Mar. and 22 Sept.	20 Apr. and 24 Aug.	21 May and 23 Jul.	21 June
0600 and 1800					9	15	18
0700 and 1700			1	10	18	25	27
0800 and 1600		2	10	19	28	34	37
1900 and 1500	6	10	17	27	37	44	46
1000 and 1400	12	15	24	34	44	52	55
1100 and 1300	15	19	28	39	50	58	61
1200	17	20	29	40	51	60	63
	Solar azimuth angle, z'						
0600				90	83	77	74
0700			108	101	94	88	85
0800		125	120	114	106	100	97
0900	139	138	133	127	120	114	110
1000	152	151	148	143	137	131	128
1100	166	165	163	161	157	153	151
1200	180	180	180	180	180	180	180
1300	194	195	197	199	203	207	209
1400	208	209	212	217	223	229	232
1500	221	222	227	233	240	246	250
1600		234	240	246	254	260	263
1700			252	259	266	272	275
1800				270	277	283	286

altitude angle is negative, the sun lies below the horizon and the azimuth angle is then of no value in calculations of solar gains.

The numerical value of direct solar radiation, I_δ, may be obtained using the altitude angle alone, from

$$I_\delta = K_1 \exp(-K_2/\sin \alpha')(\text{kW m}^{-2}) \qquad (4.113)$$

where K_1 (kW m^{-2}), is the apparent solar radiation in the absence of atmosphere, and K_2 is an atmospheric correction factor, and are empirical constants, given in Table 4.27.[12] For calculation purposes, these constants may be approximated by

$$K_1 = 1.15 + 0.09 \sin((t - 295)2\pi/365)$$
$$K_2 = 0.17 + 0.04 \sin((t - 100)2\pi/365)$$
$$K_3 = 0.09 + 0.05 \sin((t - 100)2\pi/365)$$

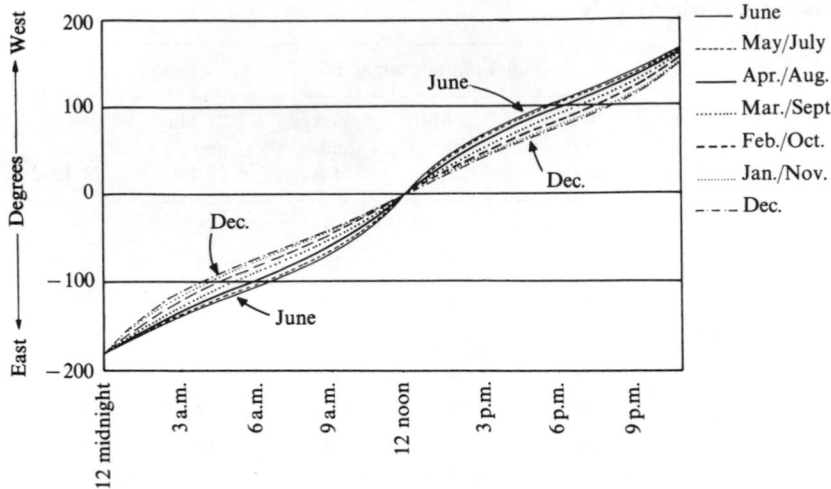

Figure 4.52 Variation of the azimuth angle of the sun with time of day for each month of the year for the latitude 51.7 °N. Note: When the altitude angle is zero from Fig. 4.51, the sun is below the horizon and the azimuth angle is meaningless.

Table 4.27 Constants for use in solar equations

Month	Jan.	Mar.	May	July	Sept.	Nov.
K_1	1.23	1.19	1.1	1.09	1.15	1.22
K_2	0.14	0.16	0.2	0.21	0.18	0.15
K_3	0.06	0.07	0.12	0.14	0.09	0.06

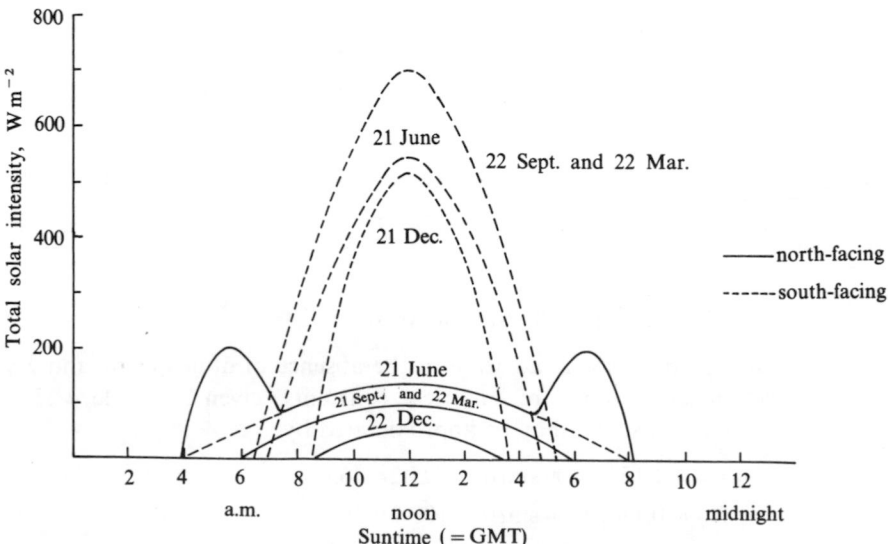

Fig. 4.53a

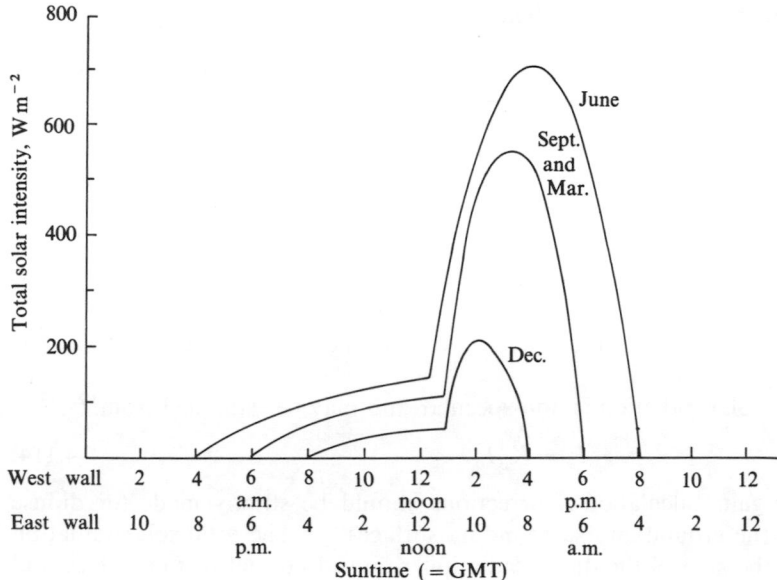

Fig. 4.53b

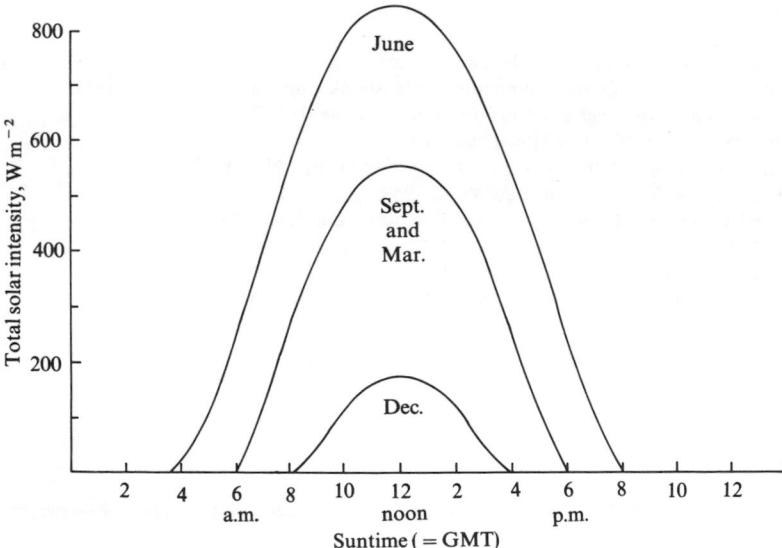

Fig. 4.53c

Figure 4.53 Annual and diurnal variations of total solar radiation (direct plus diffuse) striking vertical and horizontal surfaces: (a) for north and south-facing walls; (b) for east and west-facing walls; and (c) for a horizontal surface, for the latitude 51.7°N.

Table 4.28 Direct solar radiation intensities, $W\,m^{-2}$

Altitude angle	5	10	15	20	25	30	35	40	45	50	60	80
Normal to sun	210	388	524	620	688	740	782	814	840	860	893	920
On horizontal	18	67	136	212	290	370	450	523	594	660	773	907
On south-face	210	382	506	584	624	642	640	624	594	553	477	160

Table 4.28 lists direct solar radiation intensities with a clear sky for places below 300 m elevation.

Direct solar radiation is specular and may be resolved normal to surfaces at various inclination angles using geometrical laws.[1]

Scattered, or diffuse, solar radiation is non-specular and may be estimated from[12]

$$I_{sc} = K_3 I_\delta \qquad (4.114)$$

When performing solar gain calculations, corrections should be strictly made for diffuse radiation reflected from the ground and surrounding surfaces.[1,12] The total solar radiation received by a surface is the sum of the direct radiation resolved normal to the surface and the diffuse radiation received, the latter being typically 5 to 10 per cent of the direct radiation. Figure 4.53 show the annual and diurnal variations of total solar radiation received by north, south, east and west walls and horizontal surfaces for the latitude 51.7 N.

REFERENCES

1. O'Callaghan, P. W., *Building for Energy Conservation*, Pergamon Press, Oxford, 1978.
2. Kreith, F., *Principles of Heat Transfer*, 3rd Edition, International Textbook Company, New York, 1973.
3. Knudsen, J. G. and Katz, D. L., *Fluid Dynamics and Heat Transfer*, McGraw-Hill, Tokyo, 1958.
4. Streeter, V. L., *Fluid Mechanics*, 8th edn, McGraw-Hill, Tokyo, 1985.
5. General Electric Company, *Heat Transfer Databook*, Genium Publishing Corp., New York, 1984
6. McAdams, W. H., *Heat Transmission*, McGraw-Hill, New York, 1958.
7. CIBSE, *Guidebook*, Chartered Institution of Building Services Engineers, London, 1990.
8. Kays, W. M. and London, A. L., *Compact Heat Exchangers*, McGraw-Hill, London, 1964.
9. Hottel, H. F. and Savofin, A. F., *Radiative Transfer*, McGraw-Hill, New York, 1967.
10. Howell, J. R., *A Catalogue of Radiation Configuration Factors*, McGraw-Hill, New York, 1982.
11. Brinkworth, B. J., *Solar Energy for Man*, Compton Press, England, 1972.
12. Jones, W. P., *Air Conditioning Engineering*, 3rd edn, Arnold, London, 1985.

FURTHER READING

Badr, O., Probert, S. D. and O'Callaghan, P. W., 'Rankine Cycles for Steam Power Plants', *Applied Energy*, **36**, 1991–231, 1990.
Badr, O., Probert, S. D. and O'Callaghan, P. W., 'Air Conditioning Processes', *Applied Energy*, **37**, 13–35, 1990.

FIVE

ENERGY TECHNOLOGIES

5.1 FUELS AND COMBUSTION

Fossil fuels

Fossil fuels consist generally of the combustible elements: hydrogen, H_2, carbon, C, and sulphur, S. They often also contain the incombustible element nitrogen, N_2, moisture, H_2O, and minerals.

When combusted, the carbon oxidizes to carbon monoxide, CO (if insufficient air is present), and carbon dioxide, CO_2, the hydrogen oxidizes to water (steam), H_2O, and the sulphur (in coals and liquid fuels) becomes sulphur dioxide, SO_2. The carbon and sulphur dioxides dissolve in water to form carbonic, H_2CO_3, and sulphuric, H_2SO_4, acids, which contribute to acid rain.

When combustion temperatures are high, nitrogen in air can combine with oxygen to form the NOx gases, e.g. nitrous oxide, N_2O, which dissolved in water forms nitric acid, HNO_3. The action of sunlight on exhaust fumes can also form nitrous oxides, the main contributors to photochemical smog.

Moisture present in a fuel absorbs sensible heat from the flame to supply the latent heat needed to change phase from liquid to vapour.

Any minerals present form the residual ash.

Smoke is a dispersion in air of small solid particles of uncombusted fuels (carbon and sulphur) with some airborne ash.

Solid fuels The typical composition of solid fossil fuels is shown in Table 5.1.

Liquid fuels These are petroleum oils which are mixtures of hydrocarbons, for example:

- Paraffins, C_nH_{2n+2}

Table 5.1 Composition of solid fuels

Coal	% C	% H	% O	% N + S	% Ash	% Moisture
Anthracite	90.3	3.0	2.3	1.4	2.8	1.0
Bituminous coal	74.0	6.0	13.0	2.3	4.8	2.0–10.0
Lignite	56.5	6.0	32.0	1.6	4.5	up to 15.0

Table 5.2 Composition of liquid fuels

Oil	% C	% H	% S
Motor petrol	85.5	14.4	0.1
Vaporizing oil	86.8	12.9	0.3
Kerosene	86.3	13.6	0.1
Diesel (gas) oil	86.3	12.8	0.9
Light fuel oil	86.2	12.4	1.4
Heavy fuel oil	86.1	11.8	2.1

Table 5.3 Composition of gaseous fuels

Gases	H_2	CO	CH_4	C_2H_4	C_4H_8	O_2	N_2	CO_2
Coal gas	49.4	18.0	20.0	—	2.0	0.4	6.2	4.0
Producer gas	12.0	29.0	2.6	0.4	—	—	52.0	4.0
Blast furnace gas	2.0	27.0	—	—	—	—	60.0	11.0

- Olefins and naphenes, C_nH_{2n}
- Aromatics, C_nH_{2n-6}

and typical compositions are shown in Table 5.2.

Gaseous fuels Most hydrocarbons in petroleum deposits occur naturally as liquids, but a few exist in the gaseous phase at atmospheric temperatures and pressures. Methane, CH_4, is a gaseous paraffin and is the major constituent of natural gas (Table 5.3).

Common gaseous fuels are listed below:

Hydrogen	H_2
Methane	CH_4
Ethylene	C_2H_4
Ethane	C_2H_6
Propane	C_3H_8
Propylene	C_3H_6
Iso-butane	C_4H_{10}

Combustion

The atomic weights of the elements contained in fossil fuels are listed below:

$$
\begin{array}{ll}
\text{O} & 16 \\
\text{C} & 12 \\
\text{N} & 14 \\
\text{H} & 1 \\
\text{S} & 32
\end{array}
$$

Carbon in oxygen

	$C + O_2 = CO_2$
Equation	$C + O_2 = CO_2$
mols	$1 + 1 = 1$
atoms	$1 + 2 = 3$
molecular weight	$12 + 32 = 44$
kg	$12 + 32 = 44$
kg	$1 + 2.67 = 3.67$
m^3	$0 + 1 = 1^\dagger$

Thus 1 kg of carbon requires 2.67 kg of air for complete combustion, and produces 3.67 kg of CO_2.

Hydrogen in oxygen

	$2H_2 + O_2 = 2H_2O$
Equation	$2H_2 + O_2 = 2H_2O$
mols	$2 + 1 = 2$
atoms	$4 + 2 = 6$
molecular weight	$4 + 32 = 36$
kg	$4 + 32 = 36$
kg	$1 + 8 = 9$
m^3	$2 + 1 = 2$

Thus 1 kg of hydrogen requires 8.0 kg of air for complete combustion, and produces 9.0 kg of H_2O.

Methane in air Air contains 21 per cent O_2 and 79 per cent N_2. Thus, for complete, or *stoichiometric* combustion:

	$CH_4 + 2(O_2 + (79/21)N_2) = CO_2 + 2H_2O + (2 \times 79/21)N_2$
kg	$16 + 64 + 7.52 \times 28 = 44 + 36 + 7.52 \times 28$
kg	$1 + 4 + 13.16 = 2.75 + 2.25 + 13.16$
mols	$1 + 2 + 7.52 = 1 + 2 + 7.52$
m^3	$1 + 2 + 7.52 = 1 + 2 + 7.52$

$\dagger$ because, from Avogadro's principle that all gases occupy equal volumes per mol at the same temperature and pressure, the volume of carbon dioxide produced is equal to the volume of the air supplied at the temperature and pressure of the combustion products.

Thus 1 kg of methane requires 17.16 kg of air for complete combustion and produces 18.16 kg of waste products: 2.75 kg of CO_2, 2.25 kg of water and 13.16 kg of hot nitrogen.

1 m^3 of methane requires 9.52 m^3 of air for complete combustion.

Wet exhaust products—analyses by volume

CO_2	1	9.5%
H_2O	2	19.0%
N_2	7.52	71.5%
Total	10.52	100.0%

Thus, if the percentage of CO_2 measured in the *wet* exhaust gases is 9.5 per cent, then stoichiometric combustion has occurred.

Often, the water vapour condenses out in the sampling probe, resulting in a higher apparent percentage of carbon dioxide, which could happen only if carbon dioxide were present in the inlet air mixture (i.e. as is sometimes the case if a *reducing*, or non-oxidizing, atmosphere is required).

The presence of excess air will *reduce* the percentage of carbon dioxide in the exhaust gases (see later).

If the water vapour completely condenses out, the analysis of the *dry* products of combustion is as follows:

Dry exhaust products—analyses by volume

CO_2	1	11.7%
H_2O	0	0%
N_2	7.52	88.3%
Total	8.52	100.0%

Temperature after combustion—adiabatic flame temperatures The *heat of formation* is the heat liberated or absorbed when one mole of a substance is formed from its elements. The heat released during combustion is calculated from the sum of the heats of formation of the

Table 5.4 Standard heats of formation

Substance	Heat, kJ mol^{-1}	Molar mass, grams	Heat, MJ kg^{-1}
O_2, N_2, etc	0	—	0
H_2O vapour	241.8	18	13.43
H_2O liquid	285.7	18	15.87
CO_2	393.8	44	8.95
CO	110.6	28	3.95
CH_4	74.9	16	4.68
CH_2 (kerosene)	25.4	14	1.81

combustion products minus the sum of the heats of formation of the reactants (methane, oxygen and nitrogen here).

Table 5.4 gives standard heats of formation for various gases.

Thus, heat released during combustion = sum of the heats of formation of the combustion products — sum of the heats of formation of the reactants.

=	CO_2	1 mol	393.8	kJ
+	H_2O	2 mols	2×241.8	kJ
+	H_2	7.52 mols	7.52×0.0	kJ
	Total	9.52 mols	877.4	kJ
−	CH_4	1 mol	74.9	kJ
−	O_2	2 mols	2×0.0	kJ
−	N_2	7.52 mols	7.52×0.0	kJ
	Total	10.52 mols	74.9	kJ

Therefore, the total heat released during the combustion of 1 mol of CH_4 is $877.4 - 74.9 = 802.5$ kJ. One mol of a substance is equivalent to M grams of that substance, where M is the molecular weight of the substance. Methane (CH_4) has a molecular weight of 16, so the combustion of 16 grams releases 802.5 kJ. Thus the combustion of 1 kg of CH_4 releases 50 156.25 kJ. The *net* calorific value for methane is thus $50.156 \, MJ \, kg^{-1}$.

The *net* (or lower) calorific value for a fuel is calculated when the H_2O in the combustion products is in its vapour form. The *gross* (or higher) calorific value for a fuel is calculated when the H_2O in the combustion products is in its liquid form. The latent heat of vaporization of water is $2.5 \, MJ \, kg^{-1}$. In the combustion of methane, 2 mols, or 36 grams of water vapour are present in the combustion products of 1 mol, or 16 grams of methane (2.25 kg water per kg methane). In order to vaporize this water, an *extra* 5.625 MJ of heat would have to be released from the fuel. Thus the gross calorific value for methane is $50.156 \, MJ \, kg^{-1} + 5.625 \, MJ \, kg^{-1} = 55.781 \, MJ \, kg^{-1}$ (or $892.5 \, kJ \, mol^{-1}$ (gross c.v.) $= 802.5 \, kJ \, mol^{-1}$ (net c.v.) $+ 90 \, kJ \, mol^{-1}$).

Table 5.5 Lower calorific values for various fuels

Fuel	Calorific value, $MJ \, kg^{-1}$		Molecular weight	$MJ \, mol^{-1}$
Hydrogen gas	120	H_2O	2	240
Liquid hydrogen	94	H_2O	2	186
Methane	50	CH_4	16	800
Heptane	48	C_7H_{16}	100	4800
Propane	46	C_3H_8	44	2024
Fuel Oil	44.7			
Petrol	44.0			
Carbon	32.6	C	12	391
Coal	28.3			
Methanol	15			

Table 5.6 Enthalpies of combustion products ($J\,mol^{-1}$) above 298.15 K (25 °C)

T K	O_2	N_2	CO_2	H_2O vapour	CO
Molar mass (gram)	32	28	44	18	28
300	54	54	67	63	54
400	3 029	2 971	4 008	3 452	2 975
600	9 252	8 901	12 916	10 505	8 947
800	15 847	15 060	22 822	18 007	15 185
1000	22 721	21 478	33 419	26 004	21 700
1200	29 789	28 131	44 506	34 512	28 445
1400	36 990	34 960	55 936	43 501	35 362
1600	44 309	41 931	67 617	52 925	42 408
1800	51 723	49 011	79 486	62 722	49 951
2000	59 239	56 170	91 503	72 846	56 769
2200	66 846	63 397	103 627	83 246	64 054
2400	74 542	70 682	115 849	93 885	71 381
2600	82 329	78 000	128 141	104 729	78 745
2800	90 205	86 365	140 501	115 752	86 148
3000	98 164	92 754	152 914	126 931	93 575

Table 5.7 Enthalpies of combustion products ($MJ\,kg^{-1}$) above 298.15 K

T K	O_2	N_2	CO_2	H_2O vapour	CO
Molar mass (gram)	32	28	44	18	28
300	0.001 69	0.001 9	0.001 5	0.003 5	0.001 9
400	0.095	0.104	0.091 1	0.192	0.106
600	0.289 125	0.317 892 9	0.293 545 5	0.583 611 1	0.319 535 7
800	0.495 218 8	0.537 857 1	0.518 681 8	1.000 389	0.542 321 4
1000	0.710 031 3	0.767 071 4	0.759 522 7	1.444 667	0.775
1200	0.930 906 3	1.004 679	1.011 5	1.917 333	1.015 893
1400	1.155 938	1.248 571	1.271 273	2.416 722	1.262 929
1600	1.384 656	1.497 536	1.536 75	2.940 278	1.514 571
1800	1.616 344	1.750 393	1.806 5	3.484 556	1.783 964
2000	1.851 219	2.006 071	2.079 614	4.047	2.027 464
2200	2.088 938	2.264 179	2.355 159	4.624 778	2.287 643
2400	2.329 438	2.524 357	2.632 932	5.215 833	2.549 321
2600	2.572 781	2.785 714	2.912 295	5.818 278	2.812 321
2800	2.818 906	3.084 464	3.193 205	6.430 667	3.076 714
3000	3.067 625	3.312 643	3.475 318	7.051 722	3.341 964

Measurement of calorific values The calorific value of a fuel is measured in calorimetric tests where the products of combustion are cooled to normal atmospheric conditions (Table 5.5). The total heat released in the process of cooling is measured. During this process, water vapour present in the combustion products condenses, releasing its latent heat, and resulting in more heat being liberated than would be available had no condensation occurred. The calorific value indicated by such tests is thus the gross calorific value. The net calorific value is then obtained by subtracting the latent heat of the water present from the gross calorific value.

The *adiabatic flame temperature* is obtained by dividing the total heat released during combustion by the total heat capacity of the combustion products.

Table 5.6 lists the enthalpies (over 25 °C) of combustion products over the temperature range 600–3000 K. Enthalpies, given in $J\,mol^{-1}$, are converted to $MJ\,kg^{-1}$ in Table 5.7.

Energy balance ($MJ\,mol^{-1}$)

heat released = heat in combustion products at 2400 K	(above 25 °C = 298.15 K) at 2200 K
0.8025 =		
for CO_2	0.115 849	0.103 627
for $2H_2O$	2 × 0.093 885	2 × 0.083 246
for $7.52N_2$	7.52 × 0.070 682	7.52 × 0.063 397
Totals	0.835 147 *too large*	0.746 864 *too small*

Interpolating to match the heat released with the heat in the combustion products, the temperature after combustion becomes 2326 K.

Thermal efficiency

$$\% \text{ heat lost in the flue gas} = 100\% - \text{efficiency} \%\qquad(5.1)$$

The *thermal efficiency of combustion* is usually based upon the net calorific value of the

Table 5.8 Enthalpy of the stack products

Combustion product	Enthalpy, $MJ\,mol^{-1}$
CO_2	0.012 916
$2H_2O$	2 × 0.010 505
$7.52\,N_2$	7.52 × 0.008 901
Total	0.100 862

fuel and is defined as

$$\text{efficiency} = \frac{\text{heat released} - \text{heat in flue gas}}{\text{heat released}} \qquad (5.2)$$

Efficiency based upon net calorific value = (gross calorific value/net calorific value) × efficiency based upon gross calorific value.

For methane, efficiency based upon net calorific value = 1.112 × efficiency based upon gross calorific value.

Parameters for a stack temperature of 327 °C (= 600 K) are shown at Table 5.8.

Heat released = 0.8025 MJ mol^{-1}. Therefore efficiency = 100% × (0.8025 − 0.100 862)/ 0.8025 = 87.43%$_{\text{net}}$. Repeating this analysis with varying stack temperatures gives the following results:

Stack temperature, K	Efficiency$_{\text{net}}$, %	Efficiency$_{\text{gross}}$, %
600	87.4	78.6
800	78.5	70.6
1000	69.2	62.2
1200	59.5	53.5

Excess air It is often necessary to supply a greater amount of air than the stoichiometric requirement so that complete combustion can be assured in practical furnaces. This has the effect of cooling the combustion products.

For stoichiometric combustion

$$CH_4 + 2(O_2 + (79/21)N_2) = CO_2 + 2H_2O + (2 \times 79/21)N_2 \qquad (5.3)$$

If there is a percentage of excess air present, A = a/100, then

$$CH_4 + 2(1 + A)(O_2 + (79/21)N_2) = CO_2 + 2H_2O + 2AO_2 + 2((1 + A)79/21)N_2$$

$$
\begin{array}{llllll}
CH_4 & + 2(1 + A)O_2 & + (2(1 + A)79/21)N_2 & = CO_2 & + 2H_2O + 2AO_2 & + (2(1 + A)79/21)N_2 \\
1 \text{ mol} & + 2(1 + A) & + 2(1 + A)79/21 & = 1 & + 2 & + 2A & + 7.52(1 + A) \\
1 & + 2 + 2A & + 7.52(1 + A) & = 1 & + 2 & + 2A & + 7.52(1 + A)
\end{array}
$$

$$(5.4)$$

Molar fraction = volume fraction of the exhaust products.
Thus, the wet analysis of the exhaust products is as follows:

CO_2 1
H_2O 2
O_2 2A
N_2 7.52(1 + A)

Figure 5.1 shows how the percentage constituents of the wet exhaust products vary with percentage excess air.

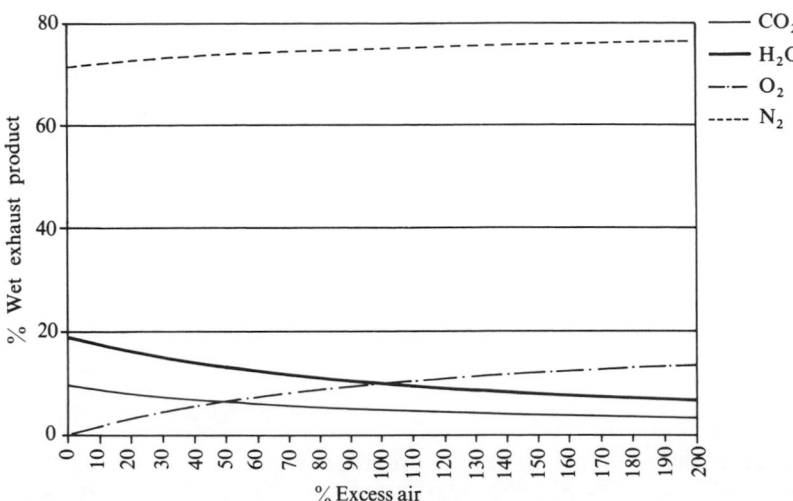

Figure 5.1 Combustion of methane—variations of wet exhaust product constituents with excess air.

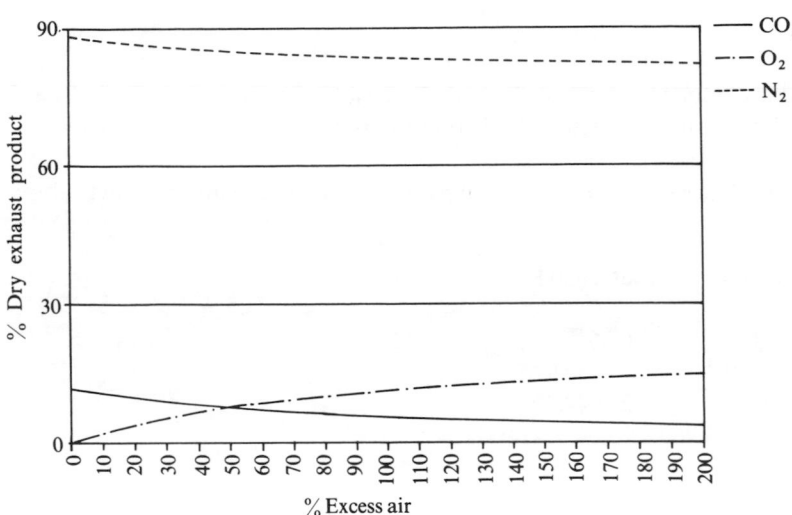

Figure 5.2 Combustion of methane variations of dry exhaust product constituents with excess air.

The dry analysis of the exhaust products is as follows:

CO_2 1
O_2 2A
N_2 7.52(1 + A)

Figure 5.2 shows how the percentage constituents of the dry exhaust products vary with percentage excess air.

Temperature after combustion

$$CH_4 + 2(1+A)O_2 + (2(1+A)79/21)N_2 = CO_2 + 2H_2O + 2AO_2 + (2(1+A)79/21)N_2$$
$$16\,kg + 64(1+A) + 210.6(1+A) = 44 + 36 + 64A + 210.6(1+A)$$
$$1 + 4(1+A) + 13.16(1+A) = 2.75 + 2.25 + 4A + 3.16(1+A)$$

The heats of formation of the combustion products of CH_4 ($kJ\,mol^{-1}$) are as follows:

CO_2	393.8
$2H_2O$ vapour	2×241.8
$2AO_2$	$2A\,0.0$
$7.52(1+A)N_2$	$7.52(1+A) \times 0.0$
Total	877.4

The heats of formation of the reactants ($kJ\,mol^{-1}$) are as below:

CH_4	74.9
$2O_2$	2×0.0
$7.52N_2$	7.52×0.0
Total	74.9

Total heat released accompanying the combustion of 1 mol (16 grams) of $CH_4 = 877.4 - 74.9 = 802.5$ as for stochiometric combustion. This represents $50.156\,MJ\,kg^{-1}$ of methane.

Energy balance (kJ) For 100% excess air, heat released = heat in combustion products above 25 °C (298.15 K):

	at 1400 K	at 1600 K
$802.5 =$		
CO_2	55.936	67.617
$2H_2O$	2×43.501	2×52.925
$2O_2$	2×36.99	2×44.309
$7.52(1+1)N_2$	15.04×34.960	15.04×41.931
Totals	742.716	892.727
	too small	*too large*

Interpolating, the flame temperature required to balance the equation is 1520 K ($= 1247$ °C). Repeating the analysis:

% Excess air	Combustion temperature, K
0	2326
40	1630
60	1440
80	1317
100	1247

Figure 5.3 shows the variation of combustion temperature with percentage excess air.

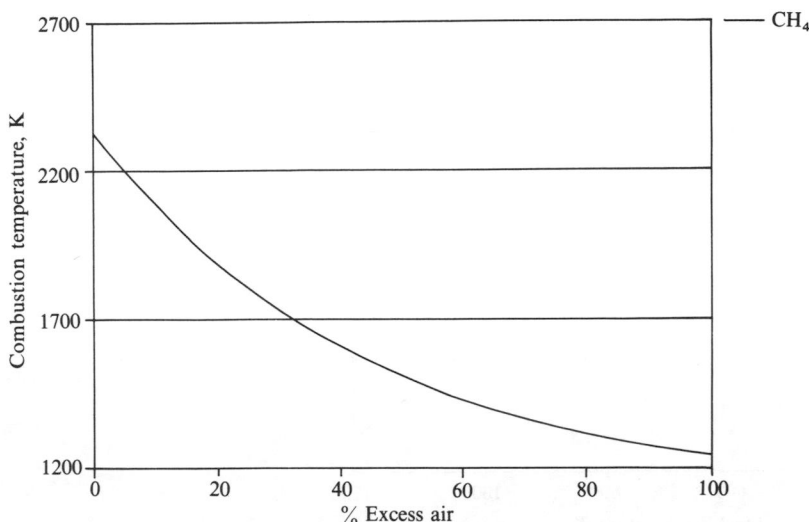

Figure 5.3 Effect of percentage excess air on the combustion temperature of methane.

Effect of percentage excess air on efficiency

$$\text{efficiency} = (\text{heat released} - \text{heat in stack})/\text{heat released} \qquad (5.5)$$

For a stack temperature of $327\,°C\,(= 600\,K)$, the enthalpy ($kJ\,mol^{-1}$) of the stack products is as follows:

Product	Enthalpy
CO_2	12.916
$2H_2O$ vapour	2×10.505
$2O_2$	2×9.252
$7.52(1 + A)N_2$	$7.52(1 + A) \times 8.901$

For 100% excess air

Product	Enthalpy
CO_2	12.916
$2H_2O$ vapour	2×10.505
$2O_2$	2×9.252
$7.52(1 + 1)N_2$	$7.52(1 + 1) \times 8.901$
Total	186.301

Heat released $= 802.5\,kJ\,mol^{-1}$, and thus efficiency $= 100\% \times (802.5 - 186.301)/802.5 = 76.78\%$.

Repeating this analysis for various stack temperatures results in the characteristic given in Fig. 5.4 and Table 5.9.

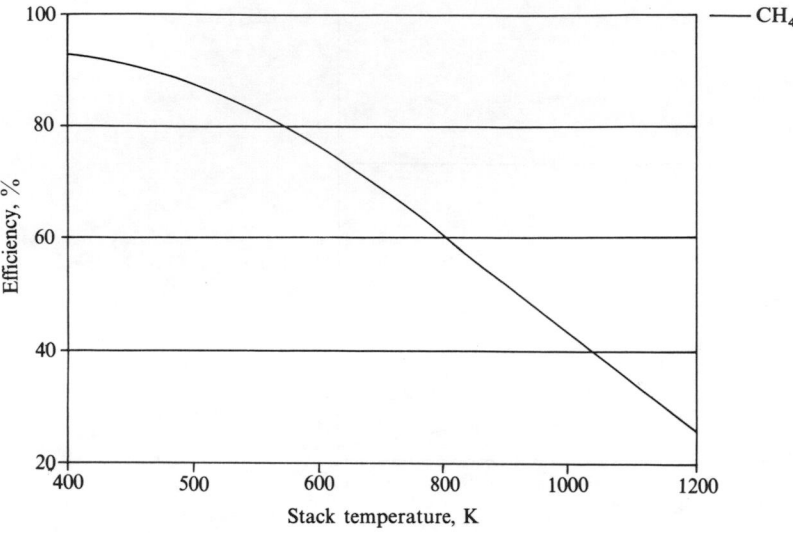

Figure 5.4 Effect of stack temperature on efficiency in the combustion of methane.

Table 5.9 Relations between stack temperatures, excess air and efficiencies

Stack temperature, °C	% Excess air	Efficiency %
1000	0	69
	50	56
	100	43
800	0	79
	50	70
	100	60
600	0	87
	50	82
	100	77

Data for the combustion of some other common gases is given at Table 5.10.

If gross calorific values are used, losses due to the enthalpies of water vapour in the exhaust gases are approximately 4 per cent if the fuel is coal, 6 per cent if the fuel is oil and 11 per cent if the fuel is natural gas.

5.2 BOILERS

Boilers are usually used for steam raising or for heating hot water for space heating, to provide domestic hot water, or for other industrial purposes. Steam is used for heating, cleaning, autoclaving and for driving turbines to provide electrical power.

Table 5.10 Combustion data for common gases

Hydrogen	$2H_2$	$+O_2$	$= 2H_2O$	
kg	4	+ 32	36	
mol(volume)	2	+ 1	=	2

Ethylene	C_2H_4	$+ 3O_2$	$= 2CO_2$	$+ 2H_2O$
kg	28	+ 96	= 88	+ 36
mol(volume)	1	+ 3	= 2	+ 2

Ethane	C_2H_6	$+ 3.5O_2$	$= 2CO_2$	$+ 3H_2O$
kg	30	+ 112	= 188	+ 54
mol(volume)	1	+ 3.5	= 2	+ 3

Propane	C_3H_8	$+ 5O_2$	$= 3CO_2$	$+ 4H_2O$
kg	44	+ 160	= 132	+ 72
mol(volume)	1	+ 5	= 3	+ 4

Propylene	C_3H_6	$+ 4.5O_2$	$= 3CO_2$	$+ 3H_2O$
kg	42	+ 144	= 132	+ 54
mol(volume)	1	+ 4.5	= 3	+ 3

Table 5.11 Boiler losses

Type of boiler	% loss at MCR
Modern Package	3.0
Water Tube	3.5
Economic Wet-back	4.0
Economic dry-back	5.0
Sectional Boiler	5.0
Cornish and Lancashire	6.5
Vertical Boiler	6.5
Sectional Boiler with no heat transfer under the furnace	8.5

MCR $\equiv$ maximum continuous rating.

Table 5.12 Pro-forma for boiler efficiency report

Type of boiler   Hot Water/Steam/Other
Make
Maximum continuous rating, MCR (kW)
Working pressure ($N m^{-2}$)
Working temperature (K)
Fuel type and grade
Type and make of stoker/burner

Test data:
Loading, % MCR

Fuel:
Chemical analysis
Gross calorific value ($MJ kg^{-1}$)

Flue gas exit temperature (K)
Boiler house temperature (K)
CO_2 or O_2 in flue gas, %
CO concentration in flue gas

Analyses:
Sensible heat of dry flue gas %
Enthalpy of water vapour in flue gas %
Unburnt gas (CO) losses %
Radiation, convection and miscellaneous %
Blow-down Losses %
Total losses %

Efficiency (100 − total losses)

Estimate of annual plant efficiency %

Small- to medium-sized space heating installations generally use low pressure, low temperature (82 °C flow, 70 °C return) hot water in sectional cast iron boilers. Larger space-heating systems may use steam or high pressure, high-temperature (200 °C flow, 100 °C return) hot water. Shell (fire tube) boilers are used for steam and high-temperature hot water. Steam systems can need steam from 2 to 17 atmospheres pressure. Values for efficiency calculated from analysing stack losses must be further discounted for losses due to casing losses and blow-down losses. These are typically of the order of 7 per cent for a steam boiler.

Radiation, convection and miscellaneous losses are approximately as shown at Table 5.11.

For a boiler operating at loads below the maximum continuous rating, the losses (Table 5.11) may be multiplied by the turn-down ratio based upon metered fuel supplied. The blow-down loss is taken as zero for hot water boilers.

A form to enable estimates of boiler efficiency is shown at Table 5.12, and typical efficiencies for various types of boiler at Table 5.13.

Energy conservation in boilers

Heat losses from the flue gases can be minimized by running boilers efficiently.

The amount of combustion air should be limited to that necessary to ensure complete combustion of fuels at all times, with a slight margin of excess air to suit the installation. Excess air increases flue losses. Too little air results in incomplete combustion of the fuel. Heat transfer surfaces should be kept clean and soot formation should be avoided at all times.

If the combustion mixture is set correctly and heat transfer is uninhibited, the addition of fuel additives and other 'fuel-saving' devices are unlikely to produce further improvements. An *economizer* can save up to 5 per cent of the fuel used by using waste flue gases to pre-heat the feed water via a recuperative heat exchanger. The introduction of *baffles* to bring combustion gases into closer contact with water or steam tubes is likely to be effective only on old boilers.

A recent development sprays water through the flue gases to recuperate some heat for use elsewhere. A rotary regenerator may be used to recover some of the heat in exhaust gases to heat air for space heating or for pre-heating the combustion air. *Flue dampers* might be used to close down flues when boilers sharing the same chimney close down intermittently.

Table 5.13 Typical efficiencies for different boiler types (based on gross calorific value)

Boiler type	Fuel	Efficiency %
Industrial water tube boiler with economizer	Oil	84–87
	Gas	80–83
Shell boiler without economizer	Oil	81–84
	Gas	77–80
	Coal	77–81
Sectional boilers low temperature hot water	Oil	75–80
	Gas	71–76
	Coal	71–76

Radiation and convection losses may be reduced by proper insulating techniques. These become appreciable especially under low loads.

Boiler *blow-down* is necessary to remove sludge from precipitated salts, and so prevent scale forming as an extra thermal resistance on the tubes. Frequencies and durations of blow-down should be just enough to achieve this—and no more. A heat exchanger might be installed to recover heat from blow-down steam or water, to pre-heat feed water.

Correct chemical water treatment is essential to prevent scale formation and to reduce or eliminate corrosion in the boiler. In steam systems, as much condensate as possible should be recovered and returned to the boilerhouse to recover heat, make-up water and chemicals used for water treatment. Continuous logs of boiler performance should be kept.

Fuel Efficiency Booklets 14 (Oil), 15 (Gas) and 17 (Coal), produced by the UK Energy Efficiency Office, contain some useful charts for estimating boiler efficiencies rapidly.[1] Energy management checklists for furnaces, boilers, boilerhouse auxiliaries, indirect heated vessels and driers are given in Appendix 2.

5.3 INSULATED PIPEWORK SYSTEMS

All pipework and valves carrying hot water, steam and condensate should be properly lagged to economic optima.

Consider a metal pipe containing a hot fluid:

Inside pipe temperature		100 °C
Annual mean outside pipe temperature		10 °C
Pipe thermal conductivity k		60 W m^{-1} K^{-1}
Length of pipe L		100 m
Inner Diameter of pipe $2r_i$		0.1 m
Outer Diameter of pipe $2r_o$		0.11 m
Pipe thermal resistance R_p	$\ln(r_o/r_i)/2[\pi]kL$	0.000 002 5 K W^{-1}
Lagging Thermal Conductivity k_1		0.04 W m^{-1} K^{-1}
Lagging outer diameter $2r_1$		0.33 m
Lagging thermal resistance R_1	$\ln(r_1/r_o)/2[\pi]k_1L$	0.043 712 4 K W^{-1}
Total thermal resistance R_t		0.043 714 9 K W^{-1}
Unit cost of lagging		100 £ per m^3
Volume of lagging		69.11 m^3
Cost of lagging		£6 911.5
Heat lost from uninsulated pipe		355 987 16 W
Cost of energy		0.03 £ per kWh
Cost per year of heat lost from uninsulated pipe		£935 534 3
Heat lost from insulated pipe		2 058.794 W
Cost per year of heat lost from insulated pipe		£541.0
Annual savings by insulating		£935 480 2
Straight payback period		6.47 hours

Thus the addition of lagging 11 cm thick to the pipe of outer diameter 11 cm costs £ 6911, but pays back the investment in a matter of hours. Economic lagging thicknesses on existing insulated pipework should be reviewed regularly.

5.4 BUILDING HEAT BALANCE

Figure 5.5 shows a schematic view of a building. Energy enters in fuels, electricity, as sundry gains from people and solar gains. Heat energy leaves the system boundary as stack losses, fabric transmission losses and ventilation losses. The annual heat balance equation is as follows:

energy supplied in fuels and electricity + sundry gains

= fabric transmission heat losses + ventilation losses + stack losses

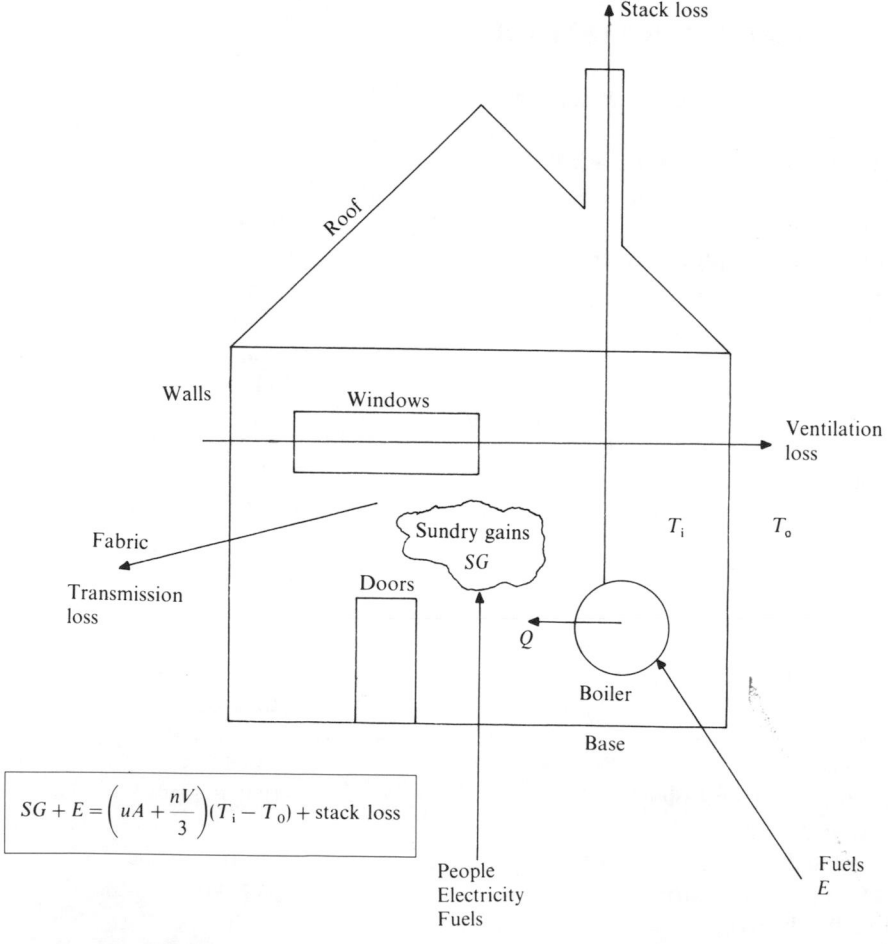

$$SG + E = \left(uA + \frac{nV}{3} \right)(T_i - T_0) + \text{stack loss}$$

Figure 5.5 Building heat balance.

or, from Chapter 2,

$$E + SG = (UA + nV/3)(T_i - T_o) \times \text{(number of hours per year)}/1000 + (1 - \eta)E(\text{kWh}) \qquad (5.6)$$

where E = total energy supplied in one year, kWh

$\quad SG$ = sundry gains from electrical devices, people and solar influx, kWh

$\quad U$ = overall U-value for the building, $\text{W m}^{-2} \text{K}^{-1}$, i.e.

$$(UA)_{\text{total}} = (UA)_{\text{walls}} + (UA)_{\text{glazing}} + (UA)_{\text{roof}} + (UA)_{\text{base}} \qquad (5.7)$$

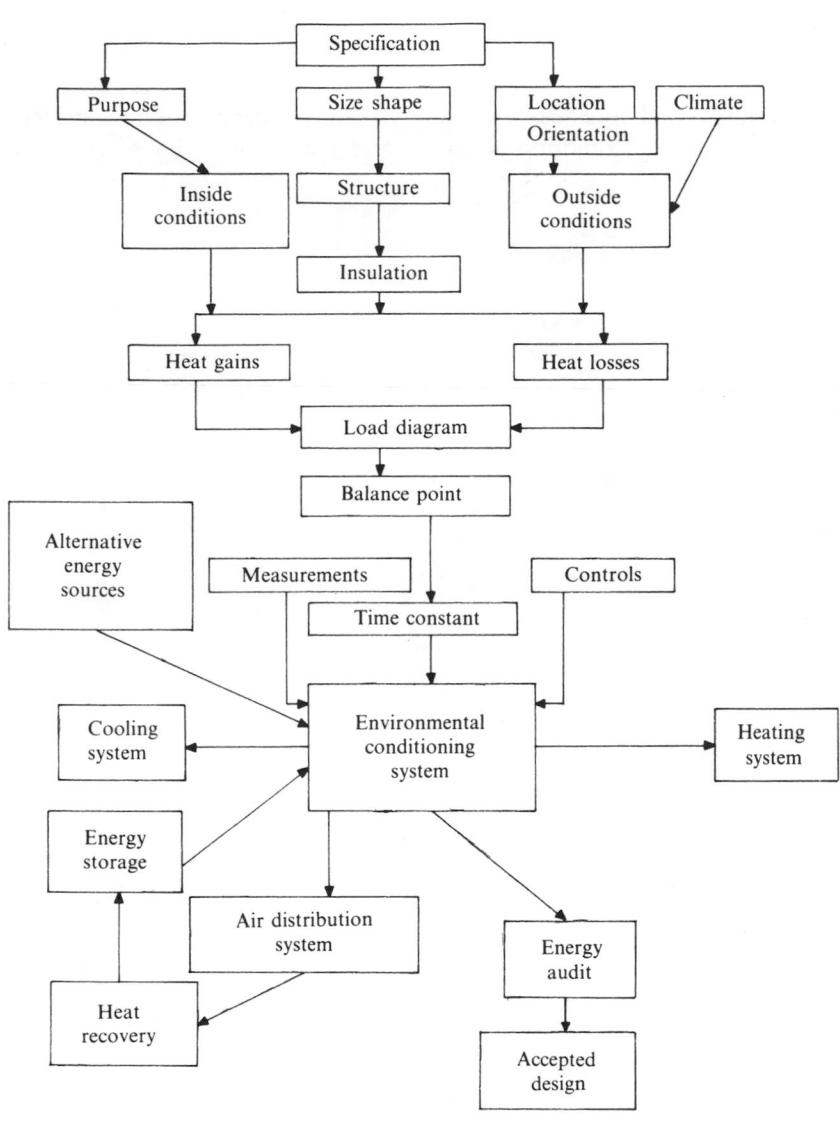

Figure 5.6 Building design chart for energy conservation.

n = average number of ventilation air changes/hour
V = enclosed volume of the building, m^3
T_i = annual mean inside air temperature, °C
T_o = annual mean outside air temperature, °C
η = annual furnace/boiler efficiency.

Figure 5.6 summarizes the considerations needed to solve this equation and so produce an energy audit or specify a heating or cooling system for the building. Also included in the chart are the energy conservation options: controls, insulation options, heat recovery, use of alternative energy sources and energy storage.

Comfort and climate

The specifications for the inside and outside temperatures, T_i and T_o, depend upon considerations of the inside thermal comfort conditions required and the outside climate. A typical

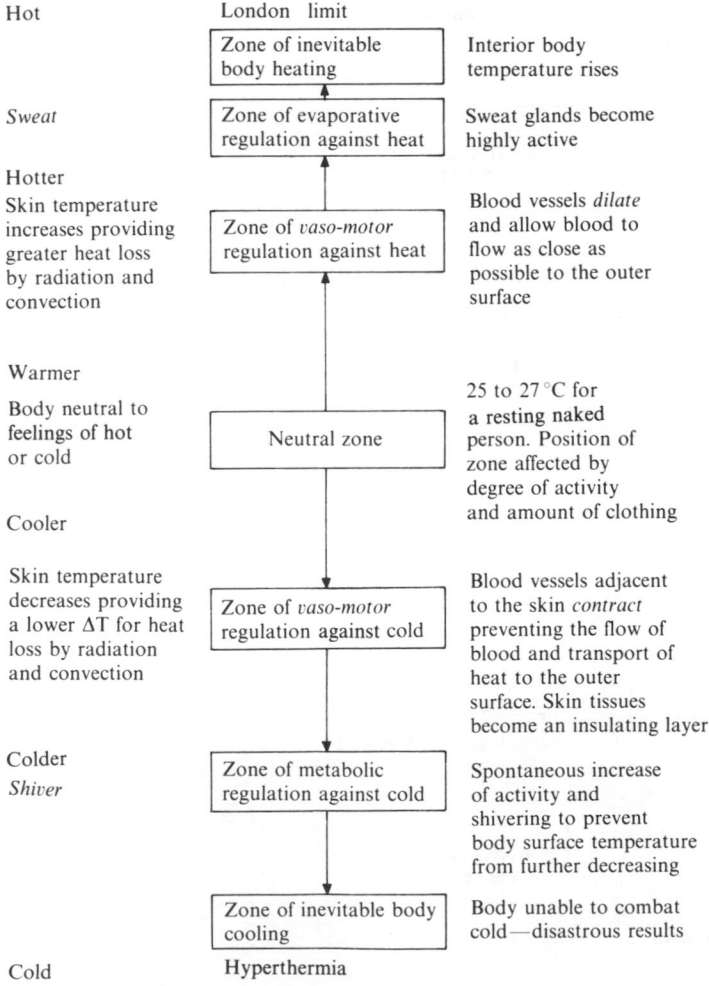

Figure 5.7 Human body regulatory processes against heat and cold.

building in the UK will be maintained at around 20 °C, whilst the mean annual outside temperature (at 51.7 °N) is about 11 °C. A reduction in mean annual temperature of 2 °C (to 18 °C) would result in a saving of 2/9 (22 per cent) of the annual energy bill. Clearly, inside air temperatures should be specified accurately and controlled precisely. Hence the importance of thermal comfort studies.

Thermal comfort An *ideally comfortable environment* is one in which the occupants experience no *heat stress* or *thermal strain*. A comfortable condition is in the neutral zone (Fig. 5.7), where a body needs to take no particular action to maintain its proper heat balance. The deep body temperature of a human being is 37 °C, and the skin surface temperature is 33 °C in a condition of thermal comfort. The body is a heat engine which converts energy in food to work and rejects heat in the process. In order to avoid overheating, this heat must be transferred from deep in the body to the external environment. The temperature of a thermally comfortable environment lies within the range 25 °C–27 °C for a naked resting human being. This temperature reduces with clothing level and increased metabolic activity (Fig. 5.8). The shaded rectangle represents the ASHRAE comfort zone.

In the comfort state, the body loses 75 per cent of its heat by natural convection to the

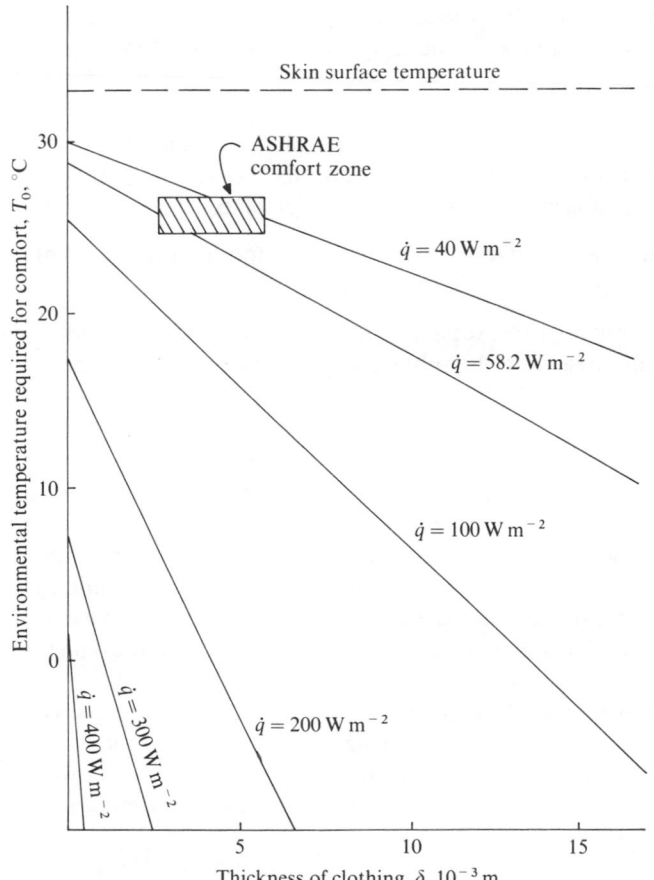

Figure 5.8 Effects of clothing levels and metabolic activities on the environmental temperatures required for human thermal comfort.

surrounding air and by radiation to surrounding surfaces, whilst the remaining 25 per cent is transferred to the environment accompanying the evaporation and respiration of water vapour. The passage of liquid water through the pores is an essential life process to prevent the body components from drying out and to ensure the correct operation of bodily organs.

When the environmental temperature *increases* from that of the comfort state, thermal comfort might be restored by loosening collars and cuffs, removing clothing and/or increasing the surface area of the body by, for example, stretching the arms.

If the skin temperature increases, heat is lost at a greater rate by convection and radiation, blood vessels dilate to allow blood flow as close as possible to the outer surface. This is known as *vasodilation*, or the zone of *vasomotor regulation against heat*. The thermal resistance between the blood vessels and the outside environment decreases, allowing a greater rate of heat flow.

If the environmental temperature increases further, the sweat glands are activated, heat loss from the body is then enhanced by the evaporation of liquid water from the skin's surface. This is known as the zone of *evaporative regulation against heat*.

Eventually, a limit to this mechanism is encountered when the skin's surface becomes 100 per cent wet and no further increased evaporation can occur. The body dehydrates, overheats, and, if the condition is prolonged, the interior body temperature rises leading to heat stroke. An external environmental temperature of 40 °C accompanied by a relative humidity of 50 per cent (the *London Limit*) could cause this condition (see Fig. 5.9).[2]

The comfort zone shown in the figure has been located by eliciting subjective responses to various environments. It applies for a basal metabolic rate, a low level of clothing and low ambient air speed.

Heat exhaustion arises from failure of the normal blood circulation. Its symptoms are fatigue, headache, dizziness, vomiting, abnormal mental reactions and fainting. It causes no permanent injury and recovery is rapid when the subject is removed to a cooler place.

Heat cramps result from loss of salt due to an excessive rate of evaporative loss. Painful muscle spasms may be avoided by the proper use of salt tablets.

Heat stroke ensues when the body is exposed to excessive heat. Then the body temperature may climb to 41 °C or higher. Sweating ceases and the subject may enter a coma, with damage to the brain and death imminent.

When the environmental temperature *decreases* from that of the comfort state, thermal comfort might be restored by tightening collars and cuffs, putting on more clothing and/or decreasing the surface area of the body by, for example, folding the arms.

If the skin temperature decreases, heat is lost at a lesser rate by convection and radiation, the skin contracts and thickens, causing the hairs on the surface to stand out providing an insulating layer. The blood vessels contract, slowing the flow of blood and the transport of heat to the outer surface of the body. This is known as *vasoconstriction*, or the zone of *vasomotor regulation against cold*.

If the environmental temperature decreases further, the subject may subjectively increase metabolic activity by stamping feet or waving arms. Shivering is a spontaneous increase of metabolic activity which prevents skin temperature from further decreases. This is known as the zone of *metabolic regulation against cold*. Eventually, the body is unable to combat lower temperatures and hyperthermia ensues.

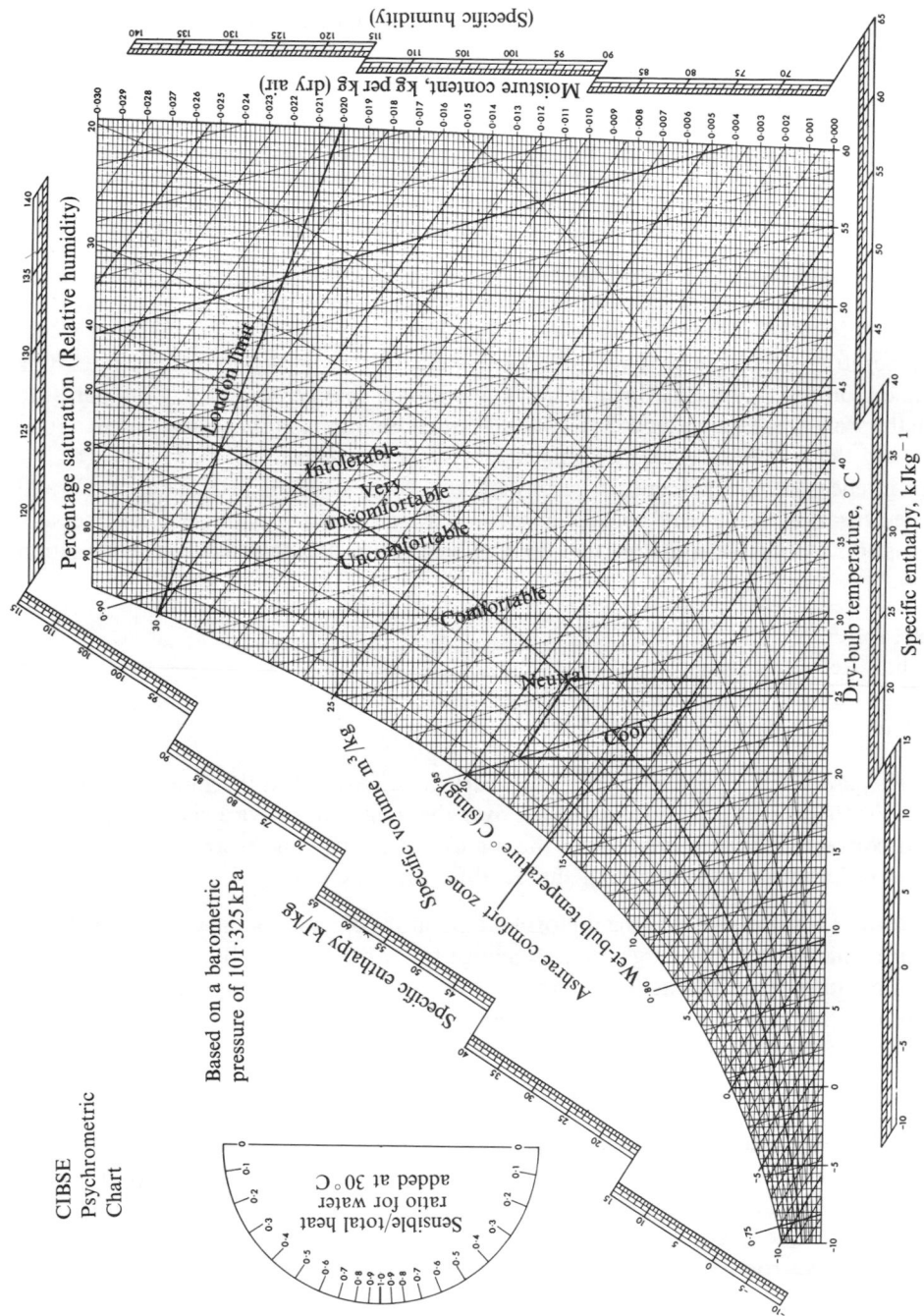

Figure 5.9 The ASHRAE thermal comfort zone.

CIBSE Psychrometric Chart

Based on a barometric pressure of 101·325 kPa

© CIBSE LONDON 1985

205

The *Bedford Comfort Scale* is a seven point scale by which subjects are asked to assess their environment; namely

1.	Much too cool	17 °C
2.	Too cool	20 °C
3.	Comfortably cool	23 °C
4.	Comfortable	26 °C
5.	Comfortably warm	29 °C
6.	Too warm	32 °C
7.	Much too warm	35 °C

The temperature which elicits a vote of 4 on the Bedford scale depends upon the levels of activity and clothing. Most experimental work has shown that corresponding temperature intervals on the scale are of the order of 3 °C. Thus, a naked human being with a basal metabolic rate might yield the responses corresponding to the temperatures shown above.

Human intellectual and perceptual performance is highest when in a state of thermal comfort. The sensation of warmth or cold is registered by the nerve endings when the skin temperature is above or below approximately 30 °C. The principle factors affecting this sensation are as below:

- Air temperature
- Mean radiant temperature
- Air velocity
- Relative humidity

Ambient air temperature, T_a is measured by an ordinary mercury-in-glass thermometer, thermocouple probe or thermistor. The rate of airflow over the measuring device should be high to ensure good convective heat transfer from the air to the probe. The instrument should be shielded from radiations emanating from surrounding surfaces.

The mean radiant temperature, T_r, is the uniform temperature of an imaginary thermally black enclosure with which a human being would exchange the same net heat by radiation as with the surrounding surfaces in the actual complex environment. If the temperatures, T_{1-n}, of all surrounding n surfaces of areas, A_{1-n} are known, the mean radiant temperature can be calculated from

$$T_r = \Sigma A_{1-n} T_{1-n} / \Sigma A_{1-n} \tag{5.8}$$

Otherwise, proprietary devices are available to measure T_r. These, in general, suppress convective heat transfer from the air and enhance radiative heat transfer by having a large black area for radiation interception.

In the comfort state, the human body exchanges approximately equal amounts of heat by radiation to the surrounding surfaces and by natural convection to adjacent air.

The effect of increased air velocity, v, is to increase the rate of convective heat loss to the ambient air. Heat will be transferred at a greater rate between the skin's surface and the air at temperature T_a. The subject will thus feel comfortable in higher air temperatures than those indicated with the Bedford scale. When the air temperature is below that corresponding to the comfort condition, increased air velocity will extract heat from the body at a greater rate, resulting in the environment feeling cooler than implied by the temperatures given with the Bedford scale.

Humidity effects The rate of evaporation of water from the skin's surface is proportional to the difference between the saturation vapour pressure of water corresponding to the temperature of the skin's surface and the partial pressure of the water vapour in the surrounding air, given by the relative humidity, ϕ, (expressed as a fraction) × the saturation pressure, p_v, corresponding to the air temperature.

Thus, as the relative humidity reduces, the air becomes drier, and the rate of evaporation from the skin's surface increases for an invariant air temperature. The subject thus feels cooler in à drier environment.

Conversely, as the relative humidity increases, the air becomes wetter, and the rate of evaporation of sweat from the skin's surface reduces for an invariant air temperature. The subject thus feels warmer in a wetter environment. Hence the difference in comfort sensations in Florida (at 90 per cent relative humidity) and Egypt (at 25 per cent relative humidity) even though the air temperatures may be the same.

The general heat balance equation over the human body In general:

The rate of *heat storage* within the body
   = the rate of *metabolism*
      − the rate of performing *work*
      − the rate of heat loss by *convection* to the surrounding air
      − the rate of heat loss by *radiation* to the surrounding surfaces
      − the rate of heat loss accompanying *evaporation* of water from the skin's surface
      − the rate of heat loss accompanying *breathing and respiration*

or

$$S = M - W - C - R - E - B \tag{5.9}$$

where S = the rate of heat storage in the body, W (S + ve—overheating, S − ve—body cooling)
    M = the rate of metabolism, which depends upon the rate of performing work, W (see Table 5.14)
    W = the rate of performing work, W
    C = the rate of heat loss by convection to the surrounding air, W
    R = the rate of heat loss by radiation to the surrounding surfaces, W
    E = the rate of heat loss accompanying the evaporation of water from the skin's surface, W
    B = the rate of heat loss accompanying breathing and respiration, W

The surface area of a typical body is $1.8 \, \text{m}^2$. The total rate of water loss from the body may be measured from weight loss. *Respired vapour loss* depends upon the rate of metabolic activity (see Table 5.14). *Convective losses* depend upon the air temperature. *Radiative losses* depend upon the mean radiant temperature. *Convective and evaporative losses* depend upon the ambient air velocity. *Evaporative losses* depend upon the partial pressure of the water vapour in the ambient air, and hence the temperature and relative humidity of the air.

Table 5.14 Metabolic rates for different activity levels

Activity	Rate of heat production, $W\,m^{-2}$
Sleeping	40
Seated quietly	60
Office work	60–80
Golf	80–150
Garage work	80–170
Vehicle driving	80–180
Domestic work	80–200
Teacher	90
Machine work	100–260
Carpentry	100–370
Light work	120
Shop assistant	120
Walking at 3 mph	150
Medium work	170
Foundry work	170–400
Tennis	200–270
Squash	290–420
Heavy work	300
Wrestling	400–500
Heaviest work possible	500

Effects of clothing The addition of clothing makes heat transfer paths more complex. Conduction, convection and radiation from the body to the inner surface of the clothing occurs. Conduction, convection and radiation heat transfers take place across air cavities formed within clothing layers. Air currents are set up inside clothing layers. Water loss is inhibited by the diffusion resistances of fabrics. The thermal resistance of clothing assemblages have been expressed in '*togs*' (1 tog $= 0.1\,K\,m^{2}\,W^{-1}$).

Climate Local climates depend upon variations in air temperature, air humidity and wind speeds and directions. Seasonal changes are due to the earth's revolution about the sun. Because the axis of the earth's rotation about its own axis is tilted by 23.5 degrees from the axis of the earth's rotation about the sun, the northern hemisphere is tilted away from the sun in winter and tilted towards the sun in summer (see Fig. 5.10). In winter, the path length for the sun's rays through the atmosphere is longer, more solar radiation is absorbed and reradiated by the atmosphere and the surface of the earth becomes cooler.

The steady-state energy balance for the earth is shown in Fig. 5.11. Note that each zone in the diagram is in thermal equilibrium, resulting in a mean earth surface temperature of 15 °C.

Weather patterns are caused by two basic mechanisms:

- Because the equatorial regions are hotter than the polar regions, hot air rises at the equator and descends towards the poles (Fig. 5.12). These thermal upcurrents in equatorial regions

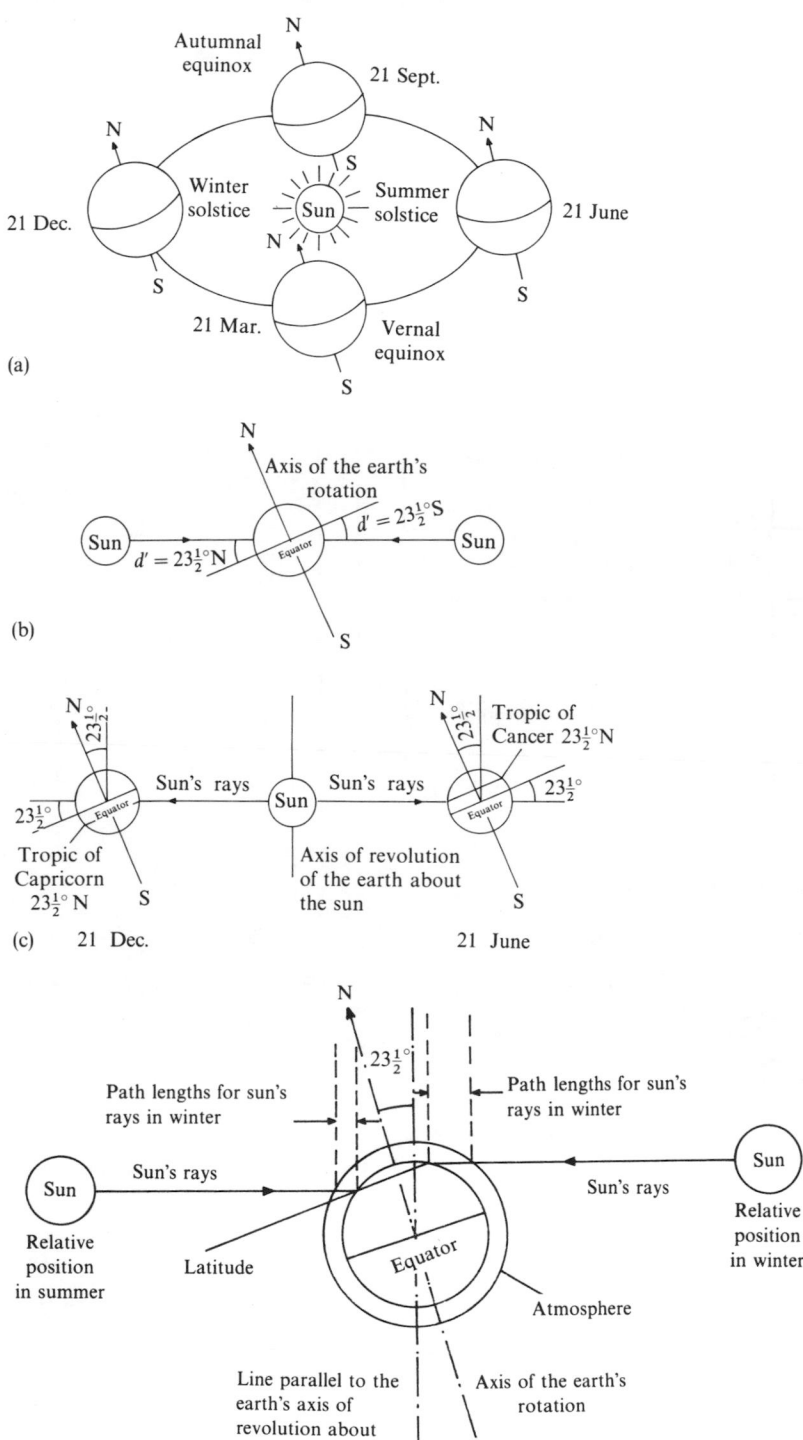

Figure 5.10 Summer and winter: (a) equinoxes and solstices; (b) declination of the sun; (c) the Tropics; (d) summer and winter path-lengths.

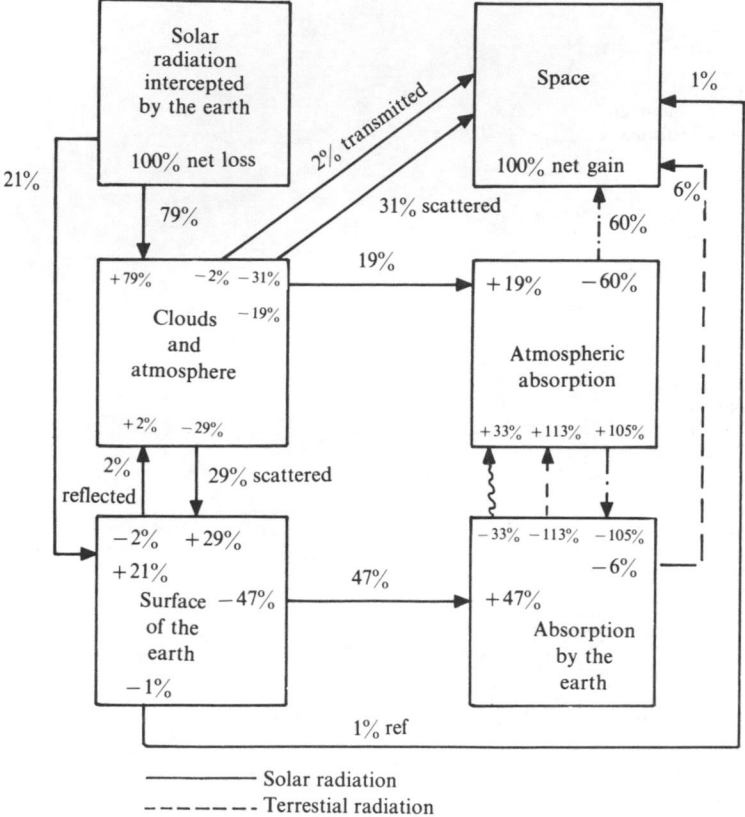

Figure 5.11 Steady-state energy balance for the earth.

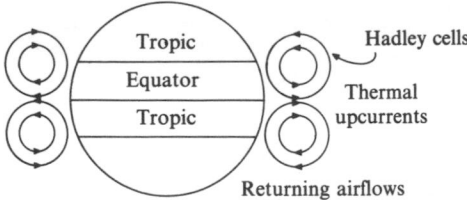

Figure 5.12 Permanent global air movements—the Hadley cells are shown, the Ferrel cells are induced near the poles.

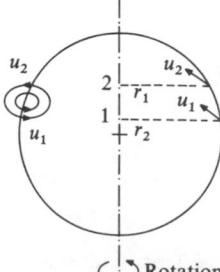

Figure 5.13 Deviation of global air movements caused by the rotation of the earth about its axis.

and down currents at higher latitudes cause the trade winds at intermediate latitudes and the doldrums in the region of the equator. They also transfer heat and water vapour from equatorial to temperate regions. For latitudes higher than 40 degrees, the earth loses more heat by radiation than it receives from the sun.

- Viscous effects cause the air mass to rotate with the rotation of the earth. This rotation is greater at the equator and so a second set of vortices are established (Fig. 5.13).

The geography of a location also effects climate, determining how much solar energy is absorbed and stored by surface materials and water, and how readily this stored energy is released to the atmosphere.

The atmosphere is comparatively transparent to insolation. Cloud forms an insulating barrier and reflects both solar energy and the low frequency infra-red radiation emitted from the earth's surface. Land masses are opaque and good absorbers. Water masses are partially transparent, energy being absorbed in depth. Since the heat penetrates deeper, the surface temperature of water does not reach as high a value as that of the land surface during the daytime. At night, land masses lose heat more rapidly than water masses, which have higher thermal capacities. Thus local differential rates of heating and cooling due to local terrain and water masses cause further localized weather patterns (Fig. 5.14). It follows that much more extreme variations in dry-bulb temperature occur in the middle of land masses. The presence of water masses form high capacity thermal stores to ballast temperature and solar flux variations.

Hills and mountains The comparative absence of dust particles and water vapour on hills and mountains result in unimpeded radiation to space at night. Thus the surfaces of high ground cool more rapidly. The air in contact with the ground cools, becomes denser and

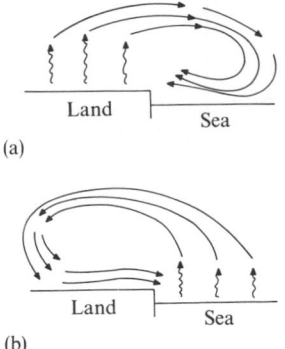

(a)

(b)

Figure 5.14 Coastal winds: (a) Day-time; (b) Night-time.

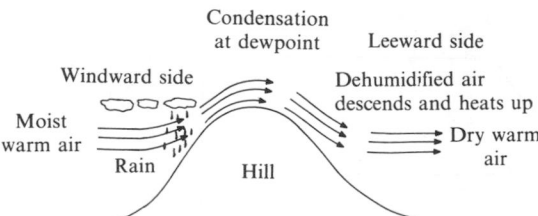

Figure 5.15 The process of natural dehumidification.

slips down into the valleys forming the so-called *katabatic* winds. As air rises, it expands and cools. Water vapour in the air condenses when the dew-point is reached. When the cold dry air flows over the mountain peaks and descends, it heats up to become warm dry air. This is the process of natural dehumidification (Fig. 5.15). One side of a mountain range may contain wet warm air whilst the other side contains dry warm air.

The combined results of all these mechanisms sum to produce the familiar weather maps (Fig. 5.16) of regions of high and low pressure, with air circulating in clockwise vortices and descending in high pressure regions and vice versa in low pressure regions.

Dew, mist and fog Water condenses from the air when the temperature falls below the dew-point, but condensation requires condensation nuclei (i.e. particles of dirt or airborne pollutants) as well as still air. These particles should have some affinity for water (i.e. they should be hygroscopic), e.g. salt, sulphur dioxide, the products of combustion. Thus mist and fog are often found over industrial areas, or at coasts where the air is wet. *Advection fog* occurs when moist sea breezes blow inland over cooler land surfaces. *Radiation fog* occurs when air in contact with the ground (this having lost heat by radiation) cools. Cloud layers associated with low pressure regions impede radiative heat loss and inhibit the formation of radiation fog. Wind rapidly disperses mist and fog.

Clouds form when warm moist air rises and is cooled adiabatically to a temperature lower than the dew-point. *Rain falls* when the droplets of condensed water become large enough so that their weights overcome frictional air resistance. *Dew* is formed when the ground loses heat by radiation to the sky. Rates of cooling differ according to the thermal diffusivities of the cooling materials. Rocks make up radiative losses by conducting heat from the ground. The result is that dew forms on low thermal conductivity grasses before rock and stone surfaces.

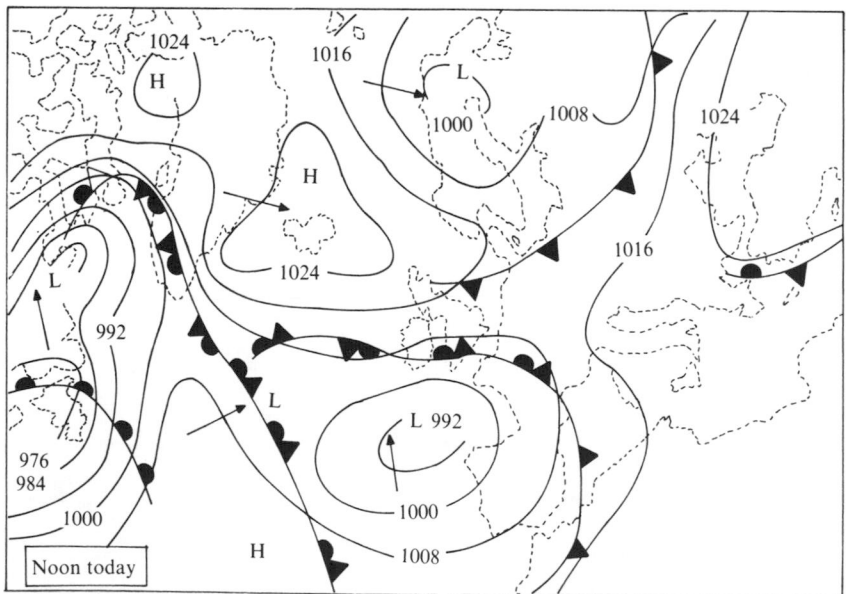

Figure 5.16 A typical weather chart. (Copyright © *The Guardian*)

Microclimates Urban conurbations affect weather locally by heat and pollution released from fossil fuel combustion, electricity usage and metabolic activities. Drainage and evapotranspiration cause higher humidity levels and smoke and waste gases reduce solar gains and give rise to fog and rain. Temperatures in cities can be 2–3 °C above those in surrounding countryside and wind speeds can be reduced by building topographies by up to 25 per cent.

Seasonal and diurnal variations Figure 5.17 shows the seasonal variations in the outside environmental psychrometric states for the UK (51.7 °N). Note that autumn conditions are wetter than spring conditions for the same dry-bulb temperatures. This difference in humidities can cause more heating to be called for in spring than in autumn, even though outside ambient air temperatures might be the same.

The earth's surface is coolest just before dawn having radiated to the sky through the night, and warmest at noon. Because of *thermal response lags*, the air temperature is highest some time after noon (2–3 pm). Temperature and humidity then follow roughly sinusoidal relationships throughout the day. Mist and dew forms when the air temperature falls below the dew-point. As relative humidity is always 100 per cent when dew is present, sensible radiation loss at night leaves the specific humidity relatively constant. The specific humidity is usually measured at 1 pm and assumed constant throughout the day until dew forms, resulting in the correspondence shown in Fig. 5.18.

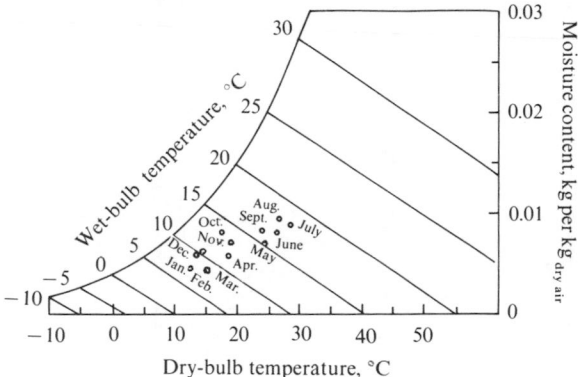

Figure 5.17 Seasonal variations in air conditions (mean monthly maximum conditions are shown, ○, for air conditioning designers).

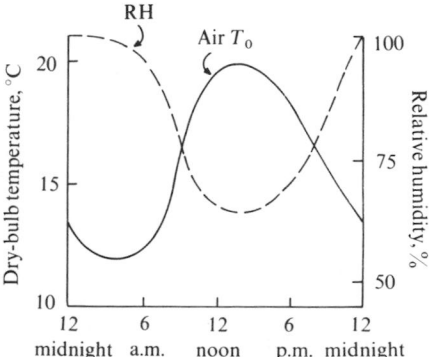

Figure 5.18 Typical diurnal variations of air conditions.

Table 5.15 Design temperatures (°C) for the United States

State	Winter dry-bulb (97.5% limit)	Summer dry-bulb (5% limit)	Summer wet-bulb (5% limit)	Summer relative humidity, %
Alabama	-6 ± 2	34	25	47
Alaska	-30 ± 13	18 ± 5	14 ± 3	64
Arizona	-4 ± 6	35 ± 4	21 ± 3	27
Arkansas	-6 ± 2	35	25	43
California	2 ± 3	30 ± 8	20 ± 2	38
Colorado	-17 ± 4	30 ± 4	18 ± 2	29
Connecticut	-16 ± 2	29	24	66
Delaware	-9	32	25	56
Florida	4 ± 3	32 ± 2	26	61
Georgia	-6 ± 2	34	26	51
Hawaii	16	30 ± 3	23	55
Idaho	15 ± 5	32 ± 2	17	19
Illinois	-17 ± 2	32 ± 1	25	55
Indiana	-16 ± 2	32	25	55
Iowa	-16 ± 2	31 ± 1	25	60
Kansas	-16 ± 1	35 ± 2	24 ± 1	39
Kentucky	-12 ± 4	33 ± 1	24	46
Louisiana	-2 ± 2	34 ± 1	26	52
Maine	-21 ± 4	27	21	57
Maryland	-10 ± 3	32	24	50
Massachusetts	-10 ± 3	28 ± 1	24	71
Michigan	-16 ± 2	29 ± 2	23	59
Minnesota	-34 ± 3	29 ± 2	22 ± 1	53
Mississippi	-4 ± 1	34	26	53
Missouri	-14 ± 2	34	25	47
Montana	-11 ± 3	29 ± 2	17	28
Nebraska	-19 ± 2	34 ± 1	23 ± 2	38
Nevada	-13 ± 6	33 ± 3	17 ± 2	17
New Hampshire	-22 ± 3	29	22	54
New Jersey	-10 ± 2	31 ± 1	24	55
New Mexico	-10 ± 5	33 ± 3	19 ± 2	25
New York	-16 ± 4	29 ± 1	26	79
North Carolina	-6 ± 2	32	26	61
North Dakota	-28 ± 1	30 ± 1	21	44
Ohio	-15 ± 1	31	23	49
Oklahoma	-10 ± 2	36	24	35
Oregon	-7 ± 6	29 ± 2	18 ± 1	32
Pennsylvania	-14 ± 2	28	23	64
Rhode Island	-12	28	23	64
South Carolina	-4 ± 1	33	25	50
South Dakota	-24 ± 2	32 ± 1	22	40
Tennessee	-8 ± 1	34 ± 1	25	47
Texas	-3 ± 4	36 ± 1	25	39
Utah	-13 ± 5	33 ± 2	18	20
Vermont	-23 ± 2	28	21	52
Virginia	-9 ± 2	31 ± 2	24	55
Washington	-7 ± 5	28 ± 4	17	31
West Virginia	-13 ± 2	30 ± 3	24	61
Wisconsin	-14 ± 3	29 ± 1	22	53
Wyoming	-21 ± 2	29 ± 2	16 ± 1	23

Arithmetic mean values indicated by the various weather stations within the State.

Figures are rounded to whole numbers.

$\pm$ indicates standard deviation.

Blank space indicates a standard deviation which is less than unity.

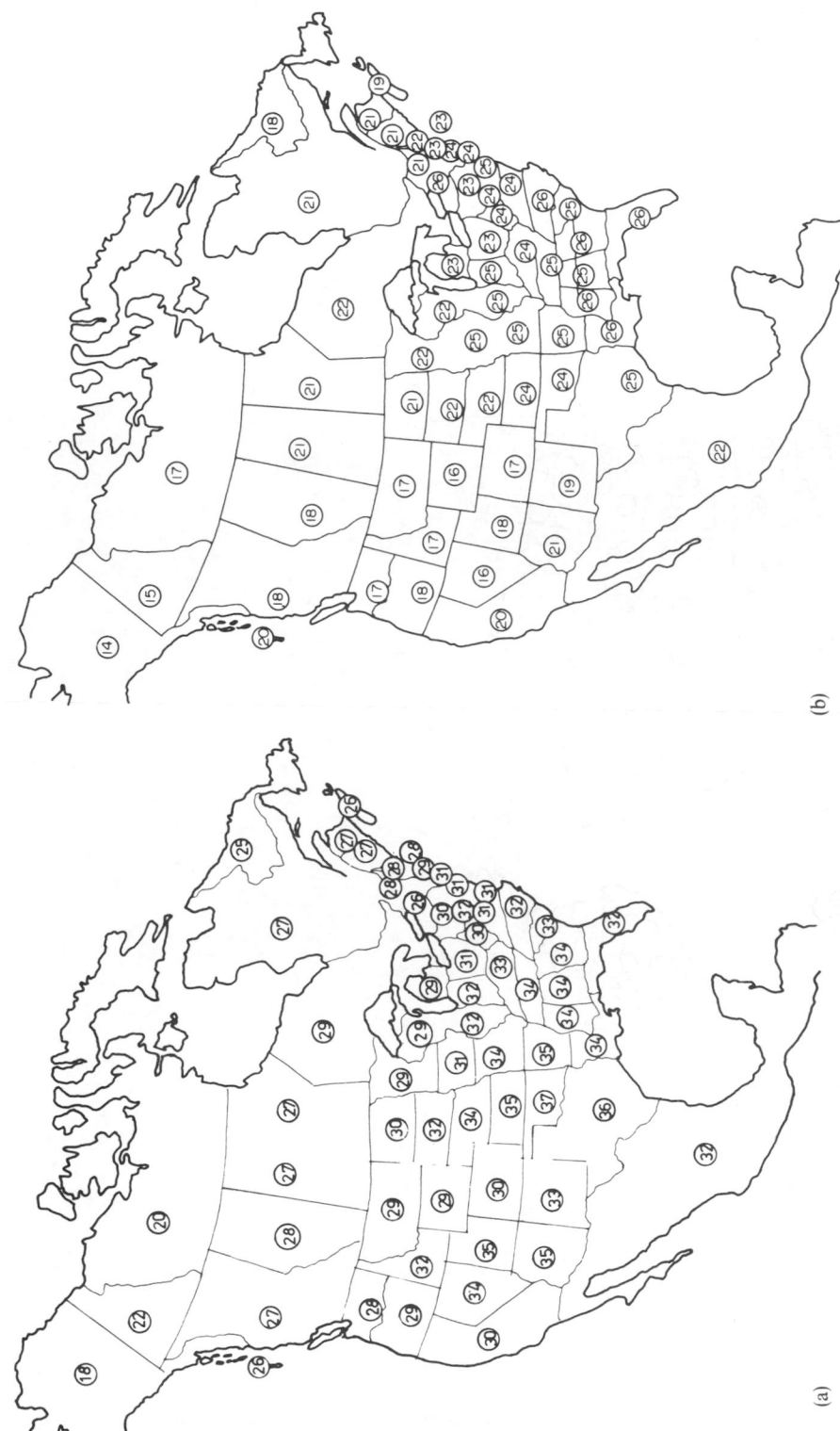

(a)

(b)

Figure 5.19 (a) Summer high (5% limit) design dry-bulb temperatures (°C) for the United States and Canada; (b) Summer high (5% limit) design wet-bulb temperatures (°C) for the United States and Canada.

215

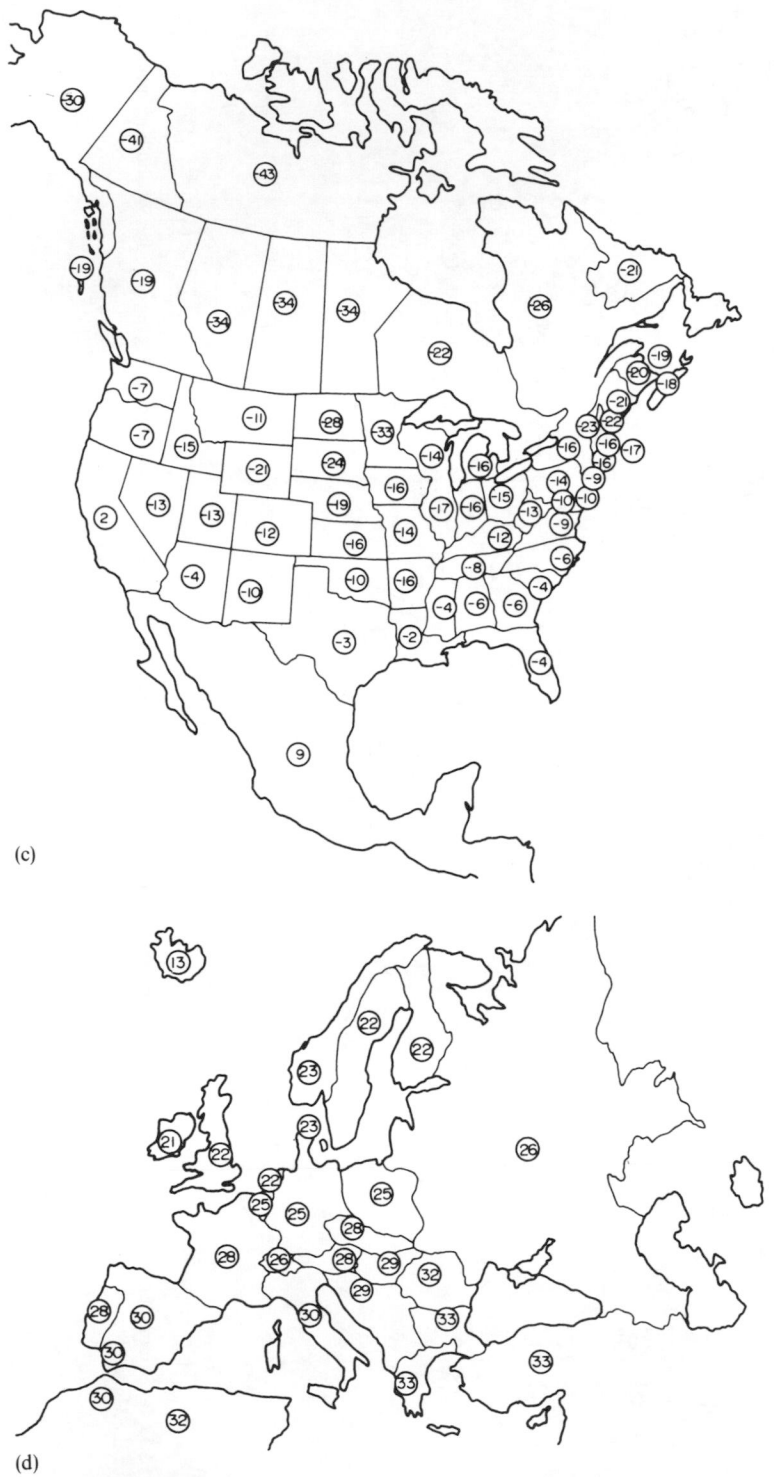

(c)

(d)

Figure 5.19 (c) Winter low (97.5% limit) design dry-bulb temperatures (°C) for the United States and Canada; (d) Summer high (5% Limit) design dry-bulb temperatures (°C) for Europe.

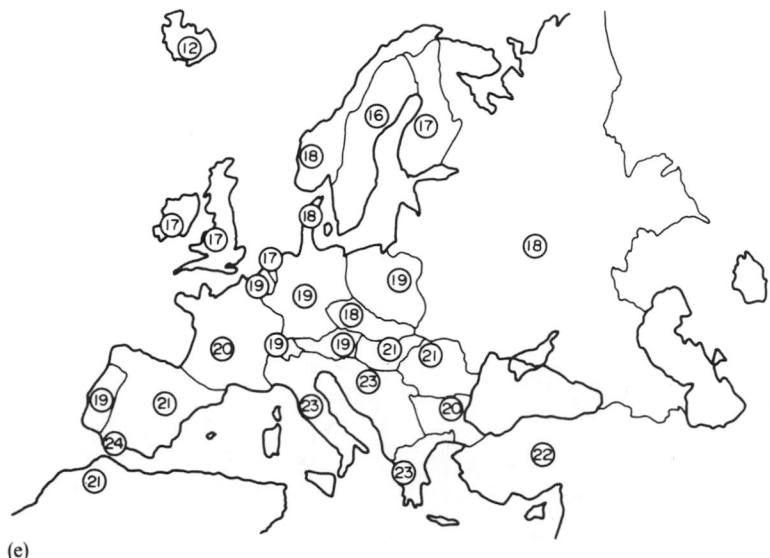

(e)

(f)

Figure 5.19 (e) Summer high (5% limit) design wet-bulb temperatures (°C) for Europe; (f) Winter low (97.5% limit) design dry-bulb temperatures (°C) for Europe.

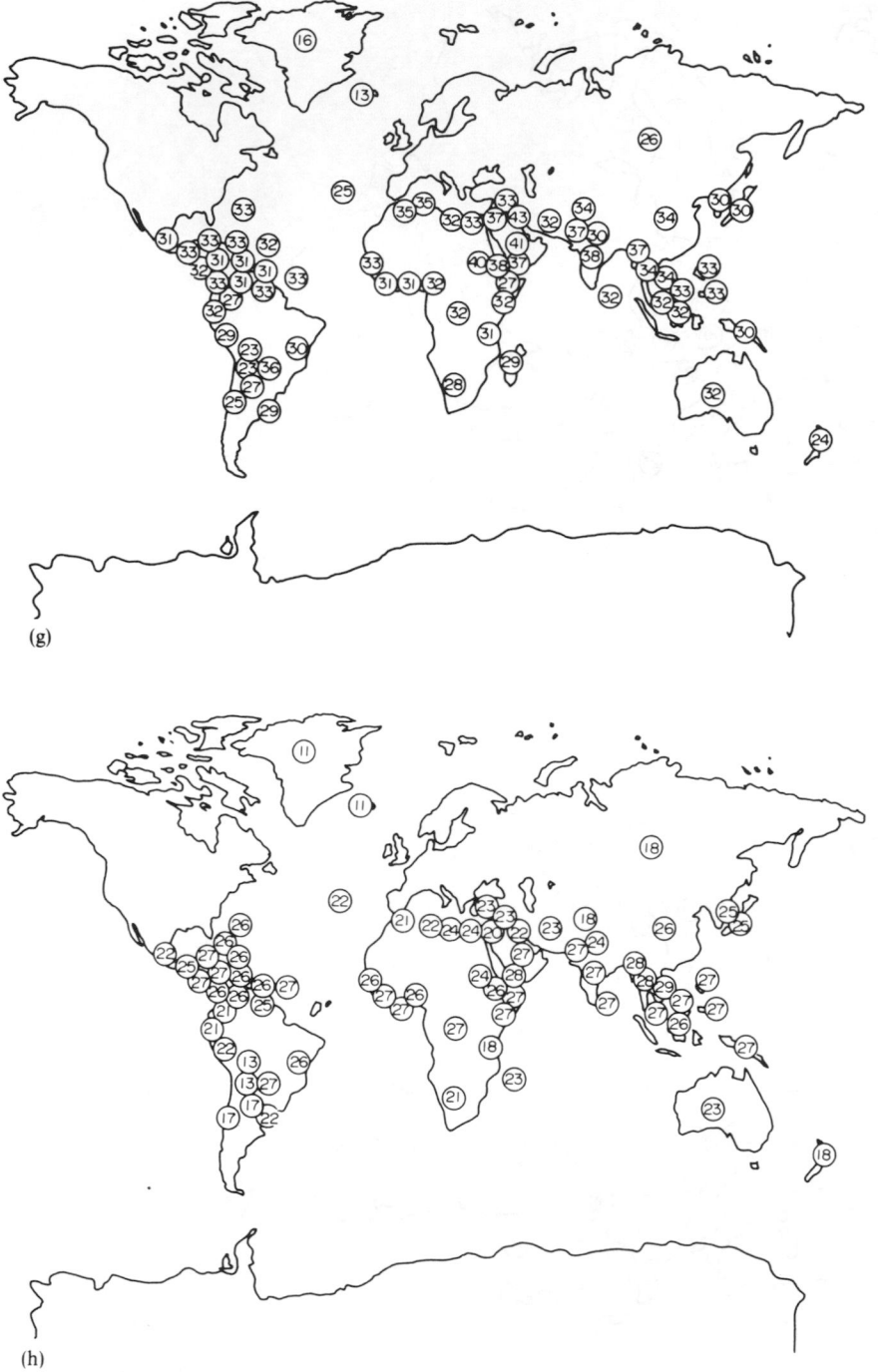

Figure 5.19 (g) Summer high (5% limit) design dry-bulb temperatures (°C) for the Rest of the World; (h) Summer high (5% limit) design wet-bulb temperatures (°C) for the Rest of the World.

Figure 5.19 (i) Winter low (97.5% limit) design dry-bulb temperatures (°C) for the Rest of the World; (j) Summer high (1% limit) design dry-bulb temperatures (°C) for the United Kingdom.

Figure 5.19 (k) Summer high (1% limit) design wet-bulb temperatures (°C) for the United Kingdom.

Table 5.16 Design temperatures (°C) for Canada

State	Winter dry-bulb (97.5% limit)	Summer dry-bulb (5% limit)	Summer wet-bulb (5% limit)	Summer relative humidity, %
Alberta	−34 ± 3	28 ± 2	18	37
British Columbia	−19 ± 13	27 ± 4	18	37
Manitoba	−34 ± 3	27 ± 2	21	58
New Brunswick	−24 ± 2	27 ± 2	21	58
Newfoundland	−21 ± 7	25 ± 1	18	50
Northern Territories	−43 ± 1	20 ± 8	17	73
Nova Scotia	−18 ± 3	25 ± 2	19	57
Ontario	−22 ± 6	29 ± 1	22 ± 1	53
Quebec	−26 ± 3	27 ± 2	21	58
Saskatchewan	−34 ± 2	29 ± 1	21	48
Yukon Territory	−41	22	15	46

Arithmetic mean values indicated by the various weather stations within the State.
Figures are rounded to whole numbers.
± indicates standard deviation.
Blank space indicates a standard deviation which is less than unity.

Table 5.17 Design temperatures (°C) for various countries

Country	Winter dry-bulb (97.5% limit)	Summer dry-bulb (5% limit)	Summer wet-bulb (5% limit)	Summer relative humidity, %
Aden	21	37	28	48
Algeria	−7	32	24	50
Argentina	0	33 ± 3	24	47
Australia	7 ± 5	32 ± 4	23 ± 3	46
Austria	−11	28	19	41
Azores	9	25	22	77
Bahamas	17	32	26	63
Belgium	−7	25	19	57
Bermuda	13	33	26	57
Bolivia	0	20	13	43
Brazil	14 ± 3	30 ± 2	25	67
British Honduras	16	32	27	67
Bulgaria	−13	33	20	28
Burma	15 ± 3	31	27	72
Cambodia	20	34	28	63
Chile	2 ± 5	25 ± 7	17 ± 4	44
China	0	34	26	52
Columbia	14 ± 6	27 ± 6	21 ± 5	57
Congo	18	32	27	67
Cuba	18	33	27	62
Czechoslovakia	−13	28	18	37
Denmark	−7	23	18	60
Dominican Republic	18	31	27	73
Ecuador	10	32	21	36
El Salvador	13	35	24	39
Ethiopia	5	27	17	34
Finland	−18	22	17	60
France	−6 ± 5	28 ± 2	20 ± 1	47
French Guinea	22	33	28	67
Germany	−10 ± 3	25 ± 2	19	56
Ghana	19	32	26	61
Gibralta	6	30	23	54
Greece	1	33	23 ± 2	39
Greenland	−22	16	11	52
Guatemala	10	27	19	45
Guyana	23	30	26	73
Haiti	19	34	27	57
Honduras	10	29	22	53
Hong Kong	10	33	27	62
Hungary	−10	29	21	48
Iceland	−8	13	12	90
India	14 ± 5	38 ± 3	27 ± 1	41
Indonesia	21 ± 1	32 ± 1	26	61
Iran	−3 ± 8	38 ± 5	23 ± 4	26
Iraq	1	43	22	13
Ireland	−4	21	17	67
Israel	4	33	31	86
Italy	−1 ± 4	30 ± 1	23	54
Ivory Coast	20	31	27	73
Japan	−6	30	25	67
Jordan	2	33	20	28
Kenya	10	26	18	45

(*Contd.*)

221

Table 5.17 (*Continued*)

Country	Winter dry-bulb (97.5% limit)	Summer dry-bulb (5% limit)	Summer wet-bulb (5% limit)	Summer relative humidity, %
Korea	−14	30	25	67
Lebanon	7	32	24	50
Liberia	20	31	27	73
Libya	9	32	24	50
Madagascar	9	29	22	53
Malaysia	22	32	27	67
Martinique	19	31	27	73
Mexico	9 ± 6	32 ± 3	22	41
Morocco	5	30	21	43
Nepal	1	30	24	60
Netherlands	−5	22	17	60
New Guinea	22	30	27	79
New Zealand	2 ± 3	24 ± 1	18	55
Nicaragua	19	33	26	56
Nigeria	20	32	27	67
Norway	−11	23	18	60
Pakistan	7 ± 5	37 ± 4	27	45
Panama	23	33	27	61
Paraguay	8	36	27	48
Peru	13	29	23	59
Philippines	23	33	27	61
Poland	−14	25	19	56
Portugal	4	28	19	41
Puerto Rico	20	30	26	72
Romania	20	32	21	36
Saudi Arabia	10	41	27 ± 2	32
Senegal	16	33	27	61
Sri Lanka	20	32	27	67
Somalia	20	32	27	67
South Africa	3	28	21	52
Soviet Union	−20 ± 6	26 ± 4	18 ± 2	45
Spain	1 ± 3	31	21	39
Sudan	13	40	24	24
Surinam	20	33	27	61
Sweden	−13	22	16	53
Switzerland	−10	26	19	50
Syria	0	37	21	21
Taiwan	9	33	27	61
Tanzania	18	31	27	73
Thailand	17	34	27	58
Trinidad	18	31	26	67
Tunisia	5	35	22	30
Turkey	−3 ± 5	33 ± 2	23 ± 3	42
United Arab Rep.	8	37	24	32
United Kingdom	−3 ± 1	22 ± 2	17	60
Uruguay	4	29	22	53
Venezuela	18	31	24	55
Vietnam	14	34	29	67

Arithmetic mean values indicated by the various weather stations within the country.
Figures are rounded to whole numbers.
± indicates standard deviation.
Blank space indicates a standard deviation which is less than unity.

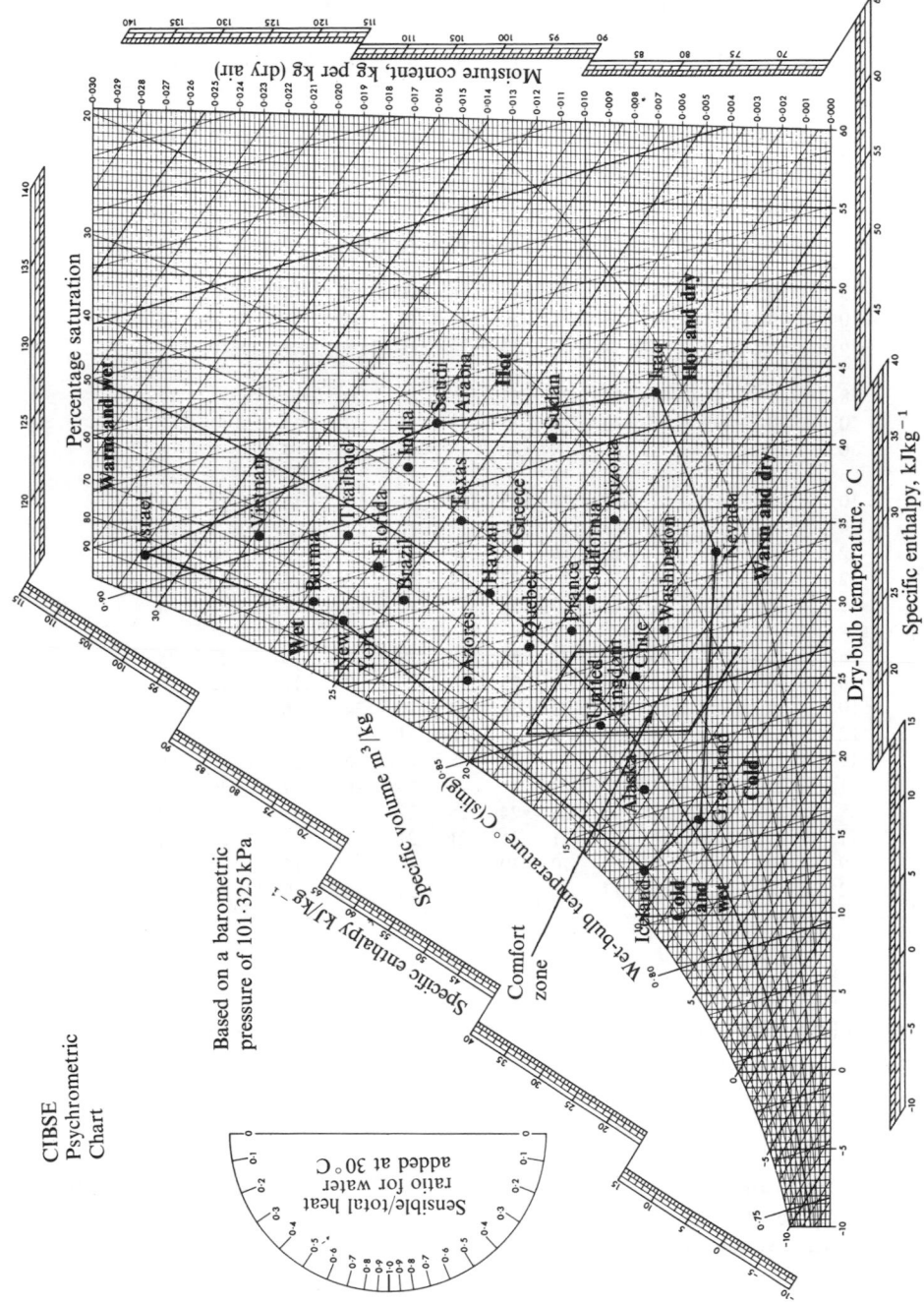

Figure 5.20 Summer design points around the globe.

223

Table 5.18 Mean monthly temperatures (°C) for various global locations

Place	J	F	M	A	M	J	J	A	S	O	N	D	Mean
United States													
New York	0	1	4	10	15	20	23	22	19	12	7	0	11
New Orleans	13	16	17	21	24	27	27	27	25	21	16	10	20
San Diego	12	12	12	14	16	17	19	20	20	18	16	13	16
Sacramento	7	9	11	13	17	21	22	22	21	18	14	7	15
Denver	−1	0	4	8	13	18	21	19	16	10	4	0	9
Canada													
Edmonton	−13	−10	−5	4	8	13	16	16	10	4	−4	10	2
Montreal	−10	−8	−3	4	13	18	20	19	15	7	0	−6	6
St John	−6	−6	0	4	7	10	15	16	10	4	−4	−10	2
Europe													
Athens	8	8	12	16	18	23	27	27	22	20	14	11	17
Bergen	0	0	3	7	10	12	16	13	11	9	5	0	7
Berlin	0	0	5	8	13	18	20	19	16	10	4	0	9
Bordeaux	6	6	7	10	13	18	20	20	16	13	10	5	12
Lisbon	10	11	13	16	17	20	22	21	20	17	15	11	16
London	4	4	7	9	13	16	19	17	14	10	7	5	11
Moscow	−12	−10	−3	3	10	16	20	16	11	5	−3	−10	4
Palermo	12	10	13	16	17	21	23	24	20	18	16	12	17
Warsaw	−5	−2	2	8	12	17	18	17	15	10	4	−1	8
Rest of World													
Aden	26	26	27	28	29	31	30	29	29	28	25	24	28
Algiers	8	8	13	15	16	21	23	24	23	19	15	12	16
Baghdad	10	10	15	21	26	30	34	33	30	24	16	8	21
Buenos Aires	22	22	21	16	10	8	8	8	11	16	18	21	15
Calcutta	19	21	27	28	29	29	28	28	28	26	22	19	25
Capetown	21	21	21	17	15	12	11	12	13	16	17	20	16
Darwin	28	29	29	27	26	25	24	26	27	29	29	29	27
Entebbe	21	21	21	21	21	21	21	21	21	21	21	21	21
Melbourne	19	19	18	16	12	10	10	8	11	13	15	13	14
Tokyo	2	5	7	12	17	21	25	24	21	16	10	5	14

Design points Building thermal designs and specifications for heating, cooling and air conditioning systems are based upon reasonable extreme heating and cooling loads corresponding to local design temperatures and humidities. (Tables 5.15–5.17 and Fig. 5.19(a)–(k)). The global range of summer design conditions are shown in Fig. 5.20 and mean monthly temperatures for various global locations are given in Table 5.18.

Further data are available from the CIBSE or ASHRAE handbooks.[2,3] Mean monthly temperatures for a given location (Fig. 5.21) may be approximated by modified sine waves and diurnal variations by nested sine waves.

The *sol-air temperature* (Fig. 5.22) may be used to take into account the effects of solar radiation gains on net transmission heat losses. Monthly air and sol-air temperatures for the location 51.7 °N are given in the *databank*. It should be emphasized however that values for sol-air temperatures are dependent upon the radiative characteristics of the building and its surroundings, as well as surface heat transfer coefficients and wind speeds (Figs 5.23).

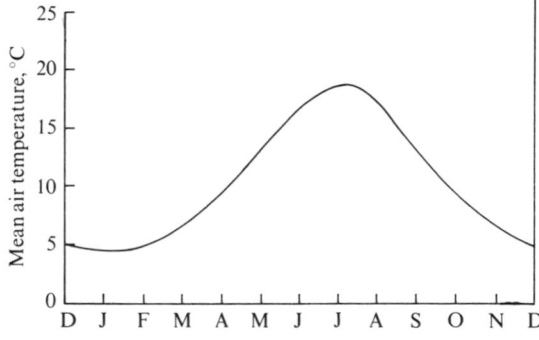

Figure 5.21 Mean monthly temperatures (°C) for the latitude 51.7 °N (UK).

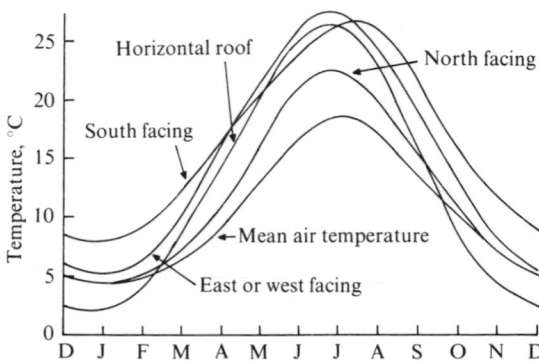

Figure 5.22 Mean representative sol-air temperatures for the latitude 51.7 °N (UK).

Heat gains to buildings

Sundry gains to buildings can sometimes comprise a high proportion of the input energy. In factories and kitchens, internal heat losses from processes often overwhelm the heat losses from the building fabric, resulting in the need for excessive fresh air ventilation to maintain the internal environment in a comfortable condition.

Solar gains Solar gains to office blocks can also be high, and, unless insolation is reduced by employing blinds, shutters or solar control glazing, such buildings may have to be air-conditioned. The mean heat gains or losses from or to the external environment through the building components can be calculated as follows:

For a wall or roof:

$$Q = UA(T_i - T_{sa}) \qquad (5.10)$$

where T_{sa} (°C) is the mean external sol-air temperature for the building component.

For glazing:

$$Q = SAI - UA(T_i - T_o) \qquad (5.11)$$

where S is the *solar gain factor* (0–1) for the glazing (see Table 5.19), A is the area (m^2) of

Table 5.19 Typical solar gain factors, SGF

Type	SGF
Single clear glass	0.76
with internal white venetian blind	0.46
with internal white cotton curtain	0.41
Double clear glass	0.64
with internal white venetian blind	0.46
with internal white cotton curtain	0.40
Single heat absorbing glass	0.45
Double heat absorbing glass	0.31
Single heat reflecting glass	0.26
Double heat reflecting glass	0.25
Double clear glass with midpane white venetian blind	0.28
Single clear glass with external canvas roller blind	0.14
Double clear glass with external canvas roller blind	0.11
Single clear glass with external white louvered sun breaker	0.14
Double clear glass with external white louvered sun breaker	0.11

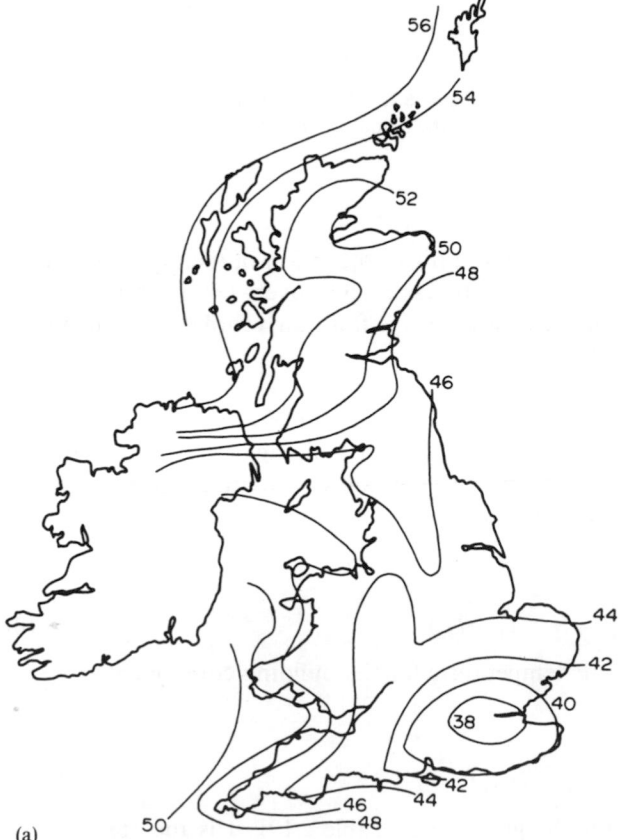

Figure 5.23 (a) Basic wind speeds (m s^{-1}) in the United Kingdom (maximum gust speeds likely to be exceeded only once in 50 years at 10 m above the ground in open country).

(a)

(b)

Figure 5.23 (b) Wind rosettes.

the glazing, and I (W m^{-2}) is the sum of direct and diffuse solar energy normal to the glazing.

For the base:

$$Q = UA(T_i - (T_o + 10\,^{\circ}\text{C}))\qquad(5.12)$$

Internal blinds and shutters absorb solar radiation and retransmit a proportion of this energy to the internal environment as heat by convection. External shutters are thus more effective in excluding solar radiation gains.

Heat gains from people In lecture theatres, auditoria, leisure centres and other public meeting places, the sensible and latent heat gains from people may dominate the heat balance equation. The sensible and latent heat proportions depend upon the rates of working, the internal environmental dry-bulb temperature and the relative humidity of the air. If not known, occupation densities could be assumed as follows:

$10\,m^2$ per person in an office block
$20\,m^2$ per person in executive offices
$2\,m^2$ per person in restaurants
$0.5\,m^2$ in cinemas and theatres

Heat gains from electrical devices All electricity supplied to a building is a propensive sundry gain, which aids heating systems and combats cooling arrangements. The exceptions to this include electricity used for refrigeration, compressed air, extract fans and domestic hot water.

Electricity for lighting often constitutes a major cooling load, especially in deep plan buildings. Approximate heat dissipations from lamps are given in Table 5.20.

An 80 W fluorescent fitting needs 100 W of power supplied—the extra 20 W is liberated directly from the control gear as heat. The heat liberated from lighting is not felt immediately as a load on an air-conditioning system as the radiant energy must be converted to sensible heat in the air by convection from heated surfaces.

All the power supplied to electrical motors is eventually dissipated as heat. Heat dissipations from processes might be sensible and/or latent.

In constructing an energy audit, sundry heat gains from people, solar gains, electricity utilization and other fuel using processes must be quantified. Table 5.21 lists typical heat gains from various activities.

For a typical domestic dwelling, for example, the annual energy release for non-heating purposes can amount to an equivalent 1 kW of continuous power rating.

Table 5.20 Heat dissipations from lamps and luminaires

Illumination on the working plane (lux)	$W\,m^{-2}$ of floor	area
	Filament with reflector	Lamps with diffuser
150	19–28	28–36
200	28–36	36–50
300	38–55	50–69
500	66–88	—
	80 W white in diffusing fitting	Fluorescent lamps in louvered ceiling panel
150	8	8–11
200	11	11
300	11–16	14–19
500	22–28	22–33
1000	36–55	44–66

Table 5.21 Sundry heat gains

Activity	Hours per year	Power rating, W	Average annual rating, W
Hot water tank	8760	50	50
Hot water use	8760	50 per capita	50 per capita
Cooking heat	1000	up to 1000 per capita	up to 120 per capita
People	1665	200	40
Lights	700	100	8
Television	1000	200	65
Tape recorder	200	40	1
Record player	200	40	1
Radio	1000	40	5
Washing machine	300	1000	35
Spin drier	300	125	4
Tumble drier	300	1000	35
Iron	300	400	14
Refrigerator	8760	40	40
Freezer	8760	200	200
Cooker hood	1000	100	11
Kettle	100	3000	35
Toaster	50	1000	6
Dish washer	500	750	42
Vacuum cleaner	300	250	10
Electric blanket	300	75	0.5
Hair drier	150	700	12

Heat losses from buildings

Heat losses from buildings occur via the following means:

- Stack losses, i.e. the heat lost in the furnace flue gases, calculated from the combustion analyses
- Fabric transmission losses through the walls, windows, roof and base, calculated from U-value analyses
- Ventilation losses, calculated from the number of fresh air changes per hour

Building heat balance

A heating or cooling system must provide or extract enough heat to balance the difference between the gains and the losses throughout the year.

An annual load profile must be constructed to quantify mean and extreme thermal loads so that heating/cooling arrangements may be accurately sized. Figure 5.24 shows such an annual load profile, together with the effects of insulating the system. The *balance point* in the diagram corresponds to an outside air temperature for which no internal heating or cooling is required. This is also the *changeover* temperature, when heating should cease and cooling initiated. Heating and cooling systems should never be allowed to operate simultaneously.

For the uninsulated building, the maximum heating load (kW) is given by AB in Fig. 5.24,

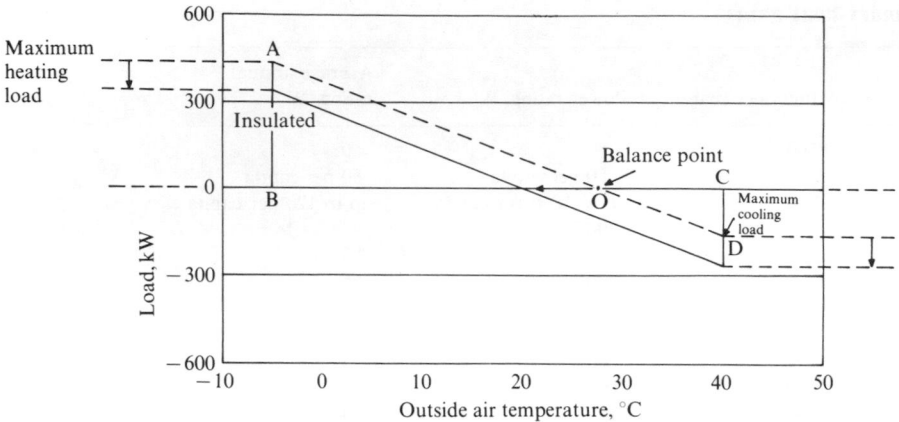

Figure 5.24 Typical load diagram for a building.

and the maximum cooling load by CD. The areas OAB and OCD are indicative of total annual heating and cooling requirements, respectively, the actual magnitudes depending upon the frequencies of occurrence of the various external air temperatures (obtainable from degree-day tables). The changeover deadspace bandwidth should be maximized to avoid excessive rates of plant cut-in.

The effect of insulating is to move the entire characteristic downwards, decreasing the heating loads and increasing the cooling loads. Thus, unless some mechanisms for varying insulation levels are included in the building design, the summer cooling requirements will increase.

Variable volume ventilation systems commonly compensate for increased cooling loads by reducing fresh air ventilation rates to a minimum during periods of heating or cooling. Around the balance point in the deadspace, when neither heating nor cooling systems are in operation, the internal air temperature is controlled by varying rates of fresh air supply.

Energy conservation in buildings

Temperature control Internal air temperatures must be exactly specified in terms of time and space and should be measured and controlled so that they exactly match these specifications. Required air temperatures vary according to the activities conducted in different internal zones, the desirable air temperature for a warehouse might be considerably less than that of an office, which, in turn will be less than that needed for a hospital ward. Ceiling voids should not be overheated and vertical *thermal stratification* should be overcome. Heat should be provided only where and when required. Time schedules should be constructed for inter-mittently occupied buildings. Depending upon the thermal responses of different zones, optimal start-up and close-down times for air-conditioning systems should be adopted. These times can be deduced from experimentation.

Ventilation control It is common that the energy requirements associated with ventilating air dominate the overall bill for heating fuels. Fresh air change rates must be exactly specified in terms of time and space, and should be measured and controlled so that they exactly match these specifications. Required air change rates vary according to the activities conducted in different internal zones. The greater the amounts of internal pollutants released, the greater the rate of air change required. Such pollutants should be removed locally (e.g. from paint spray booths, printing machines, chemical processes) using local extract arrangements. Fresh air should be provided only where and when required. Time schedules should be constructed and adhered to for intermittently occupied buildings.

The minimum amount of fresh air needed for breathing purposes is quite small, of the order of $0.2 \, \mathrm{ls}^{-1}$ per person. Greater rates of fresh air supply must however be maintained to meet the following requirements:

- Satisfaction of oxygen needs for the occupants
- Dilution of odours
- Dilution of carbon dioxide concentrations

Not less than $5.8 \, \mathrm{ls}^{-1}$ per person of fresh air is advised to prevent vitiation and palpable body odour. Recommended fresh air ventilation rates vary according to the rates of oxygen consumption within the building, and hence rates of metabolic activity, combustion in open fires, cigarette smoking and cooking, rates of odour and contaminant production and the degree of air cleanliness required (Table 5.22). Carbon dioxide is present in fresh air to the extent of about 0.03 per cent (300 ppmv), whereas concentrations up to 0.1 per cent (1000 ppmv) are acceptable in air-conditioned spaces. The rate of carbon dioxide production by a human being is about $0.0047 \, \mathrm{ls}^{-1}$.

Air movements by infiltration through a building or by controlled ventilation imposes extra loads on air-conditioning systems because outside air must be heated or cooled to the

Table 5.22 Fresh air supply requirements

Application	ls^{-1} per person	$\mathrm{ls}^{-1}\mathrm{m}^{-2}$	Occupancy for unknown air change per hr
Private dwellings	8–12	—	—
Board rooms	18–25	—	—
Bars	12–18	—	—
Stores	6–8	—	—
Factories	16–28	0.8	—
Garages	—	8.0	5
Operating theatres	—	16.0	—
Hospital wards	8–12	—	—
General offices	6–8	1.3–2.0	3–8
Private offices	8–12	1.3–2.0	3–8
Restaurants	12–18	—	5–10
Theatres, cinemas	6–8	—	5–10
Schools	14	—	—
Engine rooms	—	—	4
Baths	—	—	5–8
Lavatories	—	—	5–10
Kitchens	—	—	10–40

Table 5.23 Pressure differences ($N\,m^{-2}$) due to wind effects

Building height, m	Open country ($9\,m\,s^{-1}$)	Suburban ($5.5\,m\,s^{-1}$)	City centre ($3\,m\,s^{-1}$)
10	58	21	6
20	70	31	11
30	78	38	15
40	85	44	21
50	90	49	23
60	95	55	26
70	100	59	31
80	104	63	34

condition prevailing in the internal environment. *Infiltration* losses from a building occur when air passes through gaps in and around doors and windows. The rate of air flow is approximately proportional to the pressure difference $\Delta p\,(N\,m^{-2})$ acting across the component involved; i.e.

$$V\,(m\,s^{-1}) = K \times A(\Delta p)^n \tag{5.13}$$

where $A\,(m^2)$ is the area of the aperture, the constant $K = 0.5\text{--}0.7$ and the index n is approximately 0.5, depending upon the characteristics of the component involved.

Mean pressure differences across buildings of different heights for a free wind speed of $9\,m\,s^{-1}$ are given in Table 5.23.

The *ventilation decay equation* gives the rate at which the concentration of a contaminant, such as carbon dioxide, decays in a ventilated room under the influence of a diluting influx of fresh air.

Consider a room having a volume $V\,m^3$, in which the concentration of carbon dioxide is c ppmv. During time Δt, a small quantity of CO_2-free (for simplicity) air, Δq, enters the room. A similar quantity, Δq, of contaminated air is forced out of the room. The concentration of carbon dioxide in the room is therefore reduced by

$$\Delta c = (\Delta q/V) \times c \tag{5.14}$$

The rate of change of concentration is then given by

$$\Delta c/\Delta t = -(\Delta q/V)/\Delta t \times c/V \tag{5.15}$$

Now, $\Delta q/\Delta t = $ air change rate $ = $ constant $ = Q$, say. Therefore

$$\Delta c/\Delta t = -cQ/V \tag{5.16}$$

Integrating,

$$dc/c = -(Q/V)\,dt$$
$$\ln c = -Qt/V + \ln A$$

where A is a constant of integration.

$$\ln c - \ln A = -Qt/V$$
$$\ln(c/A) = -Qt/V$$

$$c/A = e^{-Qt/V}$$
$$c = A e^{-Qt/V} \qquad (5.17)$$

If the initial concentration in the room is c_0 at $t = 0$, the concentration at time t is given by

$$c = c_0 e^{-Qt/V} \qquad (5.18)$$

The quantity, Qt/V, is the number of air changes, n, and so

$$c/c_0 = e^{-n} \qquad (5.19)$$

This is plotted in Fig. 5.25, from which it may be seen that the concentration decay curve is exponential; 63.2 per cent of the contaminant is removed with one air change and 95 per cent with two air changes.

Alternatively, $n = \ln(c_0/c)$. Thus concentration decays of contaminants, such as carbon dioxide or steam, may be measured to ascertain internal air change rates.

Draughtproofing Air infiltration is a random occurrence and, as such, confounds any attempt to control ventilation rates and internal air conditions (temperatures and humidities). The system should be isolated from its surroundings by applying adequate weather-stripping, repairing structural defects and by controlling door and window openings, local air extracts, dampers and flues. An energy management checklist for ventilation is provided in Appendix 2.

Thermal insulation Only after the built system is adequately isolated from its surroundings and proper measurement and control systems for maintaining specified internal air temperatures and fresh ventilation rates have been installed, should attention be given to altering the thermal structure of the building by introducing roof insulation, attic insulation, dry-linings, cavity wall insulation, and double glazing. The cost-effectiveness of each of these options should be evaluated and a ranking should be produced to aid investment decisions. An energy management checklist for insulation is provided in Appendix 2.

Artificial lighting and daylighting The optimal economic balance between the use of artificial lighting and daylighting should be sought. Lighting levels should be specified, measured and

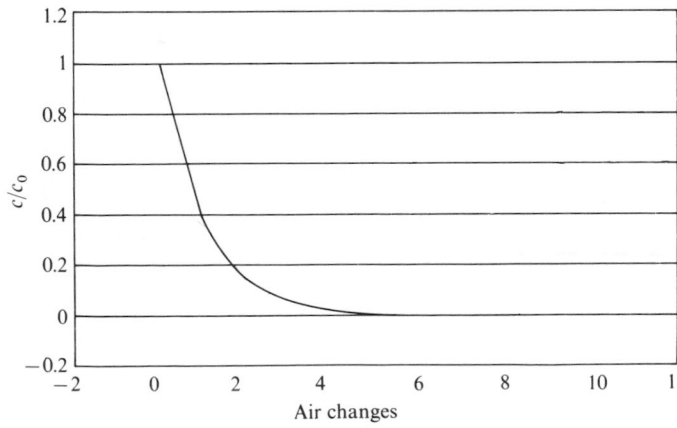

Figure 5.25 Concentration decay as a means of estimating the number of air changes in a built environment.

controlled to match zonal requirements throughout occupied periods. Lighting should be provided only when and where required. The reduction of window areas reduces the amounts of transmission losses via glazing, but may lead to the need for extra artificial lighting. Windows and luminaires should be kept clean and walls and ceilings should be painted in light colours. An energy management checklist for lighting systems is provided in Appendix 2.

Waste heat recovery When minimum fresh air change rates have been attained, any further energy savings with respect to insulation loads can only be accomplished by introducing a heat recovery system to transfer heat from the stale exhaust air to the incoming fresh air stream. Since the temperature difference between the two streams is small, this is likely to result in an expensive retrofit measure.

Heat recovery from exhaust flue gases to pre-heat combustion air, or the reclaim of heat from hot liquid effluents may prove to be cost-effective options. All heat recovery options should be carefully evaluated and costed.

A heat pump allows heat to be reclaimed and upgraded thermally for introduction back into a process. Applications for heat pump heat recovery should be sought and evaluated.

An energy management checklist for waste heat recovery is provided in Appendix 2.

Thermal storage Reject heat is often available when a need does not exist for it. The introduction

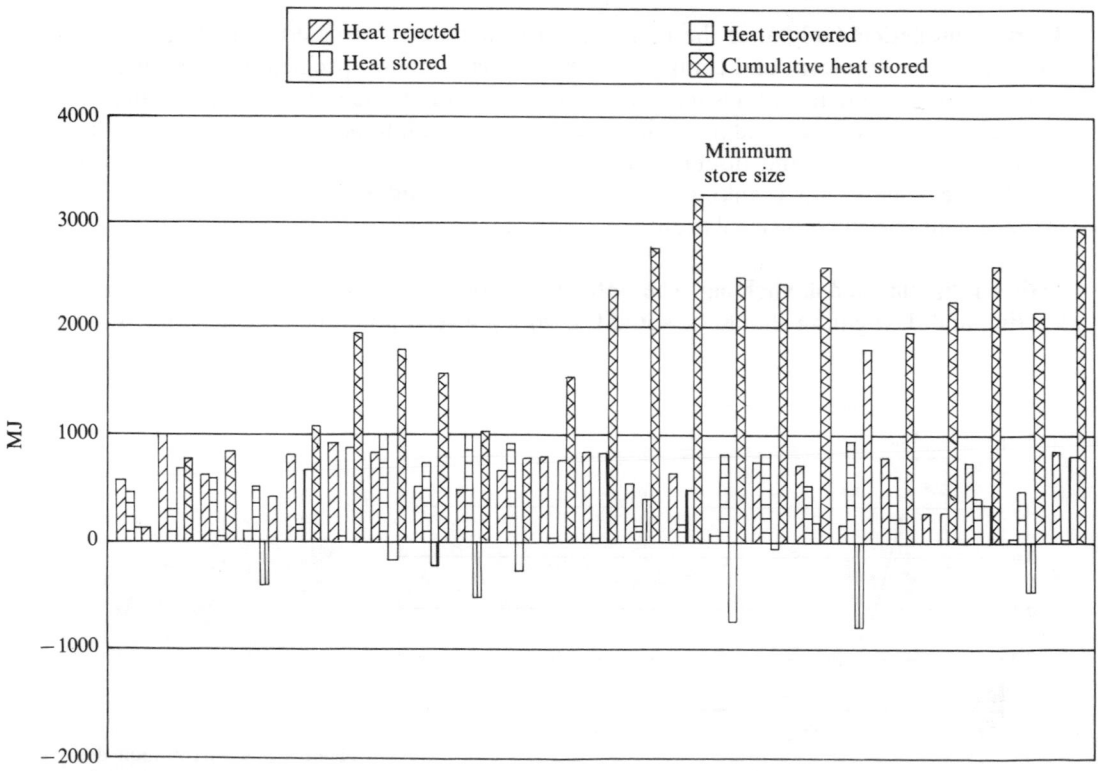

Figure 5.26 Thermal store sizing.

of a thermal store then allows matching between supply and demand. Thermal storage media include ceramics and metal matrices for high temperature thermal storage, and water for low temperature thermal storage. In order to size such a store, cumulative supply and demand schedules should be constructed and peak thermal storage requirements identified (Fig. 5.26) (see Section 5.7).

Combined heat and power The introduction of a combined heat and power plant allows the waste heat from power production to be utilized for space heating or other low-grade energy-using activity. In order to size plant, identify control strategies and to assess the economic cost-effectiveness of such a system, annual and diurnal electrical and heating loads, and heat-to-power ratios should be constructed and examined.

5.5 REJECT HEAT RECOVERY

There are many technical opportunities for the recovery of reject heat from a high-grade energy process for use by a low-grade energy operation. For example, heat in flue gas may be used to pre-heat combustion air using a recuperative heat exchanger, or directly for drying stock where a reducing furnace atmosphere is required; warm moist exhaust ventilation air can be used to pre-heat fresh ventilating air; exhaust hot water can be used to pre-heat fresh water; reject heat from refrigeration condensers or air compressors may be redirected for space heating, etc.

In each application, the reject heat must be matched to the heat load served in the following manner:

- *Where* it is required (heat exchangers, run-around coils, heat pipes)
- *When* it is required (thermal storage)
- At the appropriate *grade* (heat pumps)

Run-around coils

Where waste heat exhaust and inlet ducts or pipes are in close proximity, recuperative or regenerative (e.g. a thermal wheel) heat exchangers may be introduced to accomplish the heat

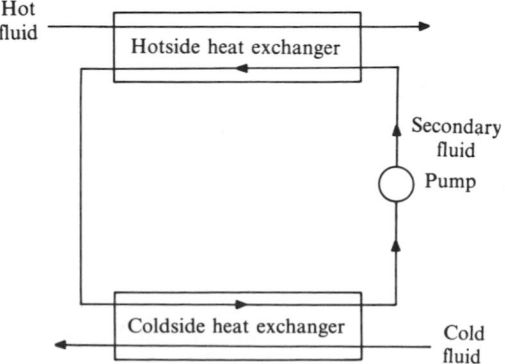

Figure 5.27 A run-around coil.

recovery. Where the waste heat ducts or pipes are located remote from inlet ducts or pipes, a *run-around coil* (Fig. 5.27) containing a secondary heat transport fluid might be employed.

Heat pipes

Very high rates of heat transfer across small temperature gradients may be accomplished when liquid-to-gas phase changes occur, as in the gravity-return (Fig. 5.28(a)) or wicked-return (Fig. 5.28(b)) heat pipe. Because the mode of operation of a heat pipe depends upon partial pressure differences, no temperature difference across the system is required theoretically. In practice, the rate of heat transfer is limited by the conductive resistances at the ends of the heat pipe. The collecting surface, or evaporator can be positioned remote from the delivery surface, or condenser. The addition of a vapour pump to aid movement then converts the heat pipe to a pumped heat pipe, or simple heat pump.

5.6 HEAT PUMPS AND REFRIGERATORS

A *heat pump* is a refrigerator system, where the emphasis is upon using the heat rejected from the condenser to serve a useful purpose. The use of heat pumps allows heat (cold) to be supplied at the *appropriate thermal grade*.

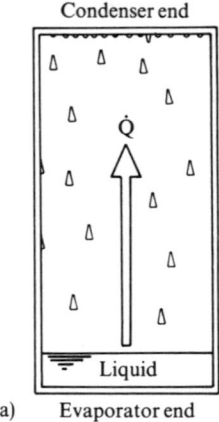

(a)   Evaporator end

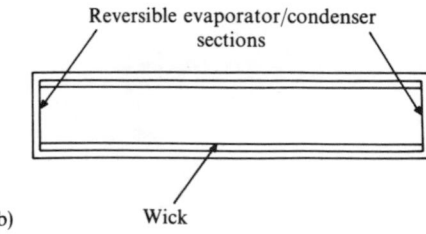

(b)   Wick

Figure 5.28 (a) A gravity-return heat pipe; (b) A wicked heat pipe.

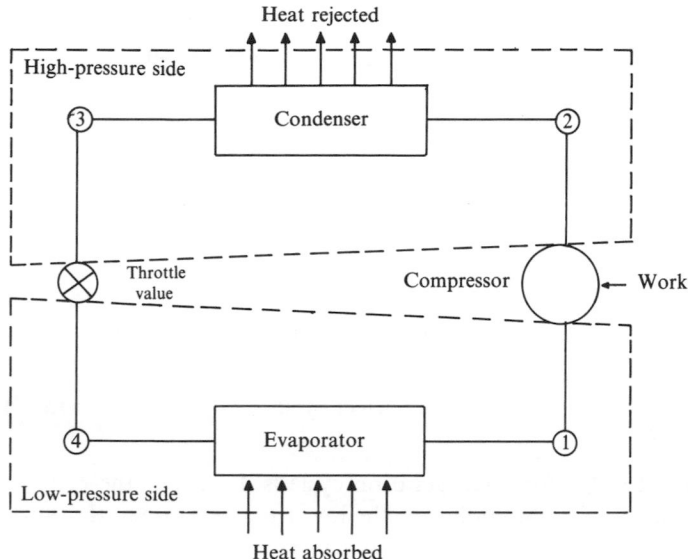

Heat rejected

Figure 5.29 Vapour compression heat pump/refrigerator.

Vapour compression systems

Figure 5.29 shows a schematic arrangement for an elementary vapour compression refrigeration, or heat pump, system.

Mode of operation

State point 1 to 2 Refrigerant vapour is compressed isentropically from a low temperature and pressure to a high temperature and pressure.

State point 2 to 3 The vapour is cooled and condensed at constant high temperature and pressure via a recuperative heat exchanger—a condenser.

The cooling medium might be fresh ventilating air or supply water to a domestic hot water in the heat pump operation. In the refrigeration operation, the heat is usually rejected to the cooling medium (water or air).

State point 3 to 4 The low temperature high pressure liquid is throttled at constant enthalpy to the original low pressure.

State point 4 to 1 The low temperature, low pressure, liquid passes through an air- or water-heated heat exchanger—the evaporator—where it evaporates, absorbing heat from the heating medium. This heating medium might be exhaust ventilating air or liquid effluent, a river or lake, or outside environmental air in the heat pump operation. In the refrigeration operation, the heat is taken from the stock refrigerated, or from air to be cooled for air-conditioning, or from water to create ice.

Coefficient of performance

The heat pump, or refrigeration, cycle is a reversed heat engine. The Carnot efficiency for a heat engine is expressed as

$$\eta = \frac{T_1 - T_2}{T_1} \tag{5.20}$$

where T_1 and T_2 are respectively the absolute temperatures (K) at which heat is added to and rejected from the cycle.

The *coefficient of performance*, COP_{hp}, for a heat pump cycle is defined as the ratio of the heat rejected at the higher temperature to the work input to the vapour compressor, i.e.

$$COP_{hp} = \frac{Q_h}{W} = \frac{T_1}{T_1 - T_2} \qquad \text{for a perfect cycle} \tag{5.21}$$

The *coefficient of performance*, COP_{ref}, for a refrigeration cycle is defined as the ratio of the heat received (i.e. taken from the stock to be cooled) at the lower temperature to the work input to the vapour compressor, i.e.

$$COP_{ref} = \frac{Q_c}{W} = \frac{T_2}{T_1 - T_2} \qquad \text{for a perfect cycle} \tag{5.22}$$

Note that $COP_{hp} = COP_{ref} + 1$.

Because of thermodynamic irreversibilities and other losses, practical COPs are lower than the Carnot COP. Tables 5.24 and 5.25 list typical practical COP_{hp} for various cycles. Electrical resistance heating has a coefficient of performance of unity.

Table 5.24 Coefficients of performance for various heat pumps assuming no mechanical losses, superheating or subcooling

Cycle details	Condenser temperature, °C	Evaporator temperature, °C	Coefficient of performance
Ideal Carnot	50	10	8.1
	50	4	7.0
	50	−1	6.3
	30	−15	5.7
Freon 12	50	10	7.2
	50	4	6.2
	50	−1	5.4
	30	−15	4.8
Ammonia	30	−15	4.7
Carbon dioxide	30	−15	2.6
Methyl chloride	30	−15	4.6
Sulphur dioxide	30	−15	4.7

Table 5.25 Coefficients of performance for practical Freon 12 heat pump cycles

Cycle details	Condenser temperature, °C	Evaporator temperature, °C	Coefficient of performance
Freon 12 Ideal	50	10	5.5
	50	4	4.7
	50	−1	4.1

Operational classifications

Heat pumps arrangements may be classified as follows:

- Air-to-air systems, which use outside environmental or exhaust ventilating air as the heat source to heat inside environmental air
- Air-to-water systems, which use outside environmental air as the heat source to heat domestic or heating water
- Water-to-air systems, which use outside or reject water sources to heat ventilating air
- Water-to-water systems, which use outside or reject water sources to heat domestic or heating water

Frost or ice accumulation on and around external heat exchanger matrices can be problematical.

Absorption systems

Figure 5.30 shows a schematic representation of an absorption refrigeration system. Comparing this with that for the vapour compression system (Fig. 5.29), it may be seen that the compressor of the latter has been replaced with an absorption/generation arrangement, in which the low pressure, low temperature vapour leaving the evaporator at state-point 10 [state-point 1 in the vapour compression cycle (Fig. 5.29)] is absorbed into a liquid absorbent to form a liquid mixture. This liquid mixture at state-point 1 is then pumped to the higher pressure at state-points 2 and 3 where it enters a heat exchanger, termed the *generator*, to which heat is supplied at constant high temperature to evaporate the refrigerant out of the high pressure liquid mixture. The high pressure, high temperature, refrigerant vapour then continues to the condenser as for the vapour compression system.

The liquid absorbent is returned through a throttle valve to the absorber, via a high pressure liquid-to-liquid heat exchanger which recoups some of the heat in the returning absorbent to pre-heat the absorbent/refrigerant mixture on its journey from the absorber to the generator.

Suitable combinations of refrigerant/absorbent combinations include water/lithium bromide and ammonia/water mixtures respectively, the solubilities of the refrigerant in the absorbent being higher the lower the temperature.

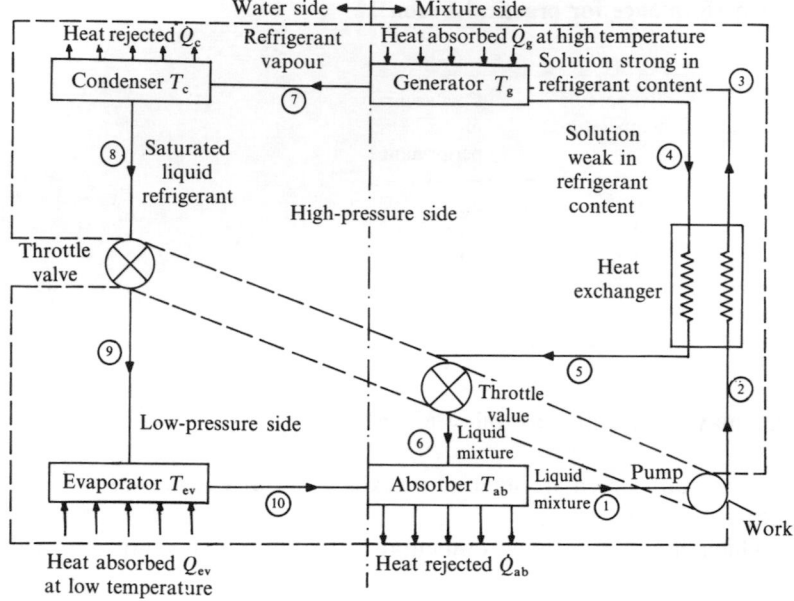

Figure 5.30 Lithium bromide/water absorption heat pump/refrigeration system.

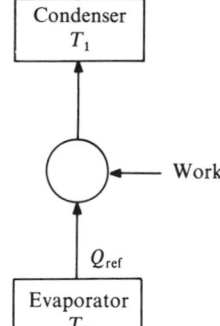

Figure 5.31 Schematic representation of the function of a vapour compression heat pump/refrigeration system.

The advantage of the absorption system over the vapour compression system is that the work required at the liquid pump in the former is considerably less than the work required to compress the vapour in the latter. In the absorption cycle, the bulk of the energy supplied is in the form of heat, which can be reject heat or that available from a solar collector.

Figure 5.31 shows a schematic representation of the function of a vapour compression refrigeration system. Work is supplied to '*pump*' heat from the evaporator at T_2 to the condenser at T_1.

The ideal coefficient of performance, *COP*, for an absorption refrigeration system may be evaluated by considering the system to be comprised of two thermodynamically reversible machines, which, taken together, perform the function of the absorption plant (Fig. 5.32):

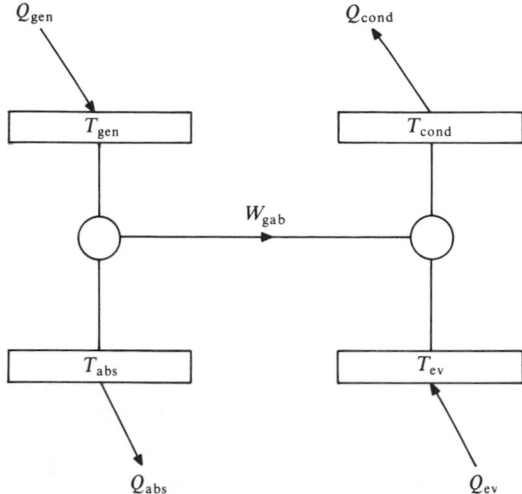

Figure 5.32 Schematic representation of the function of an absorption heat pump/refrigeration system.

- A reversible heat engine which receives heat, Q_{gen} at T_{gen}, and rejects heat, Q_{abs} at T_{abs}, whilst producing a quantity of work, W_{gab}
- A reversible heat pump which receives heat, Q_{ev} at T_{ev}, and rejects heat, Q_{cond} at T_{cond}, whilst absorbing a quantity of work, W_{gab}

This system is identical in function to the heat-engine-driven vapour compression heat pump system (Fig. 5.33).

The *COP* of the combined plant acting as a refrigerator is defined by

$$COP_{ref} = \frac{\text{cooling effect}}{\text{heat supplied}} \tag{5.23}$$

$$= Q_{ev}/Q_{gen}$$

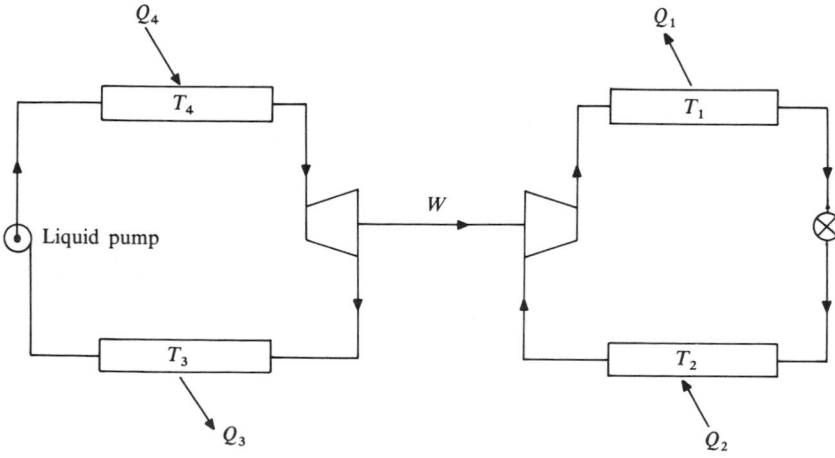

Figure 5.33 A heat-engine-driven heat pump cycle.

The Carnot efficiency, η, of the heat engine is given by

$$\eta = \frac{W_{gab}}{Q_{gen}} = \frac{T_{gen} - T_{abs}}{T_{gen}} \tag{5.24}$$

and the *COP* of the refrigerator is given by

$$COP = -\frac{Q_{ev}}{W_{gab}} = \frac{T_{ev}}{(T_{cond} - T_{ev})} \tag{5.25}$$

from which

$$COP_{ref} = Q_{ev}/Q_{gen} = \left(\frac{T_{ey}}{(T_{cond} - T_{ev})}\right) \bigg/ \left(\frac{T_{gen} - T_{abs}}{T_{gen}}\right)$$

or

$$COP_{ref} = \frac{T_{ev}(T_{gen} - T_{abs})}{T_{gen}(T_{cond} - T_{ev})} \tag{5.26}$$

When the absorber and condenser are maintained at the same temperature:

$$COP_{ref} = \frac{T_{ev}(T_{gen} - T_{abs})}{T_{gen}(T_{abs} - T_{ev})}$$

which is identical to

$$COP_{ref} = \frac{T_2(T_4 - T_3)}{T_4(T_1 - T_2)}$$

for the heat engine-driven refrigeration system.

The *COP* of the combined plant acting as a heat pump is defined by

$$COP_{hp} = \frac{\text{heat delivered}}{\text{heat supplied}} \tag{5.27}$$

$$COP_{hp} = \frac{(Q_{cond} + Q_{abs})}{Q_{gen}} \tag{5.28}$$

and when $T_{cond} = T_{abs}$

$$COP_{hp} = \frac{T_{abs}(T_{gen} - T_{ev})}{T_{gen}(T_{abs} - T_{ev})} \tag{5.29}$$

Once again, it may be proved that $COP_{hp} = COP_{ref} + 1$.

Typically, $T_{ev} = 4\,°C$, $T_{gen} = 95\,°C$ and $T_{abs} = T_{cond} = 40\,°C$, because of the chemical stabilities of the candidate materials for the refrigerant and absorbent. The ideal COP_{hp} then becomes

$$\frac{313(368 - 277)}{368(313 - 277)} = \frac{313 \times 91}{368 \times 36} = 2.15$$

Thus an ideal absorption heat pump cycle might yield over twice the heat supplied to it (at 95 °C) for a lower temperature activity (at 40 °C).

The heat-engine-driven heat pump (Fig. 5.32) could have $T_{ev} = 4\,°C$, $T_{gen} = 200\,°C$ and $T_{abs} = T_{cond} = 40\,°C$. The ideal COP_{hp} then becomes

$$\frac{313(473 - 277)}{473(313 - 277)} = \frac{313 \times 196}{473 \times 36} = 3.6$$

nearly quadrupling the heat supplied.

Components of refrigeration systems

Condensers Heat transfers from condensing or evaporating fluids are very high, so that the overall heat transfer coefficient across the walls of condensers (or evaporators) are dominated by the external heat transfer coefficients.

Refrigeration condensers for domestic applications take the form of natural convective air-cooled grids placed at the back of the appliance. Manufacturers usually recommend that a 100 mm gap be allowed between the grid and the wall to promote natural air movement. Small commercial applications may have similar condensing arrangements but finned tubes may be provided to enhance the external heat transfer coefficient. Larger installations will require a fan, or fans, to provide forced convective cooling. Water cooling is necessary for systems which must transfer heat in excess of 1 kW. Water-cooled condensers take the form of shell and internal coil (easier to manufacture) or the more expensive shell and tube (easier to maintain) arrangements. Evaporative condensers may be of the shell-and-coil or shell-and-tube type. High heat transfer rates are forced by spraying water over the external finned heat transfer surfaces and simultaneously blowing air over the surface to evaporate the water.

Evaporators Air-heated finned evaporators are frequently used for direct air cooling in air conditioning units. Care must be taken to prevent frost formation on the airside, clogging the fins. Water- or brine-heated shell and coil evaporators are common in small installations with the refrigerant passing through the coil. Shell and tube arrangements are used in large installations.

Expansion valves These are either manually set needle valves, automatic float valves or thermal expansion valves. In the latter, a temperature sensor relays signals to a spring-mounted solenoid valve to control operation.

Refrigerants The suitability of a refrigerant for a particular application depends upon economic, environmental and practical factors, as well as physical, chemical and thermodynamic properties. Substances include halocarbon compounds, such as carbontetrafluoride, CF_4 (refrigerant 14), dichlorofluoromethane, CCl_2F_2 (refrigerant 12, Freon 12, Arcton 6), hydrocarbon compounds, such as methane, CH_4, ethane, CH_3CH_3, propane, $CH_3CH_2CH_3$, or inorganic compounds, such as ammonia, NH_3, water, H_2O and carbon dioxide, CO_2.

Desirable properties of refrigerants include:

- Positive evaporating pressure to prevent possible leakage of air into the system
- Moderately low condensing pressures to allow lightweight equipment and piping on the high pressure side

- Relatively high critical temperature to prevent unduly high power requirements
- Low freezing temperature so that solidification cannot occur
- High latent heat of vaporization to give a high refrigerating effect per kilogram of refrigerant
- Inertness and stability so that is does not react with other materials in the system
- Non-corrosivity
- High dielectric strength of vapour—this is important in hermetically sealed compressors where the vapour may come into contact with electrical windings
- High heat transfer coefficients to reduce areas of heat exchangers
- Satisfactory oil solubility as oil carried from the compressor into other parts of the system may reduce heat transfer coefficients at the evaporator and condenser. The pressure-temperature characteristics of the refrigerant-oil mixture will be different from that of the refrigerant alone. The lubrication of the compressor will be affected as the viscosity of the oil reduces as it is diluted by the refrigerant. Most refrigerants are excellent degreasers
- Low water solubility. Water in the refrigerant has two principle effects
  - it causes corrosion in the system
  - it freezes in the evaporator
  Almost all refrigerants form corrosive acids or alkalis with water. These are destructive to gaskets, seals, valves and all metallic parts
- Non-toxicity
- Non-irritability
- Non-flammability
- Easy leakage detection
- Low cost
- Environmental friendliness

Typical refrigerants are detailed in Table 5.26.

Ammonia Has the following characteristics:

- Has a high latent heat of vaporization
- Has good heat transfer characteristics
- Is relatively inexpensive
- Attacks copper and its alloys in the presence of water and so needs ferrous metals
- Is insoluble with mineral oils
- Is soluble with water
- Is an irritant
- Is mildly toxic
- Burns feebly

Table 5.26 Typical refrigerants

Refrigerant	Type	Boiling point (atm. press), °C	Critical temperature, °C	Freezing temperature, °C
Ammonia	Inorganic	−30	120	−70
Carbon-tetrafluoride	Halocarbon	−115	−40	
Freon 12	Halocarbon	−30	108	−110

Carbontetrafluoride Serves as an ultra-low temperature refrigerant for use in cascade systems.

Freon 12 Exhibits the following characteristics:

- Is a clear colourless liquid
- Has a low latent heat of vaporization
- Is inert and stable
- Is miscible with mineral lubricating oil
- Is insoluble with water
- Was the first Freon to be developed and is most widely used in household refrigerators, small commercial systems and small air-conditioning systems
- *Is a most dangerous greenhouse gas*

An energy management checklist for refrigeration is provided in Appendix 2.

5.7 THERMAL RECTIFICATION AND STORAGE

Thermal rectification

A *thermal rectifier* is a device which has a high resistance to heat flow in one direction, together with a low resistance for the reversed direction of heat flow. Such devices are useful in

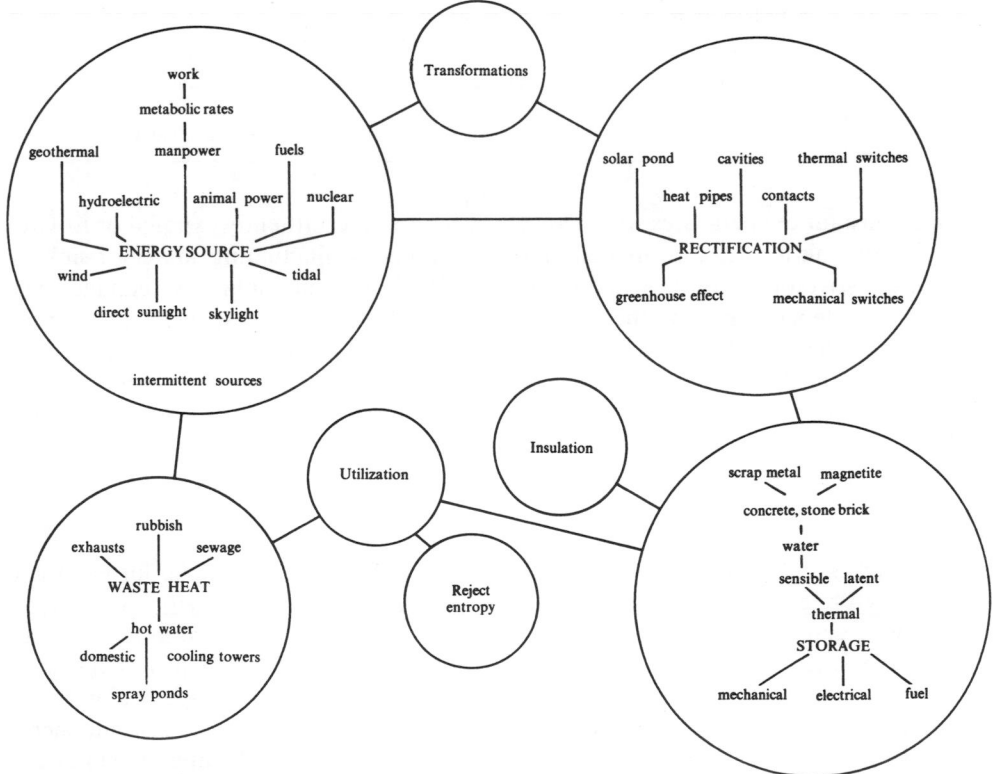

Figure 5.34 Energy flow chart.

Table 5.27 Inhibition and enhancement of heat and mass transfers

Mode of transfer	Inhibition	Enhancement	Type of rectifier
Solid conduction	Eliminate solid bridges	Provide continuous paths	Contacts subjected to thermal distortions or varying applied normal loads
	Reduce areas for heat flow	Increase areas for heat flow	
	Increase lengths of thermal paths	Decrease distance between source and sink	
	Decrease solid conductivities	Increase solid conductivities	
Convection	Low Gr number	High Re number	Damped cavities
	Evacuate	High flow rates	
	Hold gases in pockets	Pressurize	
	Reduce areas for heat flow	Reduce viscosities	
		Increase areas for heat flow	
		Promote turbulence	
Radiation	Low emissivities	High emissivities	Selective absorbers
	Shields	Mirrors	Filters
	Prevent 'seeing'		'Greenhouse effect'
	Reduce areas	Increase areas	
Mass transfer	Prevent phase changes	Promote phase changes	Heat pipe
			Heat pump
			Solar pond
			Diffusion barriers

maintaining constant temperatures within systems with are intermittently exposed or isolated from a source or sink for heat, or are continuously exposed to fluctuating source or sink for heat. Various passive or active devices are possible (Fig. 5.34) which inhibit or enhance heat flows (Table 5.27) depending upon their magnitudes and directions. These include heat pipes (Fig. 5.35), solar collectors (Fig. 5.36), special cavities (Fig. 5.37), distorting contacts (Fig. 5.38) and thermomechanically-switched arrangements. The *greenhouse effect* is a thermal rectification phenomenon, as are the body regulatory processes against heat and cold described in Section 5.4.

Energy storage

Where mismatches occur between the supply of energy, from, for example, furnaces and boilers, recovered reject heat, refrigeration systems, solar energy, ambient energy, wind, wave and tidal power, and the load to be served, it is necessary to provide some form of ballast or storage to balance supply and demand. Whereas uncombusted fossil fuel is the most compact and durable storage medium, combusted energy may be stored as superheated steam, or in thermal or chemical accumulators. Mechanical energy may be stored as strain energy in springs, as kinetic energy in flywheels, as potential energy in pumped water systems, or as

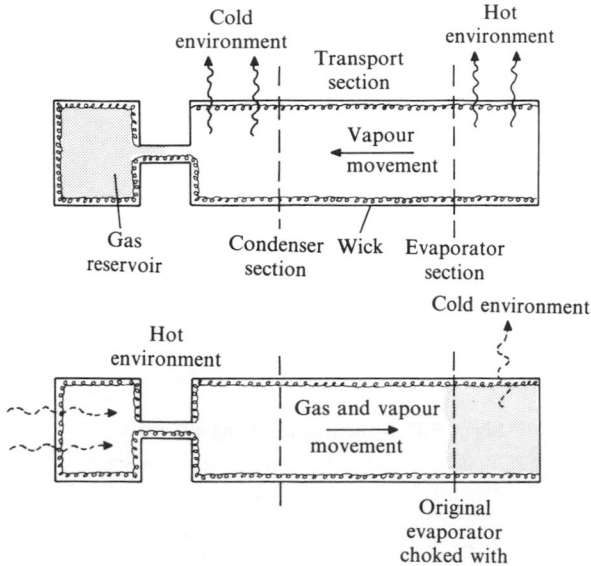

Figure 5.35 A heat pipe thermal diode (using gas to choke the evaporator for one direction of heat flow).

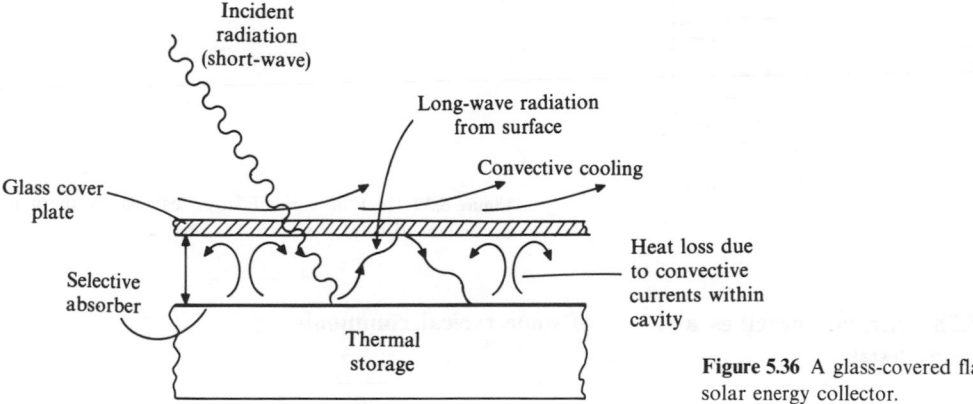

Figure 5.36 A glass-covered flat plate solar energy collector.

pressure energy in compressed air accumulators. Electrochemical storage is accomplished in batteries (Fig. 5.39).

Thermal energy storage as sensible heat in solids or liquids (Table 5.28), latent heat in phase change materials (Table 5.29), or in the solution heat capacities of compounds (Table 5.30), may be used to peak lop and smooth demand curves (Fig. 5.40) so that boilers may be operated at maximum efficiency (Fig. 5.41).

Liquid stores exhibit thermal stratification effects (Fig. 5.42) which should be encouraged so that, to retain exergetic potential, cold return liquid is fed to the base of the accumulator whilst hot liquid is drawn from the top, as in the domestic hot water tank. The material in phase change accumulators should pass through the phase change during the operating cycle (Fig. 5.43).

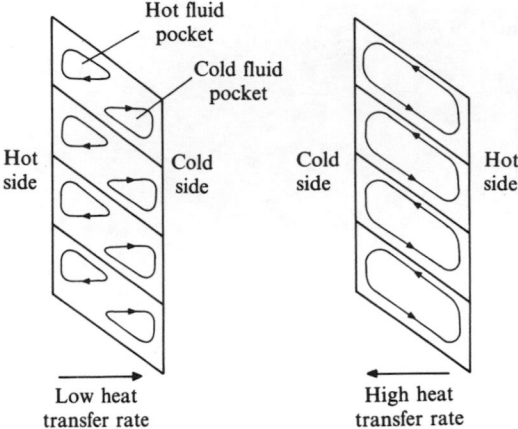

Figure 5.37 A thermal-rectifying air cavity.

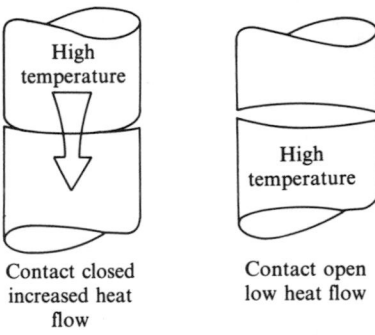

Figure 5.38 Heat flow rectification caused by thermal distortions at solid contacts.

Table 5.28 Thermal capacities at 20 °C of some typical commonly available materials

Material	Density, ρ, $kg\,m^{-3}$	Specific heat, c_p, $J\,kg^{-1}\,K^{-1}$	Volumetric thermal capacity, ρc_p, $10^6\,J\,m^{-3}\,K^{-1}$
Aluminium	2710	896	2.43
Clay	1458	897	1.28
Common brick	1800	837	1.51
Concrete	2000	880	1.76
Glass	2710	837	2.27
Gravelly earth	2050	1840	3.77
Iron	7900	452	3.57
Magnetite	5177	752	3.89
Sandstone	2200	712	1.57
Steel	7840	465	3.68
Water	988	4182	4.17
Wood	700	2390	1.67

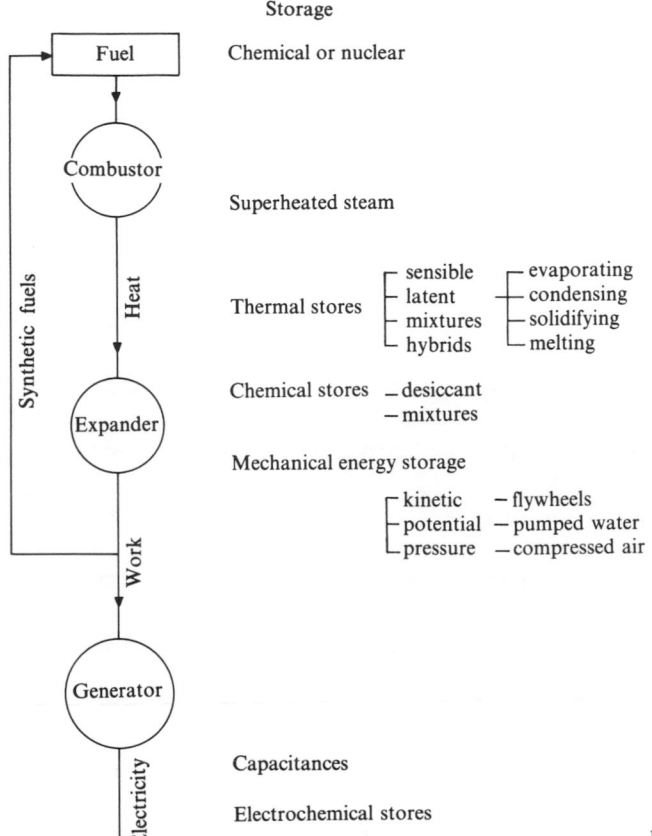

Storage

Chemical or nuclear

Superheated steam

Thermal stores ⎡ sensible ⎡ evaporating
　　　　　　├ latent ├ condensing
　　　　　　├ mixtures ├ solidifying
　　　　　　└ hybrids └ melting

Chemical stores ─ desiccant
　　　　　　　 ─ mixtures

Mechanical energy storage

⎡ kinetic ─ flywheels
├ potential ─ pumped water
└ pressure ─ compressed air

Capacitances

Electrochemical stores

Figure 5.39 Where and how to store energy.

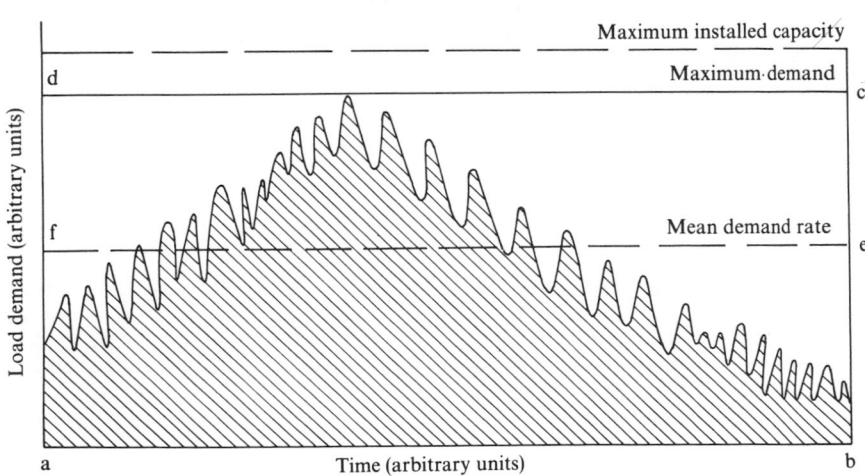

Figure 5.40 Typical load factor curve.

Table 5.29 Materials for latent heat storage

Material	Melting point, °C	Density, $kg\,m^{-3}$	Heat of fusion, $10^5\,J\,kg^{-1}$	Volume basis, $10^8\,Jm^{-3}$	Volume compared with water at 50 °C
Ice	0	913	3.36	3.07	0.65
Zinc chromate decahydrate $NA_2CrO \cdot 10H_2O$	16	1900	1.59	3.02	0.66
Calcium chloride hexahydrate $CaCl_2 \cdot 6H_2O$	29–39	1620	1.75	2.84	0.70
Sodium sulphate decahydrate $Na_2SO_4 \cdot 10H_2O$ (Glaubers salt)	32	1517	2.34	3.55	0.56
Sodium carbonate decahydrate $Na_2CO_3 \cdot 10H_2O$ (washing soda)	32–36	1425	2.70	3.85	0.52
Sodium hyperphosphate $NA_2HPO_4 \cdot 12H_2O$	40	1600	2.79	4.46	0.45
Calcium nitrate tetrahydrate $Ca(NO_3)_2 \cdot 4H_2O$	41	1819	2.10	3.82	0.52
Zinc nitrate tetrahydrate $Zn(NO_3)_2 \cdot 4H_2O$	48	1700	1.59	2.70	0.74
Sodium nitrate hexahydrate $Na(NO_3)_2 \cdot 6H_2O$	53	1680	1.52	2.55	0.78
Iron nitrate hexahydrate $Fe(NO_3)_2 \cdot 6H_2O$	60	1600	1.25	2.00	1.00
Magnesium nitrate hexahydrate $Mg(NO_3)_2 \cdot 6H_2O$	90	1630	1.78	2.90	0.69
Sodium, Na	98	956	1.14	1.09	1.83
Magnesium chloride hexahydrate $MgCl_2 \cdot 6H_2O$	115	1570	1.65	2.59	0.77
Lithium, Li	180	495	6.7	3.31	0.6
Sodium hydroxide, NaOH (caustic soda)	322	2130	2.09	4.45	0.45
Aluminium, Al	660	2600	4.02	10.50	0.19
Sodium chloride, NaCl	800	2119	4.86	10.30	0.19

Table 5.30 Solution heat capacities, SHC, of compounds

Material	Heat of solution, $MJ\,mol^{-1}$	Solubility change, $10^3\,mol\,kg^{-1}\,K^{-1}$	SCH $kJ\,kg^{-1}$ at 20 °C	at 100 °C
Ammonium nitrate	27.09	1.31	12.1	3.5
Potassium nitrate	36.13	0.27	7.4	2.8
Calcium nitrate hexahydrate	33.45	0.18	2.0	1.3
Sodium phosphate	54.43	0.08	3.7	2.0

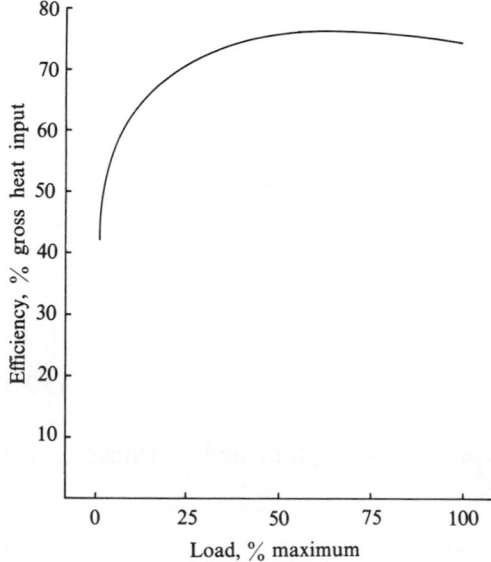

Figure 5.41 Variation in thermal efficiency for a typical oil boiler with load.

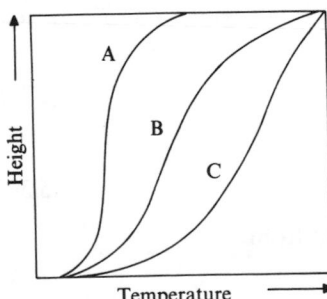

Figure 5.42 Thermal stratification characteristics.

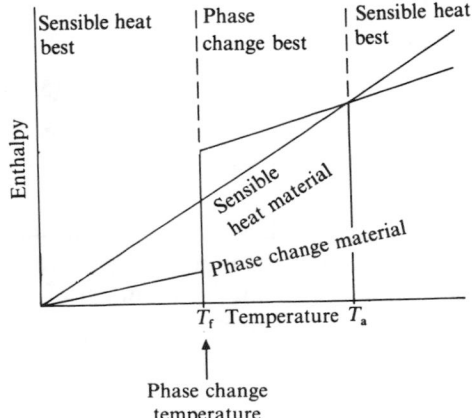

Figure 5.43 Energy storage in phase change and sensible heat materials.

Time constants

Sensible heat stores During thermal discharge, the mean temperature and the instantaneous rate of heat dissipation change continuously. To simplify the analysis, it is assumed that the temperature, T, of the store is a function of time, t, only, and is uniform throughout the system at any instant of time. It is also assumed that, over the small time period dt, the rate of heat loss from the store = the rate of heat transferred through the insulation to the surroundings at T_o, i.e.

$$-\rho C_p V dT = UA(T - T_o)dt \qquad (5.30)$$

Integrating between $t = 0$, when $T = T_{t=0}$, and $t = \Delta t$, gives

$$\log_e\left(\frac{T - T_o}{T_{t=0} - T_o}\right) = -\left(\frac{UA}{\rho c_p V}\right)\Delta t \qquad (5.31)$$

The group of parameters $(\rho c_p V / UA)$ has the dimensions of time and is defined as the *time constant*, TC, for the system. When $\Delta T = TC$

$$\frac{T - T_0}{T_{t=0} - T_o} = e^{-1} = 0.368 \qquad (5.32)$$

Physically, therefore, TC is the time taken for the temperature excess, $T - T_o$, of the system over its surroundings to fall to 36.8 per cent of its initial value, which was $T_{t=0} - T_o$. Thus, at time TC, the store temperature is given by

$$T_{TC} = T_o + 0.368(T_{t=0} - T_o) \qquad (5.33)$$

and, at any time Δt

$$T = T_o + (T_{t=0} - T_o)\exp(-\Delta t/TC) \qquad (5.34)$$

The instantaneous rate of heat dissipation can be calculated from

$$Q_t = UA(T_{t=0} - T_o)\exp(-\Delta t/TC) \qquad (5.35)$$

and the total amount of heat dissipated to the environment during Δt is obtained by integrating

$$Q_{\Delta t} = \rho c_p V(T_{t=0} - T_o)(1 - \exp(-\Delta t/TC)) \qquad (5.36)$$

Thus the total amount of heat dissipated in the interval $\Delta t = TC$ is

$$Q_{TC} = 0.632\, \rho c_p V(T_{t=0} - T_o) \qquad (5.37)$$

or 63.2 per cent of the initial heat stored.

Latent heat stores Because the temperature of a latent heat store does not vary with time, the discharge rate is constant, and the latent heat time constant cannot be expressed in terms of temperatures. Instead TC is defined as the time taken for the heat content of the store to decrease to an amount equal to the heat content of an equivalent sensible heat store at time $t = TC$; i.e.

$$Q_{TC} = 0.632\,\rho V H_{\text{fg}} = U A (T_{t=0} - T_{\text{o}}) TC \tag{5.38}$$

or

$$TC = 0.632 \left(\frac{\rho V H_{\text{fg}}}{U A \Delta T} \right) \tag{5.39}$$

Mechanical storage systems These include pumped liquid, compressed gas and flywheel kinetic energy storage.

Pumped water When the energy form is mechanical power it has an exergetic potential of unity (i.e. wind, tidal or wave power). The available energy may be stored in part using either a rotational system or by pumping.

For *pumped water storage*, energy stored $= mgh$(J). For example, energy density $= 0.0018\,\text{MJ}\,\text{kg}^{-1}$ or $1.7\,\text{MJ}\,\text{m}^{-3}$ at $180\,\text{m}$ head. Exergy density $=$ energy density.

Water may be pumped to a higher level using surplus (i.e. 'off-peak') work or electricity to drive a turbine when reclaim is required. The large-scale system at Dinorwic at North Wales, UK has a power capacity of $1500\,\text{MW}$ for 5.2 hours (i.e. $\approx 3 \times 10^7\,\text{MJ}$). The overall efficiency of this mechanical store is ≈ 76 per cent. All pumped water schemes suffer mass (and hence potential energy) losses by evaporation.

Compressed air Assuming that the air mass returns to environmental temperature after compression, the energy density of the pressure energy $= P/\rho$. But $\rho = m/V = P/RT$, therefore, the energy density of the pressure energy $= RT = 287\,T\,(\text{J}\,\text{kg}^{-1})$. Exergy density $=$ energy density.

For example, at $4100\,\text{kN}\,\text{m}^{-2}$ and $T = 293\,\text{K}$, energy density $= 0.084\,\text{MJ}\,\text{kg}^{-1}$ or $4.09\,\text{MJ}\,\text{m}^{-3}$.

Assuming that the temperature after compression is maintained, the total energy density after compression includes the heat energy created by compression, and is given by energy density $=$ pressure energy $(\gamma = 1)$ + heat energy $(\gamma = (T - T_{\text{o}})/T)$ where γ is the exergetic potential. Assuming adiabatic compression from P_1, V_1, T_1 to P_2, V_2, T_2, then $P_1 V_1^{\gamma'} = P_2 V_2^{\gamma'}$ where γ' is the adiabatic index.

For a perfect gas, $PV = mRT$, therefore

$$\frac{T_2}{T_1} = \left(\frac{P_1}{P_2} \right)^{(1 - \gamma')/\gamma'}$$

For example, if $T_1 = 293\,\text{K}$, $P_1 = 100\,\text{kN}\,\text{m}^{-2}$, $P_2 = 4100\,\text{kN}\,\text{m}^{-2}$ and $T_2 = 846\,\text{K}$:

pressure energy after compression $= RT_2 = 0.24\,\text{MJ}\,\text{kg}^{-1}$

thermal energy after compression $= c(T_2 - T_{\text{o}}) = 0.40\,\text{MJ}\,\text{kg}^{-1}$

where c is the specific heat $(\text{J}\,\text{kg}^{-1}\,\text{K}^{-1})$.

total energy after compression $= 0.64\,\text{MJ}\,\text{kg}^{-1}$

density after compression $= P/RT = 16.9\,\text{kg}\,\text{m}^{-3}$

total energy density after compression $= 10.8\,\text{MJ}\,\text{m}^{-3}$

exergy of pressure energy after compression $(\gamma' = 1) = 0.24\,\text{MJ}\,\text{kg}^{-1}$

Table 5.31 Approximate maximum energy densities (E_{max}) for flywheel kinetic energy storage

	Density, $kg\,m^{-3}$	Tensile strength, $10^{10}\,MN\,m^{-2}$	E_{max}, $MJ\,kg^{-1}$
Aluminium alloys	2600	0.41	0.1
E-glass	2500	0.33	0.7
Carbon fibre	1520	0.28	0.8
S-glass	2480	0.48	1.0
Kelver	1480	0.36	1.3
Fused silica	2160	0.50	3.1

Table 5.32 Comparative energy densities (environmental temperature = 20 °C)

Substance	Energy content, $MJ\,kg^{-1}$	Exergy content, $MJ\,kg^{-1} \times ((T - T_0)/T)$	Energy density, $MJ\,m^{-3}$	Comments
Matter	9×10^{10}	9×10^{10}	—	Complete transformation of matter into energy
Uranium	8×10^7	8×10^7	1.5×10^{12}	Release via nuclear fission at 8.5% conversion of matter
Hydrogen gas	143.0	143.0	2 288	Synthetic fuel at $20\,000\,kN\,m^{-3}$
Liquid hydrogen	94.0	94.0	6 570	Synthetic fuel at -253 °C
Methane gas	50.0	50.0	37	Synthetic or natural fuel at STP (25 °C 1 atm.)
Liquid heptane	48.0	48.0	30 000	Synthetic fuel
Liquid propane	46.0	46.0	23 000	Synthetic fuel
Fuel oil	44.7	44.7	37 000	Fossil fuel
Petrol	44.0	44.0	35 000	Fossil fuel
Lipids	37.6	37.6	60 000	Body fat derived
Carbon	32.6	32.6	50 000	
Ethyl alcohol	29.7	29.7	26 000	
Coal	28.3	28.3	45 000	Fossil fuel
Proteins	20.9	20.9	—	
Ammonia	19.0	19.0	11 200	Synthetic fuel
Carbohydrates	16.7	16.7	—	Sugar and starch
Glucose	15.7	15.7	24 000	
Methanol	15.0	15.0	12 000	Synthetic fuel
Average man in good health	4.0	—	4 000	Chemical energy
Metal Hydrides	3.3	3.3	8 200	Synthetic fuel
Flywheels	up to 3.1	3.1	6 700	Kinetic energy
Silver	1.15	0.73	12 000	Sensible thermal storage at 500 °C
Ceramics	0.9	0.7	1 700	Sensible thermal storage at 1000 °C
Gravelly earth	0.92	0.57	1 900	Sensible thermal storage at 500 °C
Compressed air	0.64	0.50	10.8	Includes heat of compression

Table 5.32 (*Continued*)

Substance	Energy content, $MJ\,kg^{-1}$	Exergy content, $MJ\,kg^{-1} \times ((T - T_0)/T)$	Energy density, $MJ\,m^{-3}$	Comments
Sodium chloride	0.49	0.36	1 000	Latent thermal storage at 800 °C
Electrochemical batteries:	*Typical life*	*4000 cycles or*	*disposable*	
Silver-zinc	0.32–0.56	0.32–0.56	—	Wet—missiles
Zinc-mercury	0.32–0.40	0.32–0.40	—	Dry—medical uses
Lead-acid	0.2–0.3	0.2–0.3	(267–400)	Wet—automobile
Zinc-alkaline-manganese dioxide	0.24	0.24	—	Dry—shavers
Silver or cuprous chloride magnesium	0.2–0.44	0.2–0.4	— —	Wet—torpedoes
Silver oxide-cadmium	0.16–0.23	0.16–0.23	—	Wet—space uses
Nickel-cadmium	0.1	0.1	—	Wet—emergency lighting
Leclanche (Zn-C)	0.08–0.16	0.08–0.16	— (100–200)	Dry—flashlights
Aluminium	0.4	0.29	1 000	Latent thermal storage at 800 °C
Concrete	0.44	0.28	880	Sensible thermal storage at 500 °C
Lithium	0.67	0.24	3 300	Latent Thermal storage at 180 °C
Sodium hydroxide	0.21	0.11	450	Latent thermal storage at 322 °C
Compressed air	0.084	0.084	4.09	Assumed returned to ambient temperature
Water	0.33	0.06	330	Sensible thermal storage at 80 °C
Silver	0.23	0.05	240	Sensible thermal storage at 100 °C
Gravelly earth	0.18	0.04	380	Sensible thermal storage at 100 °C
Ammonia	0.15	0.02	87	Sensible thermal storage at 30 °C
Sodium	0.11	0.02	109	Latent thermal storage at 98 °C
Concrete	0.09	0.019	175	Sensible thermal storage at 100 °C
Various latent Heat materials see Table 5.20	0.25	0.013	370	Latent thermal storage at 32–41 °C
Liquid carbon dioxide	0.36	0.01	217	Sensible thermal storage at 10 °C
Pumped water	0.0018	0.0018	1.7	1870 m head
Ice	0.34	−0.024	300	Latent thermal storage at 0 °C

Conversion efficiencies neglected.

Datum temperature is that of the environment.

Figures in brackets assume a typical density of $1333\,kg\,m^3$.

$$\text{exergy of thermal energy after compression } (\gamma' = 0.65) = 0.26 \, \text{MJ kg}^{-1}$$
$$\text{total exergy after compression} = 0.50 \, \text{MJ kg}^{-1},$$
$$\text{effective exergetic potential after compression} = 0.78$$

Thus mechanical energy may be stored in pressure vessels or in underground caverns by pumping on compressed air. This method is particularly suitable for electricity-generating stations to ballast for off-peak loads where pumped water storage is not possible. The energy density is however low and it is beneficial to reclaim the thermal energy after compression using regenerative storage heat exchangers.

Flywheel energy storage systems Table 5.31 lists the maximum energy densities for some flywheel energy storage systems.

Maximum exergy density = maximum energy density = $\sigma/2\rho (\text{J kg}^{-1})$, where σ = maximum hoop stress (N m^{-2}) and ρ is the density (kg m^{-3}) of the material. Thus low-density/high-strength materials are appropriate.

Table 5.32 lists a range of energy storage systems, together with their specific energy and exergy contents.

It is interesting to note that petrol, costing £2 per gallon costs £440 per m^3 = £440 per $35\,000 \, \text{MJ} = 1.26$ per MJ = 4.54 p per kWh.

A common Leclanche dry battery costing £1 may have a volume of $1 \times 10^{-4} \, \text{m}^3$ and may contain 0.015 MJ. This equates to £67 per MJ, £241 per kWh or 5312 times as expensive as petrol.

REFERENCES

1. Energy Efficiency Office, UK, Department of Energy, *Fuel Efficiency Series of Booklets*, 1–20, 1989.
2. ASHRAE, *Handbook of Fundamentals*, New York, 1988.
3. CIBSE, *Guidebook*, Chartered Institution of Building Services Engineers, London, 1990.

FURTHER READING

Avallone, E. A. and Baumeister III, T.: *Mark's Standard Handbook for Mechanical Engineers*, 9th edn, McGraw-Hill, New York, 1987
Fanger, P. O.: *Thermal Comfort*, Danish Technical Press, New York 1971.
O'Callaghan, P. W.: *Building for Energy Conservation*, Pergamon Press, Oxford, 1978.
O'Callaghan, P. W.: *Design and Management for Energy Conservation*, Pergamon Press, Oxford, 1981.

INSTRUMENTATION, MEASUREMENT AND CONTROL

6.1 INSTRUMENTATION AND MEASUREMENT

This section is confined to the measurement and control of those variables commonly affecting the energy audit. These are as below:

- Temperature
- Heat flux
- Radiation
- Humidity
- Fluid velocities and flow rates
- Pressure in fluids

The basic principle of measurement is that the measuring device must not effect the characteristics of the system being measured.

6.2 TEMPERATURE

More than 60 per cent of all observations made in industry are measurements of temperature.

What is temperature?

Heat is the energy possessed by a substance in the form of kinetic energy of atomic or molecular translation, rotation or vibration (Fig. 6.1). Temperature is a *measure* of the kinetic energy of the molecules, atoms and ions of which matter is composed.

Neither heat nor work is a property of a thermodynamic system. Both are transient

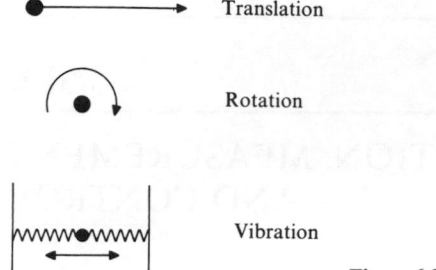

Figure 6.1 Kinetic energies in atoms and molecules.

qualities, only appearing at the boundary of a system while a change of state occurs within the system.

The heat content of a body is the product of its mass, its temperature and its specific heat.

The concept of temperature arises from the Zeroth Law of Thermodynamics—*Two systems, each in thermal equilibrium with a third are in thermal equilibrium with each other*—if two bodies are in thermal equilibrium, no net heat passes between them, and they are said to be at the same temperature. Thus the temperature defines the equilibrium state of a system and determines the rate at which heat will be transferred to or from it.

The chief observable physical effects of a change in the heat content of a body include the following:

- Rise or fall in temperature
- Expansion and contraction
- Change of state
  - from solid to liquid (melting)
  - from liquid to solid (freezing)
  - from solid to gas (sublimation)
  - from gas to solid (condensation)
  - from liquid to gas (evaporation and boiling)
  - from gas to liquid (condensation)
- Electrical effects (Peltier (voltage difference gives a cooling effect), Seebeck (thermocouple effect—temperature difference gives a voltage))

Heat cannot be detected in a system unless it is 'touched' by something, to or from which heat is transferred by one or more of the following methods:

- Conduction
- Convection
- Radiation

A measuring probe having observable temperature-dependent properties must be brought into 'contact', and should achieve thermal equilibrium, with the system to be measured. It should not, in the process, alter the equilibrium condition of the system to be measured.

Thermodynamic temperature scale

Many values of temperature are merely *indices* which cannot be used in mathematical analyses. All such scales must be converted to the thermodynamic temperature scale (i.e. from absolute zero) in order to be meaningful.

Quantitatively, temperature differences are defined to be proportional to the work output of a Carnot cycle engine, operating between a heat source and a cold sink (Fig. 6.2).

$$\text{Work done} = Q(T - T_0)/T \tag{6.1}$$

Because the efficiency of converting heat to work is always less than 100 per cent from the second law, T_0 can never be zero. From Charles' Law, $T = f(V)$ at $p = $ constant ($pV = mRT$), for perfect gases, we know the limit to be $-273.16\,^\circ\text{C} = 0\,\text{K}$.

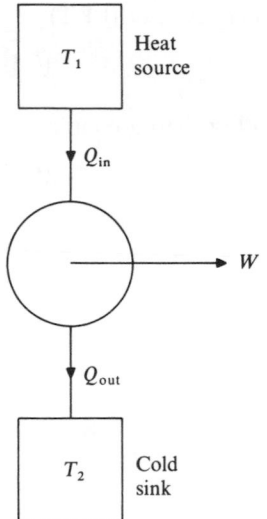

Figure 6.2 Carnot engine.

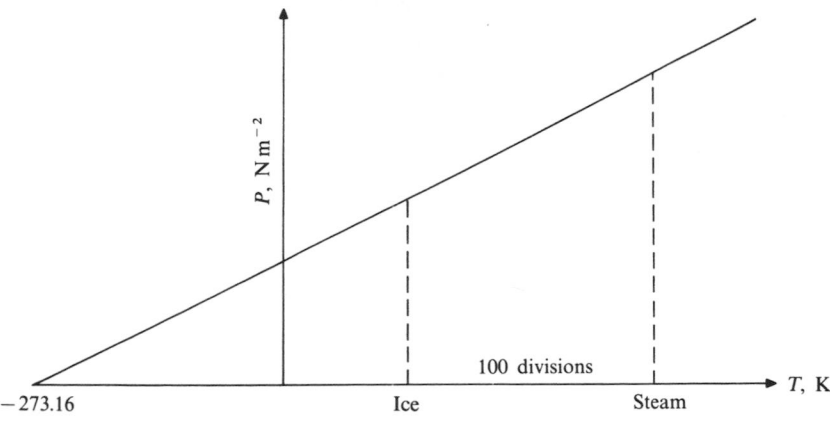

Figure 6.3 The international temperature scale.

The ideal gas thermometer reproduces the *thermodynamic temperature scale* (Fig. 6.3) since $T = PV/mR$ for a perfect gas. If this is applied at constant volume, $T = $ constant $\times P$.

The international temperature scale is based upon 100 divisions between the ice point and the steam point. Instruments are calibrated within this scale using standard primary and secondary phase change points as shown in Table 6.1.

Divisions of the thermodynamic temperature scale Standard calibration probes are used to define the variations of temperature between the fixed points, as follows:

Oxygen point to ice point Standard platinum resistance thermometer (PRT).

$$R = R_0(1 + AT + BT^2 + C(T - 100)T^3) \tag{6.2}$$

Ice point to antimony point **(630.5 °C)** Standard platinum resistance thermometer (PRT).

$$R = R_0(1 + AT + BT^2) \tag{6.3}$$

Antimony point to gold point Standard platinum–platinum/rhodium alloy thermocouple.

$$E = a + bT + cT^2 \tag{6.4}$$

Table 6.1 Standard phase change points

Point	Status	Temperature, °C	Notes
Oxygen point	Primary	−182.97	liquid to gas
Carbon dioxide	Secondary	−78.50	solid to gas
Mercury	Secondary	−38.87	freezing
Ice point	Primary	0.00	ice and liquid water
Sodium sulphate decahydrate	Secondary	32.38	transition of form
Steam	Primary	100.00	liquid/vapour
Naphalene	Secondary	218.00	condensing vapour
Cadmium	Secondary	320.90	freezing
Sulphur	Primary	444.60	liquid/vapour
Aluminium	Secondary	660.10	freezing
Silver	Primary	960.80	solid/liquid
Gold	Primary	1063.00	solid/liquid
Nickel	Secondary	1453.00	freezing
Platinum	Secondary	1769.00	freezing
Rhodium	Secondary	1960.00	freezing
Iridium	Secondary	2443.00	freezing
Tungsten	Secondary	3880.00	melting

Measurement of temperature

Temperature measuring devices are classified into *contact methods* and *non-contact methods*. In contact methods, the probe is placed in direct contact with the object under scrutiny and thermal equilibrium is established by material contact between the hot body and the testing body. In non-contact methods, the temperature measuring device does not physically touch the object being measured. There are two principle types of non-contact methods:

- Methods in which radiation emitted is observed from a distance. Thermodynamic equilibrium is then established as the radiation emitted by excited atoms and molecules is returned to the ground state. Example of such devices are the radiation pyrometer which tunes the observed radiation 'colour' with that of an electrically heated filament, and the infra-red camera.
- Methods in which density changes in transparent fluids are observed from a distance. Examples of these devices include:
  - The *interferometer*, in which the fluid temperature is sensitive directly to the fluid density, ρ
  - The *Schlieren technique*, in which the fluid temperature is sensitive to the first derivative with respect to distance of the fluid density, ρ
  - The *shadowgraph technique*, in which fluid temperatures is sensitive to the second derivative with respect to distance of fluid density, ρ.

Contact methods of temperature measurement can be broadly classified as follows:

- The homeotherm—humans as thermometers (cf. thermal comfort and the Bedford scale).
- Devices based upon the expansion of solids, such as strips and washers, cantilevers, U-shapes, spirals, helixes (used for thermostats and cut-outs).
- Devices based upon the expansion of liquids, such as liquid-in-glass thermometers, liquid-in-metal (Bourdon pressure gauge) thermometers.
- Devices based upon the expansion of gases, such as the gas thermometer (based upon Charles' law and sometimes used in electrical storage heaters).
- Devices based upon the expansion of vapours, such as vapour pressure thermometers which compensate for temperature losses along lines.
- Devices based upon changes in electrical resistivity with temperature such as resistance thermometers and thermistors. The former use platinum (PRT—platinum resistance thermometer), nickel or copper, the electrical resistance R of each of these materials being proportional to temperature, T. Thermistors utilize semiconductor materials for which R is proportional to $1/T$.
- Devices based upon the thermoelectric Seebeck effect, in which a voltage is produced by connecting hot and cold junctions.

The choice of the method to be used for measuring temperature depends upon the following factors:

- The temperature range to be examined
- The permissible time lag
- The compactness required

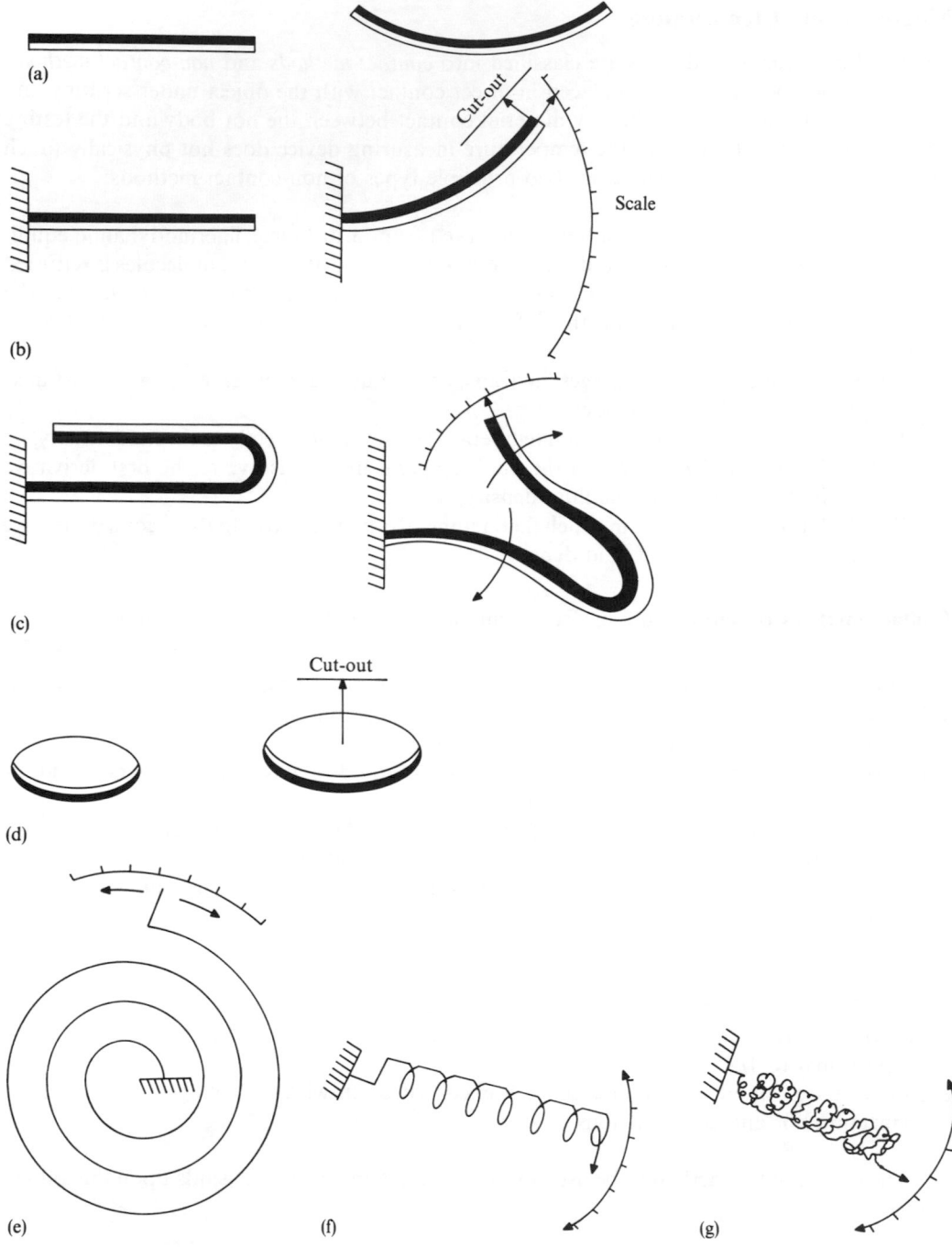

Figure 6.4 Temperature measuring devices based upon the expansions of solids: (a) Bimetallic strip; (b) Bimetallic cantilever; (c) U-shaped bimetallic cantilever (doubles response); (d) Bimetallic washer; (e) Spiral bimetallic or metallic strip; (f) Helix bimetallic or metallic strip; (g) Double helix bimetallic or metallic strip.

- The robustness required
- The ease of reading for unskilled personnel
- The position required for read-out
- The accuracy required
- The allowable cost

Types of contact sensor

Devices based upon the expansion of solids Bimetallic strip devices are usually used in passive control elements, thermostats and cut-outs. They are not particularly accurate, suffer from severe hysteresis effects and require frequent recalibration. Such devices are available to measure temperatures from $-70\,°C$ to $500\,°C$, with an accuracy, when recently calibrated, $\pm 1\%$. Their response speeds are slow but they are very compact and robust. Types include metal strips, washers, cantilevers, U-shaped cantilevers, spirals, helixes and double helixes (Fig. 6.4). Greater amplification of translational movements results in greater sensitivity but introduces longer time lags, greater hysteresis effects and lesser degrees of reproducibility due to thermal strains.

Devices based upon the expansion of liquids Liquid-in-glass thermometers, in which the expansion of a liquid is observed through movement along a capillary tube, are the commonest devices. Such thermometers are available to measure temperatures from $-200\,°C$ to $650\,°C$ with an accuracy $\pm 0.1\%$ to 1.0%. Suitable liquids are as follows:

Mercury	$-35\,°C$ to $540\,°C$
Alcohol	$-55\,°C$
Toluene	$-80\,°C$
Pentane	$-196\,°C$

Liquid-in-glass thermometers have low cost, are reliable, simple and have long lives. Because glass is not perfectly elastic, however, a hysteresis effect occurs on heating and cooling. It is not uncommon that such a device may take some days to relax after cycling between $0\,°C$ and $100\,°C$. Time lags are high because of the high thermal capacity of the liquid bulbs.

A correction must be made during calibration for the expansion of the glass. Devices are usually calibrated for either total immersion or partial immersion into the fluid being measured.

Liquid-in-metal thermometers are more robust, respond faster and suffer from less

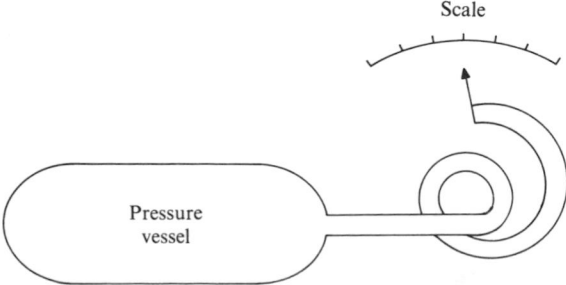

Figure 6.5 Liquid-in-metal thermometer.

hysteresis. They may be used to measure temperatures from $-200\,°C$ to $650\,°C$ with accuracies varying from $\pm0.1\,°C$ to $1.0\,°C$. They consist of a transducer bulb connected by capillary tubing to a pressure (Bourdon) gauge (Fig. 6.5), the whole being fully-filled with mercury.

The advantages of such systems are that the reading gauge can be located a considerable distance from the point of measurement and that the arrangement allows a pen recorder to be actuated. The speed of response is, however, slow and temperature losses occur along the capillary tubing. Each installation needs to be calibrated.

Devices based upon the expansion of vapours Vapour pressure thermometers usually consist of liquids in metals and operate on the principle that, when a liquid and its vapour are in equilibrium, the pressure in the system depends only on the temperature. Such devices thus compensate for *temperature losses* along the capillary tubing. They may be used to measure temperatures in the range $-240\,°C$ to $316\,°C$.

Devices based upon the expansion of gases These are the only devices which describe absolute zero temperature $(-273.16\,°C)$. They are used at constant pressure so that the volume of the gas is a constant × temperature. They may be used to measure temperatures within the range $-240\,°C$ to $538\,°C$. Zero temperature is defined by extrapolating to zero volume, when all motion is stilled.

Devices based upon changes in electrical resistivity Electrical resistance thermometers depend upon the practically linear relationship between temperature and the electrical resistances of pure platinum, nickel or copper (Fig. 6.6).

$$\text{Electrical resistance, } R = R_0(1 + aT + bT^2) \tag{6.5}$$

where a and b are calibration constants. Such devices may be used to measure temperatures in the range $-250\,°C$ to $1000\,°C$ with accuracies as good as ±0.1 per cent. Speeds of response are fast.

Thermistors use semiconductor materials which have negative temperature coefficients of resistance much higher than those positive values for resistance thermometers (Fig. 6.7). They can be made very small and robust and can measure point temperatures. They may be used to measure temperatures in the range $-50\,°C$ to $300\,°C$ with accuracies within ±1 per cent. Response times are of the order of a few seconds.

Temperature-dependent changes of resistance are measured using a *Wheatstone bridge* circuit (Fig. 6.8).

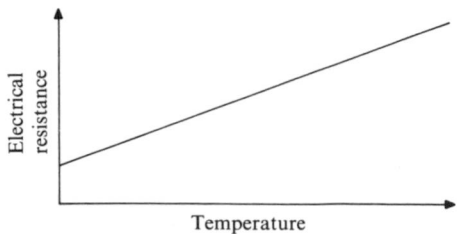

Figure 6.6 Electrical resistance thermometer.

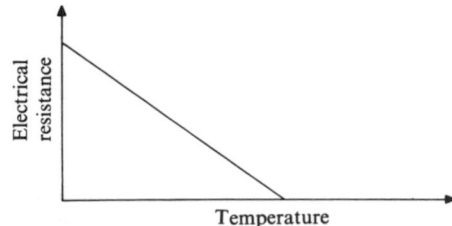

Figure 6.7 The thermistor.

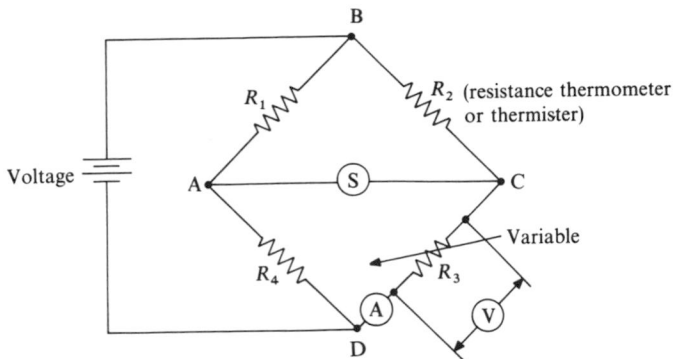

Figure 6.8 Wheatstone bridge circuit. When R_3 is adjusted so that the current flowing between A and C is zero, the galvonometer has a zero (null) reading. Then

$$\frac{R_1}{R_4} = \frac{R_2}{R_3}$$

If R_1 and R_4 are standard resistances, R_2 can be calculated.

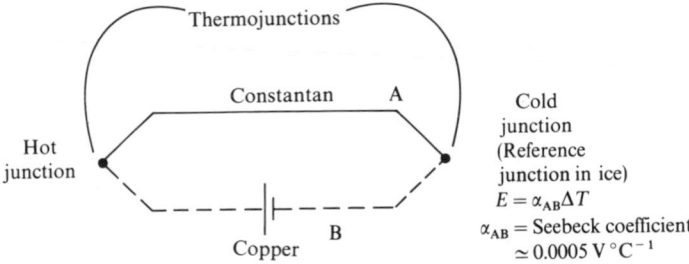

Figure 6.9 Thermocouple.

Devices based upon thermoelectric effects Thermocouple elements work on the thermoelectric *Seebeck effect*—if a closed circuit is made between two dissimilar metals, an electrical currents flows in the circuit when the two junctions are maintained at different temperatures—Fig. 6.9.

A standard thermoprobe junction should have a welded spherical joint 0.35 mm to 0.65 mm in diameter. Seebeck voltage, $E = \alpha T$, where α is the relative *Seebeck coefficient* ($\simeq 0.00005\ V\,°C^{-1}$) between the two metals involved.

Characteristics of some common thermocouples are shown in Table 6.2.[1]

Jacob lists polynomial constant coefficients for converting millivolts to °C for the above thermocouple types.[1]

Chesmond indicates that, with suitable calibration, types J, K and T may be used down to $-184\,°C$.[2]

Thermocouples are small, simple, reliable and have good reproducibility. Little external power is required, although the signals must be amplified for input to computational systems. Because such thermoprobes have low thermal capacities, speeds of response can be extremely fast. Thermocouples measure differential temperatures and so a reference junction must be employed (e.g. well-mixed crushed or shaved ice). Junctions formed using base metals (e.g.

Table 6.2 Thermocouples[1]

Type	Description	Range of calibration, °C	Accuracy	Seebeck coefficient, $\mu V\,°C^{-1}$
E	Nickel (10%) chromium/constantan	−100 to 1000	±5°C	62
J	Iron/constantan	0 to 760	±1%	51
K	Nickel (10%) chromium/ nickel (5%)(Al, Si)	0 to 1370	±0.7%	40
R	Platinum (13%) rhodium/platinum	0 to 1000	±0.5%	7
S	Platinum (10%) rhodium/platinum	0 to 1750	±1.0%	7
T	Copper/constantan	−160 to 400	±0.5%	40

copper, constantan) can measure temperatures in the range −200 °C to 1400 °C, whilst 0 °C to 2800 °C are possible for rare metals (e.g. platinum, rhodium, iridium). High accuracies are possible (of the order ±0.05 per cent).

Whereas thermocouples are cheap to construct, thermocouple circuits require expensive instrumentation to record the very small voltages produced.

Thermocouple circuits A *thermopile* consists of a number of thermocouples connected in series to amplify the response (Fig. 6.10). *Multi-junction thermocouples* (Fig. 6.11) may be used to obtain the average temperature of a number of observations.

Transient response of thermometry

The parameter which effects the rate of heating or cooling of a thermoprobe is the *Biot number, Bi.*

Bi = internal resistance to heat flow/external resistance to heat flow

$= R_i \times$ volume$/R_e \times$ surface area

$= ((1/k) \times V)/((1/h) \times A)$

$= hL/k$ (6.6)

where k = thermal conductivity of the solid probe, $W\,m^{-1}\,K^{-1}$
h = surface heat transfer coefficient, $W\,m^{-2}\,K^{-1}$
and L = volume/surface area; m.

The heat absorbed by the probe equals the net heat transferred to the probe, i.e.

$$\rho V s\,dT_p = hA(T_s - T_p)d\theta$$
$$dT_p/(T_s - T_p) = hA\,d\theta/\rho V s$$

or

$$d(T_s - T_p)/(T_s - T_p) = hA\,d\theta/\rho V s$$
$$-\ln((T_s - T_p)/(T_s - T_{p,0})) = hA\theta/\rho V s$$
$$(T_s - T_p)/(T_s - T_{p,0}) = \exp(hA\theta/\rho V s)$$ (6.7)

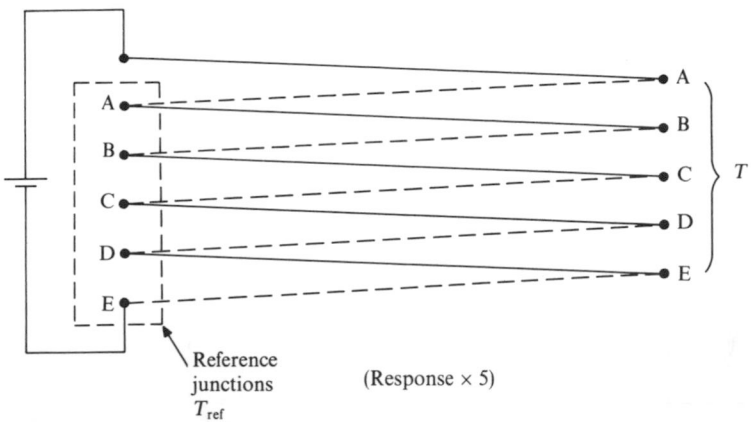

Figure 6.10 Thermopile.

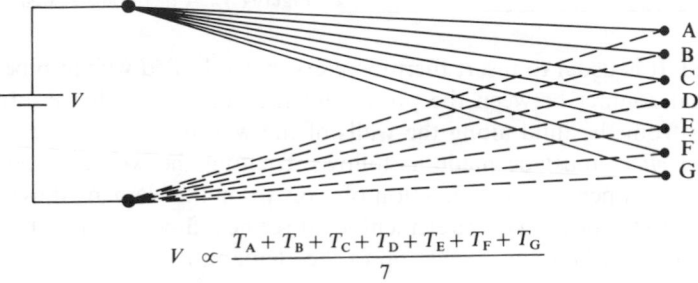

$$V \propto \frac{T_A + T_B + T_C + T_D + T_E + T_F + T_G}{7}$$

Figure 6.11 Averaging thermocouple circuit.

The quantity $\rho V s / h A$ has the dimensions of time (seconds) and is known as the *time constant, TC,* for the system. Thermoprobes having high time constants are slow to respond to changes in system temperature, whilst those having low time constants exhibit a fast response.

Letting $L = V/A$ and multiplying top and bottom of the time constant by kL

$$TC = h\theta kL / \rho V kL^2$$
$$= (hL/k) \times (k\theta / \rho V L^2)$$
$$= Bi \, Fo \qquad (6.8)$$

where Fo is the *Fourier modulus* $= \alpha\theta/L^2$ and $\alpha = k/\rho s$, the *thermal diffusivity* of the probe. Thus

$$(T_p - T_s)/(T_o - T_s) = e^{-Bi \, Fo} \qquad (6.9)$$

Measurement errors and corrections

Bulk fluid temperature in a pipe Due to boundary layer effects, the temperature is not uniform across a pipe or duct section. Thermoprobes must be traversed across the section to integrate the distribution for bulk fluid temperatures.

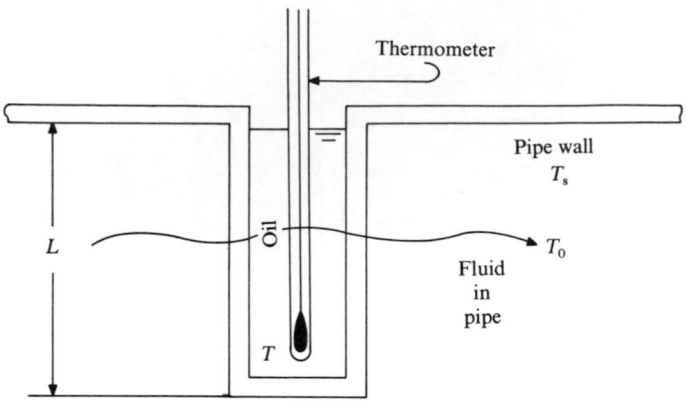

Figure 6.12 A thermometer well.

Fin effect in a thermometer well It is usual to insert thermometers into oil-filled well in pipes (Fig. 6.12) so that good thermal contact between the probe and the system is achieved. In such circumstances, heat flows from the fluid along the walls of the well to the cooler pipe walls. A correction for this *fin effect* must be made as follows: treating the well as a rod (Fig. 6.13) and assuming that the temperature is a function of x only, for the element shown, rate of heat flow by conduction into the bottom of the element + rate of heat flow by convection into the surface dx long = rate of heat flow out of the top of the element; i.e.

$$q_{in} + q_{conv} = q_{out} \tag{6.10}$$

or

$$-kA\,dT/dx + hP\,dx(T - T_o) = -kA\,dT/dx + d/dx(-kA\,dT/dx)dx \tag{6.11}$$

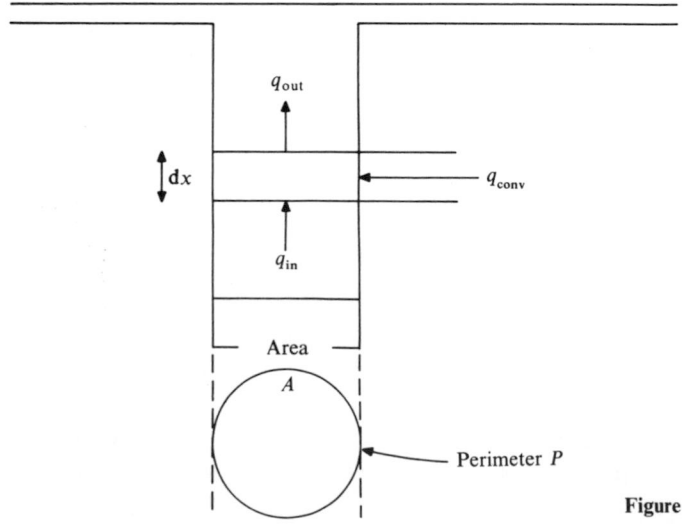

Figure 6.13 Notation for the thermometer well.

where k = effective thermal conductivity of the well, $W\,m^{-1}\,K^{-1}$
A = cross-sectional area of the well, m^2
h = surface heat transfer coefficient of the well sides, $W\,m^2\,K^{-1}$
P = perimeter of the well, m
T_o = temperature of the fluid in the pipe, °C.

This may be written in as alternative form:

$$hP/kA(-T + T_o) = d^2T/dx^2 \qquad (6.12)$$

which is a standard form of an ordinary linear differential equation, and can be solved to obtain

$$\frac{T_o - T}{T_o - T_s} = \frac{\cosh(m(L-x)) + (h/mk)\sinh(m(L-x))}{\cosh(mL) + (h/mk)\sinh(mL)} \qquad (6.13)$$

where $m^2 = hP/kA$, T is the temperature indicated at the bottom of the well, T_s is the pipe surface temperature and T_o is the true fluid temperature.

The error in the indicated reading may be reduced by inclining the well to increase the length L.

If T is the indicated temperature at the bottom of the well:

$$\frac{T_o - T}{T_o - T_s} = \frac{1}{\cosh(mL) + (h/mk)\sinh(mL)} \qquad (6.14)$$

If heat is carried away from the measuring point by the measuring device, the measured temperature will be lower than the true temperature.

Conduction and radiation from thermocouples Conduction along thermocouple probe leads similarly results in a lower temperature being measured and should be corrected for. Radiation exchanges between temperature probes and surrounding surfaces also result in spurious measurements. These may be compensated for as follows: referring to Fig. 6.14, heat loss by radiation from the probe of area A (m^2) to the wall = heat gain by convection from the fluid; i.e.

$$A\varepsilon\sigma(T_p^4 - T_w^4) = hA(T_g - T_p) \qquad (6.15)$$

from which the true gas temperature is given by

$$T_g = (\varepsilon\sigma/h)(T_p^4 - T_w^4) + T_p \qquad (6.16)$$

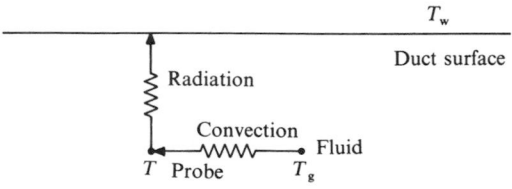

Figure 6.14 Correction for radiation in ducts.

where ε = surface emissivity of the probe
σ = Stephan–Boltzmann constant ($= 5.67 \times 10^{-8}$), $W\,m^{-2}\,K^{-4}$
T_w = temperature of the wall, °C
T_g = temperature of the gas, °C
T_p = temperature indicated by the probe, °C.

Tips for accurate temperature measurements

- Correct readings via thermal analyses.
- Use sloping oil wells.
- Place probes in turbulent regions.
- Inside solids:
  - Ensure that the ratio of the depth of the hole drilled to take a thermoprobe to the diameter of the hole, L/d, is greater than 10
  - Ensure good thermal contact between the probe and the system
  - Ground electrical thermoprobes
  - Use long thin thermoprobe leads inserted along isotherms or via guard heaters.
- Surface temperatures:
  - Use thin disk-shaped thermocouples in good contact with the surface
  - Examine a small area insulated at the outside
  - Ensure that the surface properties are not altered by the probe.
- Gas temperatures:
  - Radiation-shield thermoprobes
  - Speed-up convective flows (suction pyrometer).

Measurement of room air temperatures

The accurate measurement of air temperature is relatively straightforward, provided that the effect of thermal radiation on the detector is eliminated. This is most easily accomplished by mounting the sensor within an aspirated radiation shield, on the upstream of the fan so that fan heat does not raise the temperature of the air being measured. In general, the larger the surface area of the probe, the more radiant energy exchanges with surrounding surfaces. Thus the use of fine thermoelement wires reduce the effects of thermal radiation. The wires and the probe should also have low radiant emissivities. Even a very small thermoprobe bead with surface emissivity equal to 0.1 would produce a response proportional to

$$T_p = 0.8T_a + 0.2T_r \tag{6.17}$$

where T_p (°C) is the temperature indicated by the probe T_a (°C) is the air temperature and T_r (°C) is the mean radiant temperature of the surrounding surfaces.

Most temperature sensors will produce a measurement of temperature somewhere between the air temperature and the mean radiant temperature. Aspirated devices artificially increase the air velocity around the sensor and hence increase the convective component of the heat transfer to or from the probe relative to the radiative component.

In situ measurements are usually taken using a mercury-in-glass thermometer properly shielded from radiation, wiped dry before taking the reading to eliminate evaporative cooling, and given sufficient time to equilibrate with the temperature of the surrounding air (i.e. to settle down to give a steady-state reading).

6.3 MEASUREMENT OF HEAT FLUX

Transient techniques

Slug-type sensor for surface measurements These are used to measure heat flux exchanges from the surfaces of solids. A slug of metal is buried in, and insulated from the surface (Fig. 6.15). As for the measurement of surface temperatures, it must be ensured that the surface properties (e.g. radiant emissivity, coefficient of convection) are not altered by the insertion of the measuring device.

Neglecting heat losses through the back insulation and the thermoelement leads, heat received in time dt = energy stored in time dt; i.e.

$$qA\,dt = ms\,dT \qquad (6.18)$$

where q = uniform rate of heat flux entering the surface, $W\,m^{-2}$
 A = surface area of the slug receiving heat, m^2
 m = mass of the slug of metal, kg
 s = specific heat of the metal, $J\,kg^{-1}\,K^{-1}$
 dT = temperature rise of the slug in time dt, °C.

Thus

$$q = (mc/A)\,dT/dt \qquad (6.19)$$

For greater accuracy, a *loss coefficient* may be included in the equation, i.e.

$$q = (mc/A)\,dT/dt + K\,\Delta T \qquad (6.20)$$

where ΔT is the temperature difference (°C) across the casing, and the correction factor, K ($W\,m^{-2}\,K^{-1}$), is obtained via calibration of the system (i.e. under a known radiant flux).

Whole bodies The rates of temperature rise or fall of cylinders, spheres, or plates (Fig. 6.16), may be measured to estimate the rate of heat received or lost over entire surfaces.

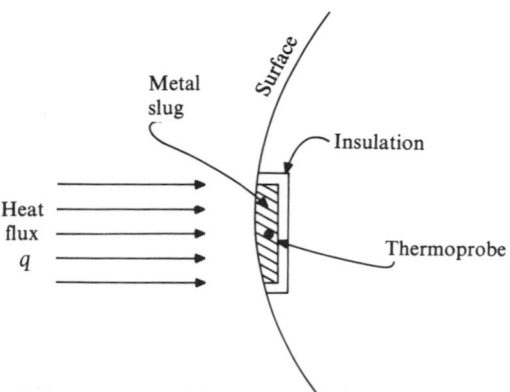

Figure 6.15 Slug-type heat flux sensor.

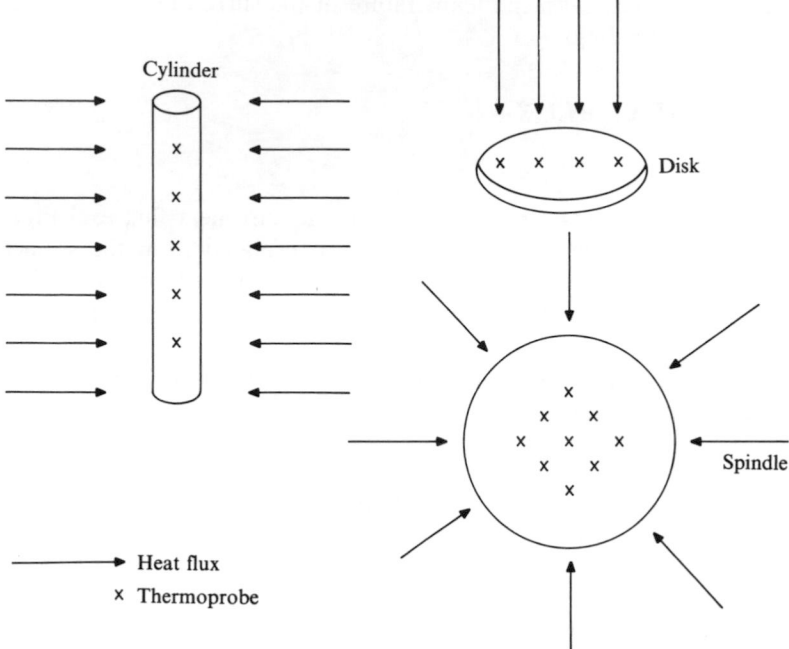

Figure 6.16 Total heat flux sensors.

Steady-state techniques

Transient sensors converge to steady-states when external conditions do not change with time. Surface thermocouples may be used to measure rates of heat flux entering the surfaces of solids.

The *Gardon gauge* consists of a thin constantan disk, which is connected to a large copper

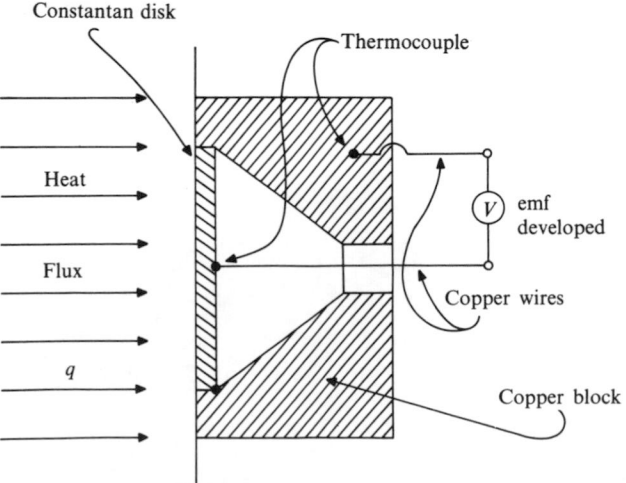

Figure 6.17 Gardon gauge heat flux sensor.

heat sink (Fig. 6.17), whilst a thin (less than 0.1 mm in diameter) is connected at the centre of the disk. This configuration forms a differential thermocouple between the centre of the disk and its edges. When the disk is exposed to a uniform and constant heat flux, the equilibrium temperature difference established is directly proportional to the heat flux being received. The system must be calibrated under a known heat flux.

When only the radiation component of the heat flux is required, the front of the sensor may be shielded against convection using a sapphire window. Alternatively, radiation shields may be installed externally to the surface when only the convective flux is required.

Heat flux rates *inside solids* may be monitored by measuring steady-state temperature distributions (using inserted thermocouple probe arrays) inside solids constructed from materials of known thermal conductivities. Armco iron is often used as a standard reference material.

6.4 MEASUREMENT OF RADIATION

Radiation pyrometers

In the *disappearing filament optical pyrometer*, high-temperature thermal radiation is viewed via a site glass which contains a heated wire. The heat to the wire is adjusted until the 'colour' temperature of the wire exactly matches that of the radiation viewed, i.e. the filament 'disappears'. The temperature of the wire then equals that of the body emitting the radiation.

Radiation fluxes may also be measured using the surface and body heat flux devices described previously.

Many other radiation detectors focus radiation on to a blackened temperature measuring probe (e.g. thermometor, thermister, or thermopile).

Measurement of mean radiant temperature

In thermal comfort studies, the mean radiant temperature (°C) refers to the shape of the human body and so is a factor which is difficult to measure precisely. One method is to measure the areas, A_i (m^2), and surface temperatures, T_i (°C) of all walls, ceilings and floors with thermocouple systems and then calculate the mean radiant temperature, T_r, from

$$T_r = \Sigma A_i T_i / \Sigma A_i \qquad (6.21)$$

i.e. weighting the surrounding surface temperatures by the areas of the surfaces. This gives a mean value for the mean radiant temperature throughout the room, whereas strictly, the mean radiant temperature is a function of position in the room, and, at a given location within the room, the contribution of each radiating area should be weighted according to the solid angle subtended at the test point by each surface i and by the emissivities of the surfaces.[3]

Devices to measure mean radiant temperatures must suppress convective heat transfers from the surrounding air and increase radiation interception from surrounding surfaces, usually by arranging that the surfaces of the sensor are thermally 'black' and that a large surface area is presented to the radiating surrounds.

Single sphere globe thermometer

Because of its simplicity, the *globe thermometer* is usually used in practice. This consists of a black spherical copper shell (i.e. a ball-cock from a cistern) at the centre of which is placed a thermal sensor (mercury-in-glass thermometer, thermocouple probe or thermistor). The larger the sphere, the greater the amount of radiative, in relation to convective, heat exchanges. In theory, any diameter sphere may be employed, provided that the device is calibrated. The globe thermometer is influenced by dry-bulb air temperature, air velocity as well as mean radiant temperature. In still air, it indicates the mean radiant temperature exactly. In using the system in practical situations, simultaneous measurements of adjacent airspeed and dry-bulb air temperature should be taken. At thermal equilibrium, heat gain by radiation = heat lost by convection or

$$\sigma\varepsilon(T_r^4 - T_p^4) = h_c(T_p - T_a) \tag{6.22}$$

where σ = Stephan–Boltzmann constant ($= 5.67 \times 10^{-8}$), $\mathrm{W\,m^{-2}\,K^{-4}}$
   ε = surface emissivity of the device ($= 1.0$)
   T_r = mean radiant temperature, °C
   T_p = temperature indicated by the probe, °C
   h_c = velocity-dependent mean convective heat transfer coefficient over the surface of the device, $\mathrm{W\,m^{-2}\,K^{-1}}$
   T_a = dry-bulb temperature of the surrounding air, °C.

Thus this heat balance equation may be solved for the mean radiant temperature T_r.

Note: For air at 20 °C flowing with an approach velocity, U ($\mathrm{m\,s^{-1}}$), over a sphere, the mean convective heat transfer coefficient, h_c ($\mathrm{W\,m^{-2}\,K^{-1}}$), over the surface of the sphere is given by

$$h_c = 14.1 U^{0.6} \tag{6.23}$$

Two-sphere radiometer

The single globe thermometer is sluggish and needs measurements of three quantities, increasing the chance of error in estimates of mean radiant temperatures.

The two-sphere radiometer uses two spheres each approximately 50 mm in diameter. The surface of one of these spheres is gold-plated and that of the other is thermally 'black'. The two spheres are electrically heated to yield identical temperatures. In the steady-state, the mean radiant temperature of the surroundings is indicated from the ratio of the rates of thermal energy needed to be supplied to each sphere to maintain temperature equilibrium.

Directional radiometers

A *Moll thermopile* measures incoming thermal radiation in a given direction. It consists of a thermopile, suitable shielded against convection (e.g. with a polythene sheet window) and having radiation guard shielding. The voltage output is proportional to the radiation exchange between the thermopile and the surfaces within its view. The reference temperature for the thermopile is usually that at the back surface of the device (i.e. the opposite surface to that receiving the thermal radiation). This can be measured using a separate thermocouple circuit.

Infra-red imaging

The infra-red camera measures infra-red radiation and, by scanning, builds an 'image' of the radiation emanating from a system of surfaces. Great care must be taken in using such devices as the image is affected by local convection rates, as well as surface temperatures and radiative properties, Reflections from other surrounding surfaces can also confound the image.

6.5 MEASUREMENT OF PSYCHROMETRIC VARIABLES

The amount of water vapour in air (see Fig. 6.18) may be assessed by measuring the following parameters:

- Dry- and wet-bulb temperatures (°C)
- Specific humidity (kg_w per kg_a)
- Dew-point
- Relative humidity

Dry- and wet-bulb temperatures (°C)

The most widely used instrument is the wet- and dry-bulb thermometer set. The combination is used in a 'sling' (whirled by hand), in a 'screen' or louvered box (for meteorological measurements) which relies upon wind speed to speed up convective flows, or in a mechanically aspirated device.

One thermometer bulb is covered with a wick which is kept wet. Providing that the vapour pressure of the water is greater than that of the moist air being measured, evaporation of the water will occur. To effect this, sensible heat will flow from the environment by convection and radiation until a steady-state situation is reached in which the heat gain to the water exactly equals the latent heat loss from it. At this balanced state, the equilibrium temperature attained is known as the *wet-bulb temperature* of the air surrounding the wetted bulb. When dry- and wet-bulb temperatures are equal, the air is fully saturated with water and its relative humidity is unity. As the air becomes drier, the wet-bulb temperature decreases. The greater the difference between the dry-bulb temperature and the wet-bulb temperature (the *wet-bulb depression*), the drier the air. Knowledge of the wet- and dry-bulb temperatures locates the state-point of the air on the psychrometric chart and hence defines all the other variables. Wet- and dry-bulb sets must be calibrated for use. It is recommended that distilled water be used for accurate measurements.

Specific humidity

The ultimate standard for calibration of humidity measuring devices is the *gravimetric hygrometer*. In this laboratory-based apparatus, the water vapour in a sample of air is absorbed by chemicals (phosphorous pentoxide, magnesium perchlorate, calcium chloride or silica gel (a form of silica formed by coagulation of sodium silicate sol—this yields a gelatinous solid which is then dehydrated to a hard granular form)) and then carefully weighed. It thus determines the specific humidity of the air sample directly. Solid carbon dioxide or other cryogenic material can alternatively be used to freeze the water out of the air.

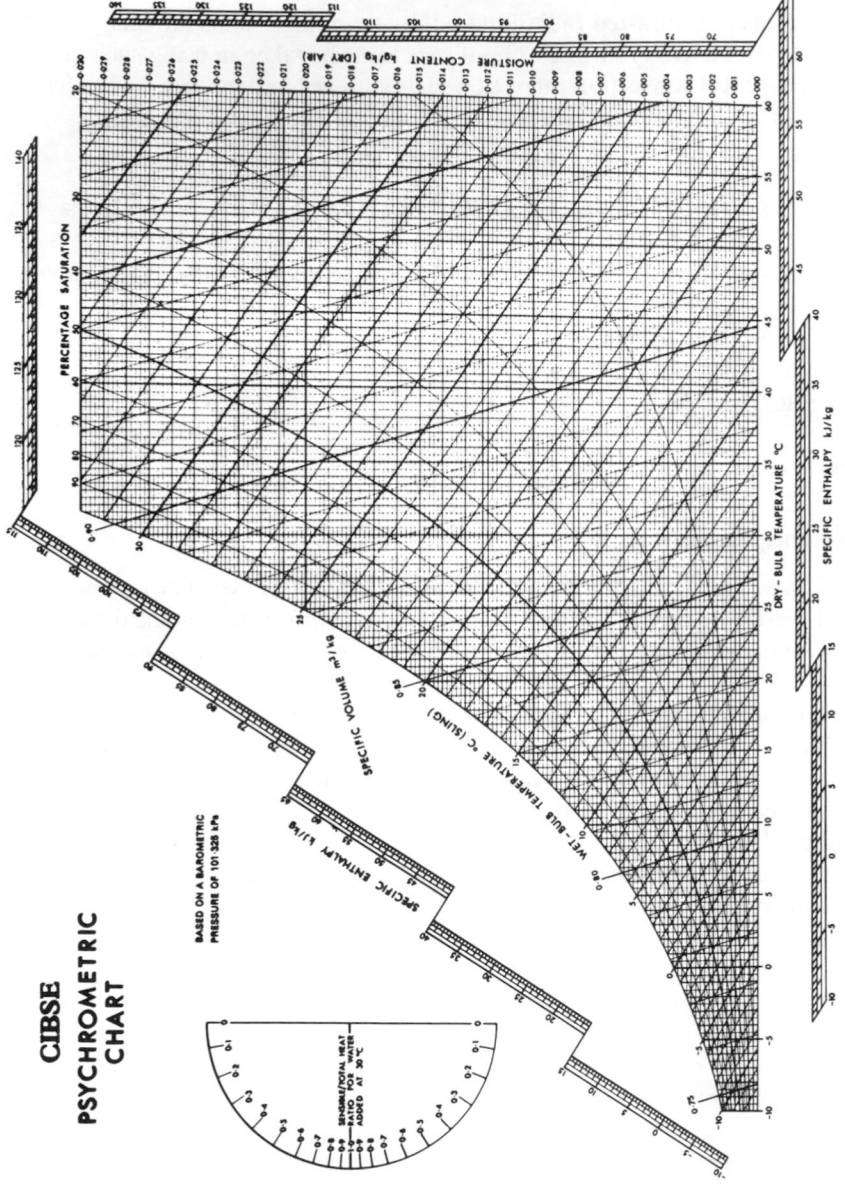

Figure 6.18 The CIBSE psychrometric chart (reproduced with the permission of the Chartered Institution of Building Services. Pads of charts size A3 suitable for permanent records are available from the Institution).

Dew-point

Dew-point hygrometers The dew-point temperature is the temperature of a surface in contact with the air at which dew would begin to form when the surface is slowly cooled. It is the saturation temperature corresponding to the partial pressure of the water vapour in the air. Dew-point temperatures can be determined by noting the temperature of a cooled polished metal surface when the first traces of condensation appear.

Heated electrical hygrometer This device consists of a tubular core covered by glass fibre with a spirally-wound winding for heating electrodes. The surface is covered with a salt solution (lithium chloride) for which its electrical resistance varies with moisture content. When moisture is absorbed, the resistance of the salt film drops, more current flows and the salt film heats up driving off moisture. When moisture is driven off, the resistance of the salt film increases, less current flows, the salt film cools down and moisture is absorbed. At equilibrium, the steady-state temperature attained is *related to* the dew-point temperature and so devices must be calibrated.

Electrolytic hygrometer In this system, a continuous supply of sample air flows through an analysis tube inside which the moisture is absorbed by a desiccant, such as phosphorous pentoxide. The water is electrolysed into hydrogen and oxygen. Consequently a measurable electrolysis current flows.

Relative humidity

Many organic materials change in dimensions with changes in relative humidity. This effect has been utilized in a number of simple humidity indicators. Materials commonly employed include human hair, animal membrane, wood and paper. No material has been found which reproduces readings continuously and accurately over a period of time, since they suffer from hysteresis effects. Nevertheless, devices using the effect are used in indicators, recorders and controllers, motion caused by dimensional changes being translated via suitable linkages to cause a pointer to move across a dial, or to actuate a pen-recorder or pneumatic or electrical control mechanism. Frequent recalibrations are needed for accurate measurements.

6.6 MEASUREMENT OF FLUID VELOCITIES AND FLOW RATES

Pitot-static tube

From the energy equation for a flowing fluid, the total energy (W) = pressure energy (W) + kinetic energy (W); i.e.

$$E = mp/\rho + mu^2/2 \tag{6.24}$$

where m = fluid mass flow rate, $kg\,s^{-1}$
p = static pressure in the flowing fluid, $W\,m^{-2}$
ρ = density of the fluid, $kg\,m^{-3}$
u = flow velocity, $m\,s^{-1}$.

If the fluid were brought to rest, from the first law of thermodynamics (energy conservation) the total pressure energy (W) in the fluid would be given by

$$PE = E = mp_o/\rho = mp/\rho + mu^2/2 \qquad (6.25)$$

where p_o ($N\,m^{-2}$) is known as the stagnation, or total, pressure of the fluid.

Cancelling m throughout:

$$p_o/\rho = p/\rho + u^2/2$$

and solving for u:

$$\text{flow velocity, } u = (2(p_o - p)/\rho)^{0.5} \qquad (6.26)$$

Thus the pitot static tube (Fig. 6.19) contains a pressure tapping at the leading edge to

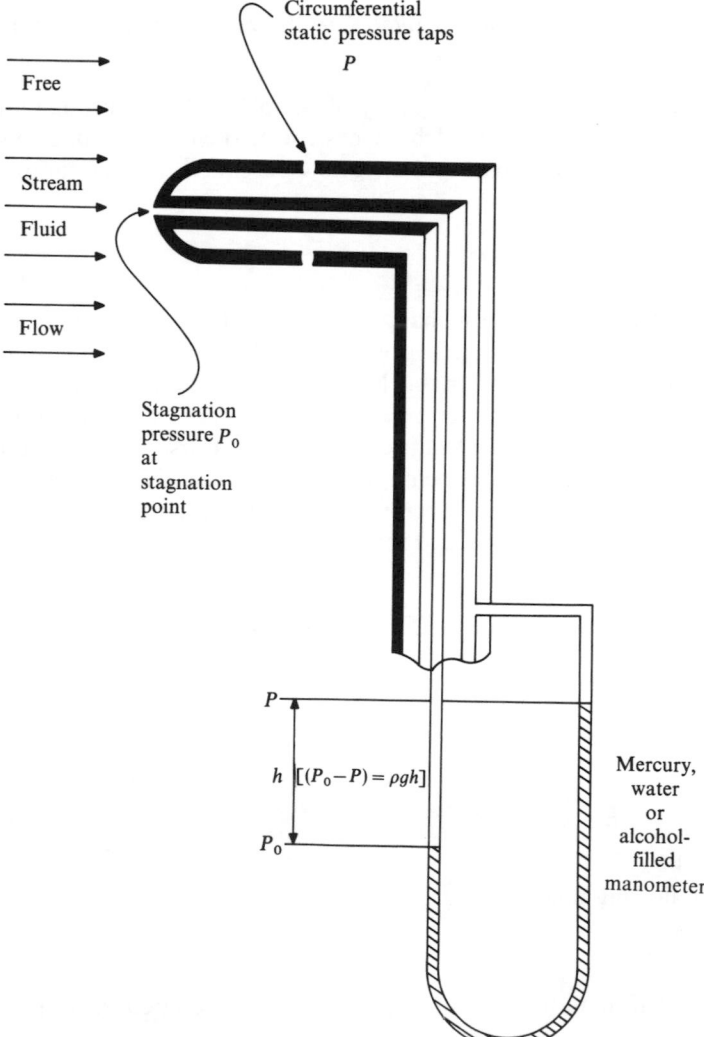

Figure 6.19 Pitot-static probe.

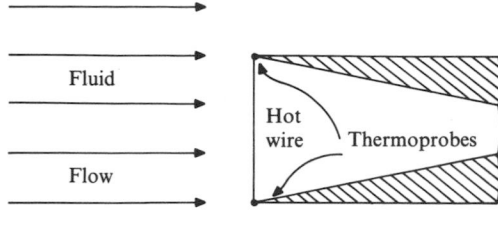

Figure 6.20 Hot-wire anemometer.

measure the stagnation pressure as the fluid is brought to rest, and circumferential pressure tappings to measure static pressure. The difference between stagnation pressure and static pressure is then indicated by a vertical or inclined liquid manometer. The fluid flow velocity may then be deduced from the difference between these two pressures.

Hot-wire anemometers

In these devices (Fig. 6.20), a thin electrically heated wire is placed so that the fluid flows perpendicularly to the axis of the wire. By equating the electrical power supplied to the wire to the heat lost from the wire to the fluid at steady-state conditions, the flow velocity can be deduced. For accurate measurements, hot wire anemometers should be calibrated in known flows.

Deflecting vane anemometers

These consist of pivoted vanes enclosed in suitable casings. As the flowing fluid exerts a pressure on the vane, the vane deflects against a spring and damping magnet combination. The movement is indicated on a scale.

Inside pipes and ducts

Orifice plates and Venturi tubes If an obstruction (such as a sharp-edged orifice or constriction) is placed in the flow inside a pipe (Fig. 6.21), the fluid flow velocity in the pipe is a function of the pressure drop across the obstructing plate and orifice. Again, devices should be calibrated in known flows for accurate measurements.

Rotameters A rotameter (Fig. 6.22) consists of a vertical tube with a tapered bore inside which a float assumes a given vertical position according to the rate of flow.

Turbine flow meters These are small water turbines placed in the flow. The rotational speed attained depends upon the flow velocity. A magnetic proximity pickup can be used to produce voltage pulses.

Measurement of low air speeds

The air flow around a person is unconfined and therefore generally turbulent. For thermal comfort studies, a single value for air speed usually suffices to describe the cooling effect of

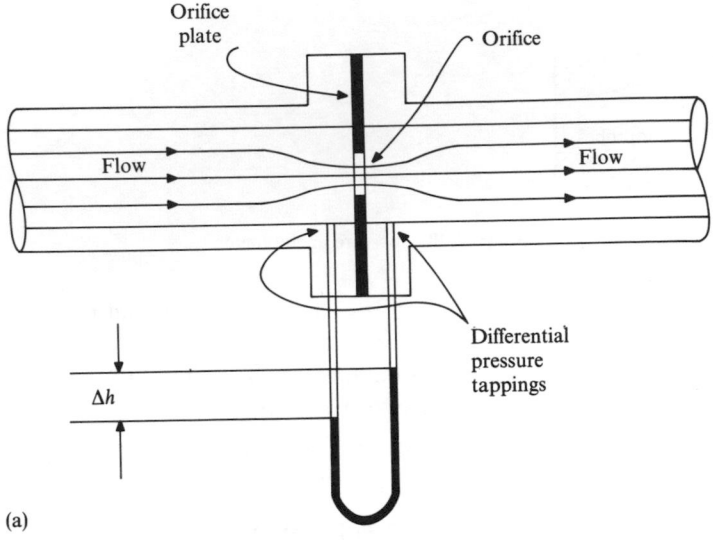

(a)

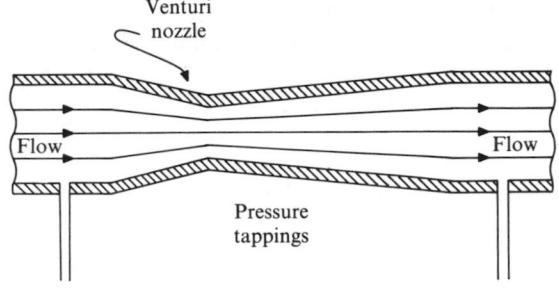

(b)

Figure 6.21 (a) Orifice plate; (b) Venturi meter.

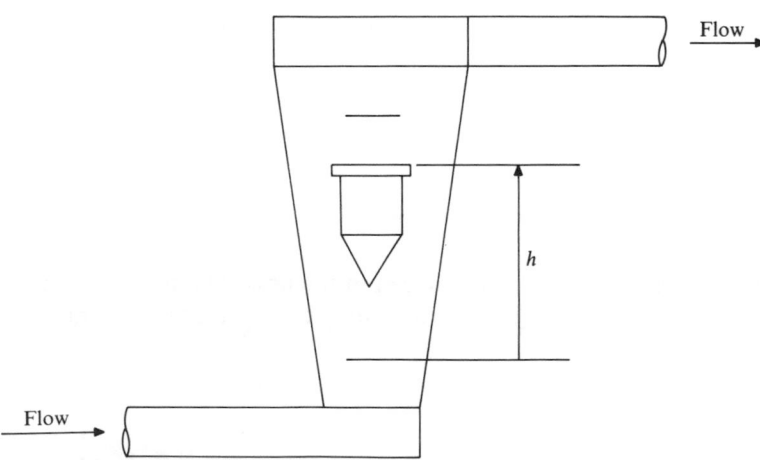

Figure 6.22 Rotameter flow meter.

the air motion. The sensible rate of heat loss (W) from the surface of the human body or clothing to the surrounding air is given by an equation of the form:

$$Q = hA(T - T_o) \qquad (6.27)$$

where h = surface-to-air heat transfer coefficient, $W\,m^{-2}\,K^{-1}$
    A = area of the surface, m^2
    T = skin or clothing surface temperature, °C
    T_o = ambient air temperature, °C.

The surface-to-air heat transfer coefficient, h, depends upon the air velocity, u ($m\,s^{-1}$). The greater this velocity, the greater the rate of heat loss for the same ambient air temperature. Hence the depressed equivalent *wind chill temperatures* often quoted by weathermen.

Airspeeds commonly encountered in internal built environments range from 0.1 to $1.0\,m\,s^{-1}$. Below $0.1\,m\,s^{-1}$ heat loss is dominated by natural convection and so is independent of air speed. Measurement of low air velocities (less than $1.0\,m\,s^{-1}$) is particularly difficult. Frequently, the level of turbulence is of the same order as the velocity. Useful data can however be obtained with any of the several instruments available, provided that they are maintained in calibration and used correctly.

The heated thermocouple anemometer In this device, a pair of heated and unheated metal–metal thermojunctions are exposed to the air stream as for the hot-wire anemometer. The air velocity is proportional to the differential voltage produced in the resulting thermocouple. Velocities as low as $0.05\,m\,s^{-1}$ have been measured successfully using such a device.

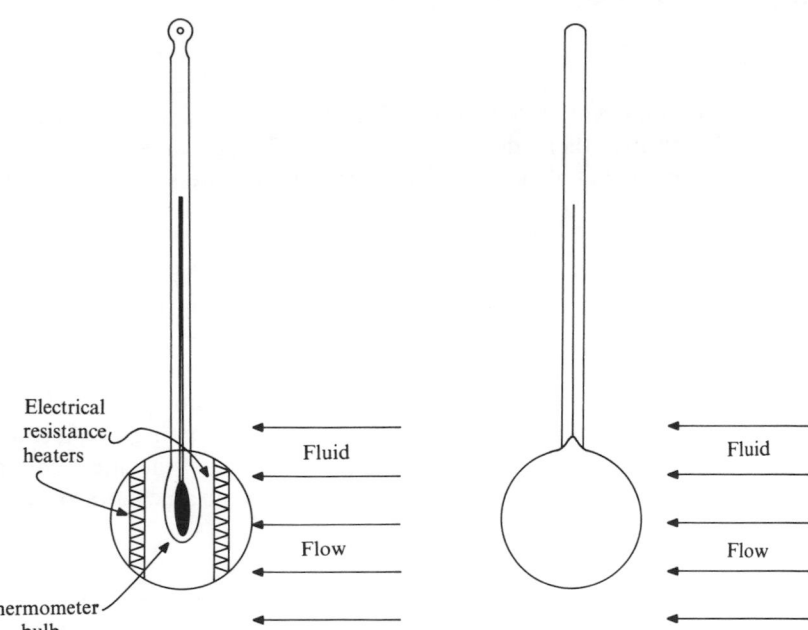

Figure 6.23 Heated bulb thermometer.   **Figure 6.24** Kata thermometer.

Heated bulb thermometer In this device, the bulb of an ordinary liquid-in-glass thermometer is inserted into a metal ball of silver or aluminium, which has a highly polished surface to reduce radiation heat exchanges (Fig. 6.23). As for the hot wire anemometer, the electrical input to the bulb equals the heat loss from the surface of the air to the sphere at steady-state. Thus the steady-state equilibrium temperature attained by the metal ball is a function of the air velocity.

Kata thermometer This is a transient device which consists of an alcohol-in-glass thermometer having a large spherical bulb (3–6 mm diameter) (Fig. 6.24). After being heated, the thermometer is placed in the airstream and the time taken for the alcohol to fall between two marks on the stem is measured. The air velocity depends upon this time for a given air temperature.

Air-borne tracers The introduction or air-borne tracer materials allow velocities (via time-lapse photography or videorecording) and directions of fluid flows to be measured. Suitable tracers include smoke, feathers, pieces of lint, radioactive and non-radioactive gases, helium filled zero buoyancy soap bubbles, and burnt metaldehyde. Measurements may be accomplished by timing the rates of movement of solid particles or by monitoring the change in concentrations of gaseous tracers. Smokes include titanium tetrachloride (noxious), fired potassium chlorate and powdered sugar (non-irritant), fumes of aqua-ammonia mixed with fumes of sulphuric acid (forms a white precipitate), air bubbled through ammonium hydroxide and then through nitric acid. Smoke tubes, candles and smoke bombs are commercially available.

6.7 MEASUREMENT OF PRESSURE IN FLUIDS

The U-tube manometer

If a tube from a pressure tapping is connected to one side of a U-tube, partially filled with a fluid of density, ρ kg m^{-3}, and the other side of the U-tube is left open to atmosphere, the pressure difference (N m^{-2}) between that of the tapping and atmospheric pressure is given by

$$\Delta p = \rho g h \tag{6.28}$$

where g is the acceleration due to gravity ($= 9.81$ m s^{-2}) and h (m) is the difference in the heads (levels) of the U-tube fluid either side of the U-tube. Thus, from Fig. 6.19

$$p_o - p = \rho g h \tag{6.29}$$

Static pressure distributions may be obtained using traversing pitot-static probes. U-tubes may be filled with low-density fluids and inclined to amplify the response. Using a light source and a photocell, a servosystem can be constructed with *tracks* the motion of the liquid column and provides an electrical signal via a potentiometer for input to an electronic data collection system.

Flexible metallic pressure transducers

Many elastic devices exist to translate pressure energy into mechanical movement. These may generally be classified as follows:

- Bourdon tubes
- Diaphragms
- Bellows

Bourdon gauges are hollow C-shaped, spiral or helical tubes which attempt to straighten out when an internal pressure is applied. They may be used for the remote measurement of very high pressures. Flat or corrugated *diaphragms* or *bellows* are often used in barometers for measuring atmospheric pressures.

6.8 DATA COLLECTION

It has been seen that suitable devices exist to convert measurements of temperature, heat fluxes, radiation, humidity, fluid velocities and flow rates to mV emfs via thermocouples or thermistors. Translational movements, such as those produced in the measurement of temperature by the expansion of solids, liquids and gases; in the measurement of relative humidity via dimensional change; in the use of U-tubes or rotameters for the measurement of pressure in fluids; and those produced by flexible metallic pressure transducers, can be converted to electrical voltage signals using potentiometer circuits (see Fig. 6.25).

For input to a computer system, these voltage signals must be amplified (or sometimes reduced) electronically to produce voltages varying within the range $-10\,\text{V}$ to $+10\,\text{V}$ dc. A 12-bit *analogue-to-digital converter* (ADC) card will convert these signals to lie in the range 0–4095 digital integers (2^{12}), which may be mathematically converted into temperatures, heat fluxes, radiation levels, humidities, fluid velocities and flow rates, or fluid pressures, with appropriate computer software.

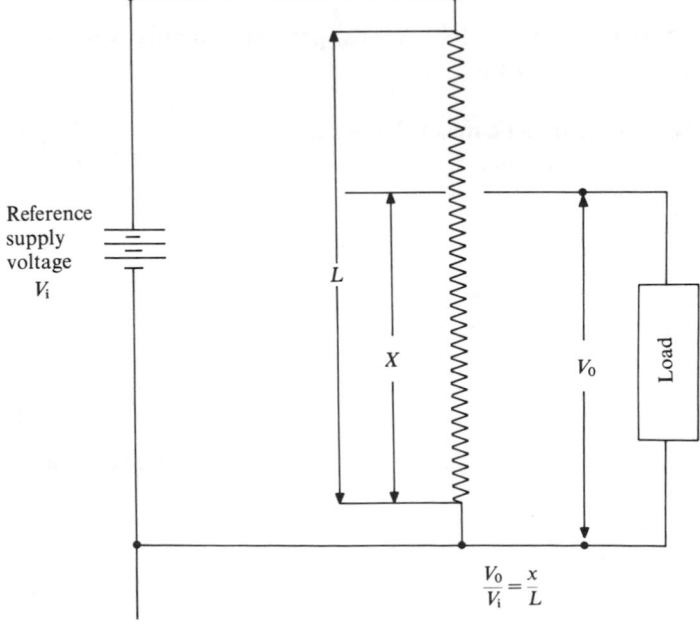

$$\frac{V_0}{V_i} = \frac{x}{L}$$

Figure 6.25 Potentiometer.

6.9 DATA ANALYSES AND PRESENTATIONS

Measurement errors should be minimized by careful attention to the measuring procedure and corrections. Degrees of reproducibility and systematic errors, often quoted by manufacturers, should be identified. The remaining random errors may be quantified by the statistical analyses of multiple samples.

Single variable

For a set of n measurements of a single variable, x_{1-n}, *the arithmetic mean value* is given by

$$x_{\text{arithmetic mean}} = \frac{\Sigma x}{n} \tag{6.30}$$

The arithmetic mean value is used to average a number of nominally-constant observations taken over space or time.

The *median value* of a set of data is the value of the variable for which half of the data set lies above it and the other half lies below.

The *modal value* of a set of data is the most commonly occurring value.

When the data is the result of a building or decaying time series, the *geometric mean value*, defined by

$$x_{\text{geometric mean}} = \sqrt[n]{x_1 x_2 \cdots x_n} \tag{6.31}$$

should be used (e.g. for the mean height of a system growing over time).

The *harmonic mean value* is calculated from

$$x_{\text{harmonic mean}} = \frac{n}{\Sigma(1/x)} \tag{6.32}$$

This should be used for averaging reciprocal variables (e.g. cost per unit quantity, where the quantities vary, or thermal conductivities in series).

Measures of the scatter of data about the arithmetic mean The *range* of the values = the heightest value − the smallest value. The *mean deviation* of the set of data is calculated from

$$\text{mean deviation} = \frac{\Sigma|x - x_{\text{mean}}|}{n} \tag{6.33}$$

where $||$ denotes the absolute values of the difference.

The *standard deviation* of the set of data about the arithmetic mean value is the root-mean-square deviation, σ, calculated from

$$\sigma = \sqrt{\Sigma(x - x_{\text{mean}})^2/n} \tag{6.34}$$

The *variance* of the data is equal to σ^2 and the *coefficient of variation, v*, is calculated from

$$v = 100\sigma/x_{\text{mean}} \tag{6.35}$$

The *standard error, S*, of the mean of n observations is defined by

$$S = \sigma/n^{0.5} \tag{6.36}$$

Random data errors

If the errors are truly random, the distribution of data about the mean value is described by the *normal probability* (*random* or *Gaussian*) *distribution* (Fig. 6.26). The probability density y is given by

$$y = \frac{1}{\sqrt{2\pi}} e^{-t^2/2} \qquad (6.37)$$

where $t = (x - x_{mean})/\sigma$.

Normalizing the differences by σ, results in the total area under the frequency distribution curve being equal to unity. Then, 68 per cent of the data will lie within 1 standard deviation either side of the arithmetic mean, 94 per cent of the data will lie within 2 standard deviations either side of the arithmetic mean; and 99.73 per cent of the data will lie within 3 standard deviations either side of the arithmetic mean (see Fig. 6.27).

For a purely random distribution, arithmetic mean, modal and median values coincide. If the data contains a bias, a skewed distribution, which has differing mean, model and median values, results (Fig. 6.28).

The standard deviation, the variance and the coefficient of variation of the data are indications of the error inherent in the data set and the level of confidence that the data provided by the measurement are accurate.

Two variables

Suppose that a set of values for a dependent variable, y, are related to a set of values for an independent variable, x. A first-order correlation between y and x may be deduced using *linear regression*. The procedure adopted is termed *the method of least squares*, which states that 'the sum of the squares of the discrepancies between observed values for y, corresponding to given values for x, and the values predicted from a linear equation should be minimized.

The set of values for the independent variable, y_1 to y_n, correspond to values for the independent variable, x_1 to x_n. It is assumed that $y = ax + b$, thus, for any value x_s:

$$y_s = ax_s + b + \text{error} \qquad (6.38)$$

Transposing,

$$\text{error} = ax_s + b - y_s \qquad (6.39)$$

The method of least squares *minimizes the sum of the squares of the errors for all the measured combinations of y and x*. Thus if $Z = (ax_s + b - y_s)^2 = $ a minimum, the condition for this is that the first-order differentials $dZ/da = 0$ and $dZ/db = 0$, or

$$\Sigma x_s 2(ax_s + b - y_s) = 0 \quad \text{and} \quad \Sigma 2(ax_s + b - y_s) = 0 \qquad (6.40)$$

Denoting summations by [ ]:

$$a[xx] + b[x] = [xy] \quad \text{and} \quad a[x] + bn = [y] \qquad (6.41)$$

Solving these two equations for a and b:

$$a = \frac{n[xy] - [x][y]}{n[xx] - [x][x]} \qquad (6.42)$$

$$y = \frac{1}{\sqrt{2\pi}} e^{-t^2/2}$$

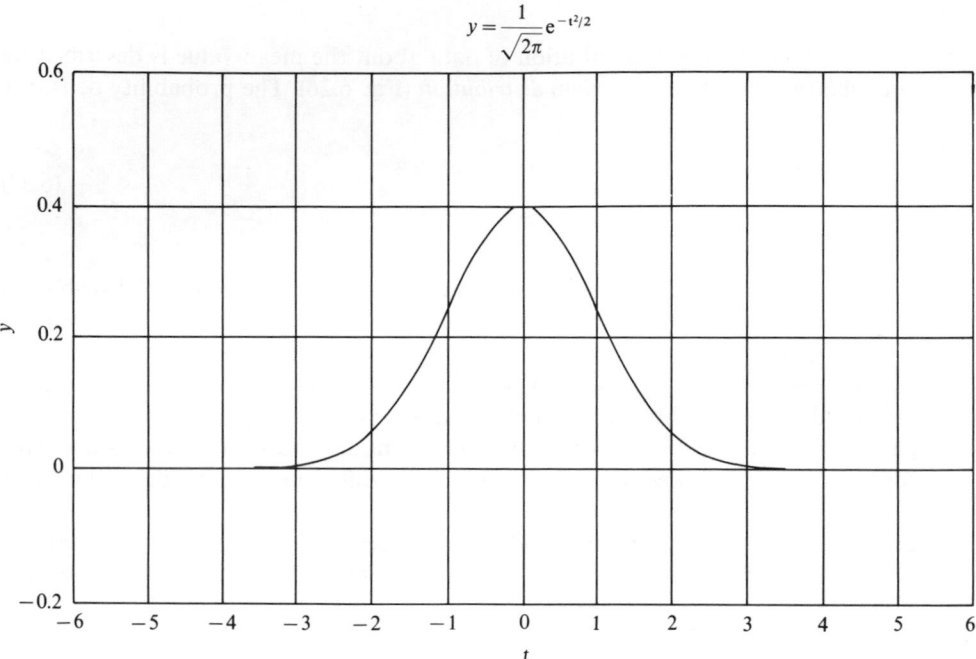

Figure 6.26 Normal probability distribution.

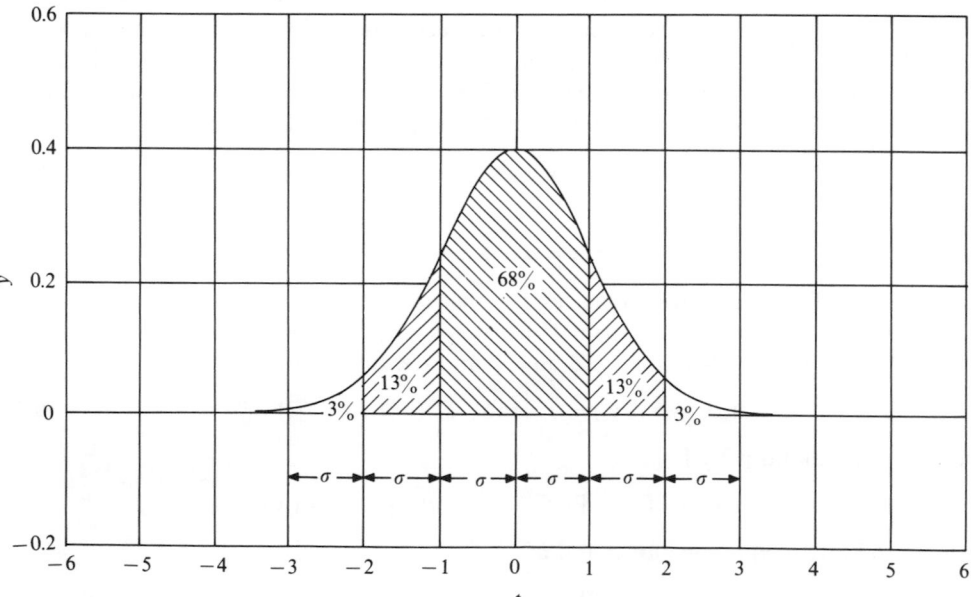

Figure 6.27 Standard deviations.

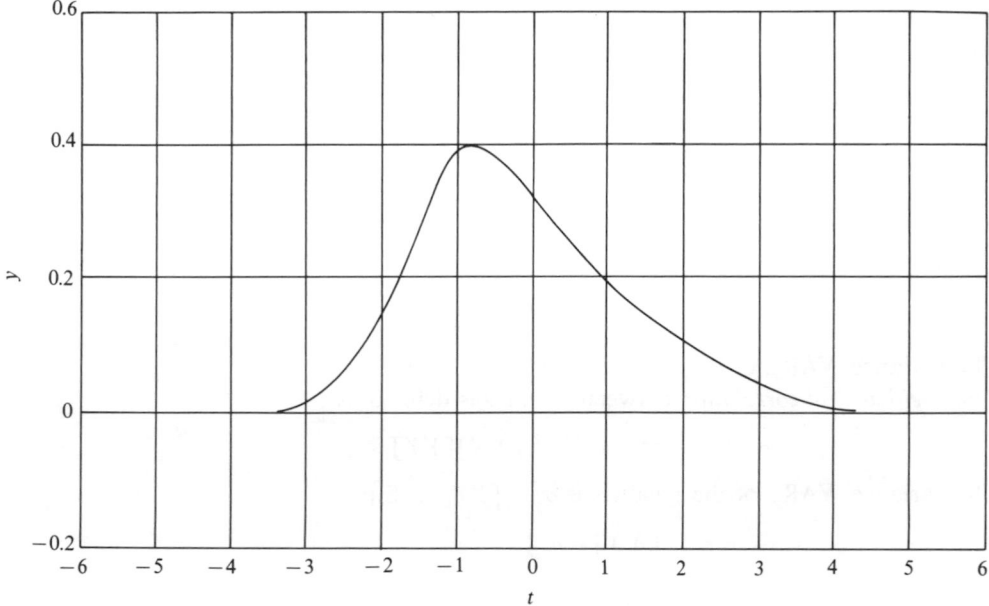

(a)

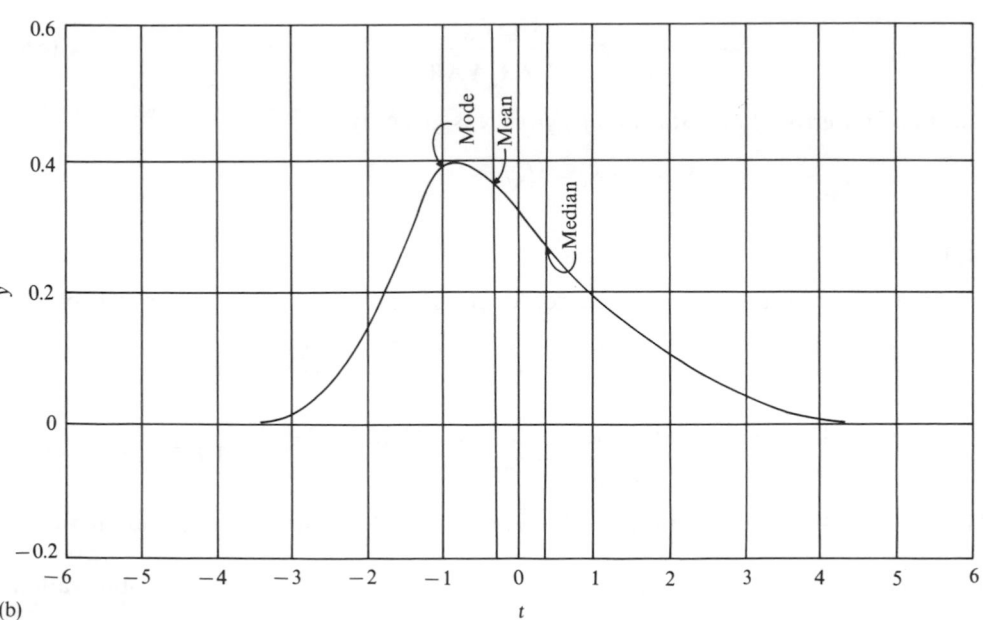

(b)

Figure 6.28 (a) Skewed probability distribution; (b) Modal, mean and median values.

and

$$b = \frac{[y][xx] - [x][xy]}{n[xx] - [x][x]} \tag{6.43}$$

If we denote $X = x - x_{mean}$ and $Y = y - y_{mean}$, the standard deviation of the x values is given by

$$\sigma_x = \sqrt{[XX]/n} \tag{6.44}$$

and the standard deviation of the y values is given by

$$\sigma_y = \sqrt{[YY]/n} \tag{6.45}$$

The variance, VAR_x, of the x values is $\sigma_x^2 = [XX]/n$.
The *coefficient of correlation* between x and y is defined as

$$r^2 = [XY]^2/[XX][YY] \tag{6.46}$$

The variance, VAR_y, of the y values is $\sigma_y^2 = [YY]/n$. Since

$$[XX] = n\sigma_x^2$$

and

$$[YY] = n\sigma_y^2$$
$$r^2 = [XY]^2/n\sigma_x^2 n\sigma_y^2$$

or

$$r = [XY]/n\sigma_x\sigma_y \tag{6.47}$$

The quantity $[XY]/n$ is known as the *covariance*, COV_{xy}, between x and y. Thus

$$r = \frac{COV_{xy}}{\sqrt{VAR_x\,VAR_y}} \tag{6.48}$$

The *standard error of the estimate of y from x* is given by

$$S_y = \sigma_y\sqrt{1 - r^2} \tag{6.49}$$

Example

Suppose that 10 observations of y are recorded for 10 values of x ($n = 10$), as shown in Table 6.3.

$$a = \frac{10 \times 818.2 - 55 \times 125.3}{10 \times 385 - 55 \times 55} = \frac{8182 - 6891.5}{3850 - 3025} = \frac{1290.5}{825} = 1.564$$

$$b = \frac{125.3 \times 385 - 55 \times 818.2}{10 \times 385 - 55 \times 55} = \frac{4820.5 - 45001}{3850 - 3025} = \frac{3239.5}{825} = 3.927$$

Thus $y = 1.564x + 3.927$ and the resulting predictions for y are given in column 4 of the table.

To determine measures of scatter, Table 6.4 is constructed from the observations, Table 6.3.

Table 6.3 Example observations

x	y	xy	xx	Predicted
1	5	5	1	5.490 909
2	8	16	4	7.055 152
3	9	27	9	8.619 394
4	10	40	16	10.183 64
5	11.7	58.5	25	11.747 88
6	12.8	76.8	36	13.312 12
7	14	98	49	14.876 36
8	16.3	130.4	64	16.440 61
9	18.5	166.5	81	18.004 85
10	20	200	100	19.569 09
Sums 55	125.3	818.2	385	
Means 5.5	12.53	81.82	38.5	

Table 6.4 Determination of scatter

x	X	XX	y	Y	YY	XY
1	−4.5	20.25	5	−7.53	56.7009	33.885
2	−3.5	12.25	8	−4.53	20.5209	15.855
3	−2.5	6.25	9	−3.53	12.4609	8.825
4	−1.5	2.25	10	−2.53	6.4009	3.795
5	−0.5	0.25	11.7	−0.83	0.6889	0.415
6	0.5	0.25	12.8	0.27	0.0729	0.135
7	1.5	2.25	14	1.47	2.1609	2.205
8	2.5	6.25	16.3	3.77	14.2129	9.425
9	3.5	12.25	18.5	5.97	35.6409	20.895
10	4.5	20.25	20	7.47	55.8009	33.615
Sums 55	0	82.5	125.3	0	204.661	129.05
Mean value 5.5	0	8.25	12.53	0	20.4661	12.905

Standard deviations are given by $VAR_x = 2.87$, $VAR_y = 4.52$, and the coefficient of correlation of y on x is given by

$$r^2 = \frac{(XY)^2}{(XX)(YY)} = \frac{129.05^2}{82.5 \times 204.661} = 0.986$$

$$r = 0.993$$

$$COV = [XY]/n = 129.05/10 = 12.905$$

$$r = \frac{COV}{\sqrt{VAR_x \, VAR_y}} = \frac{12.905}{\sqrt{8.25 \times 20.47}} = \frac{12.905}{12.995} = 0.993$$

(Note the degree of calculation accuracy required.)
Standard error of the estimate of y from x is given by

$$S_y = \sigma_y \sqrt{1 - r^2} = 4.52(1 - 0.993^2) = 0.529$$

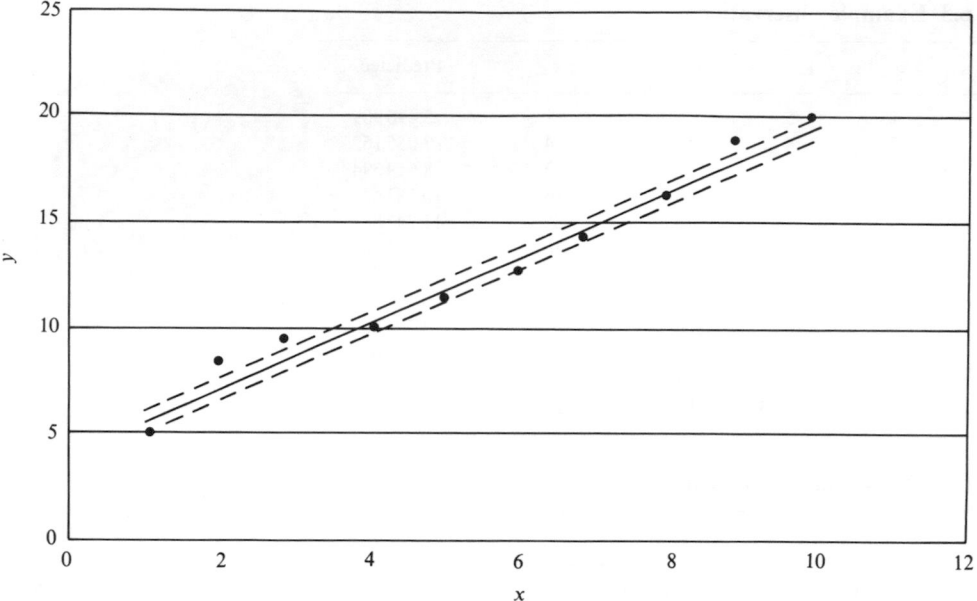

Figure 6.29 Regression line of y on x.

Therefore

$$y = 1.564x + 3.927 \pm 0.529$$

The original data, the regression line and the scatter bands (confidence limits) are shown in Fig. 6.29.

Compound errors

If the standard errors, α_1 and α_2, are known for two quantities, m_1 and m_2, then:

the standard error for the sum, $m_1 + m_2$, is $(\alpha_1^2 + \alpha_2^2)^{0.5}$

the standard error for the difference, $m_1 - m_2$, is $(\alpha_1^2 + \alpha_2^2)^{0.5}$

the standard error for the multiple km_1 is $k\alpha$

the standard error for the product, $m_1 m_2$, is $(m_1^2 \alpha_2^2 + m_2^2 \alpha_1^2)^{0.5}$

the standard error for the product $m_1 m_2 m_3$ is α

where

$$(\alpha/m_1 m_2 m_3)^2 = (\alpha_1/m_1)^2 + (\alpha_2/m_2)^2 + (\alpha_3/m_3)^2$$

the standard error for the quotient, m_1/m_2, is α

where

$$\alpha^2 = (1/m_2)^2 \alpha_1^2 + (-m_1/m_2^2)^2 \alpha_2^2$$

the standard error for the power m_1^p is α

where

$$\alpha/m_1^p = p\alpha_1/m_1 \text{ or } \alpha = (pm_1^{p-1})\alpha_1$$

the standard error of any function of m_1 to m_n is α

where

$$\alpha^2 = (\partial f/\partial m_1)^2\alpha_1^2 + (\partial f/\partial m_2)^2\alpha_2^2 + \cdots(\partial f/\partial m_n)^2\alpha_n^2$$

Curve fitting

The procedure for fitting a curve to bivariate data is essentially similar to the procedure for linear regression. Firstly, a suitable function type should be chosen according to the physics of the process being measured (e.g. a Fourier series for multiple sine waves, an exponential function for ventilation decay or transient heating). The polynomial function is general purpose but should not be used indiscriminately without considering the application under investigation.

It is assumed that

$$y = c_1 + c_2 x + c_3 x^2 + \cdots + c_m x^{m-1} \tag{6.50}$$

The degree of the polynomial chosen depends upon the number of maxima/minima expected in the data, but must be less than the number of data pairs. For example, a third order polynomial, such as

$$y = c_1 + c_2 x + c_3 x^2 + c_4 x^3 \tag{6.51}$$

will exhibit 2 maxima/minima.

In general, as for linear regression, for each data pair

$$y_{\text{meas}} - (c_1 + c_2 x + c_3 x^2 + \cdots + c_m x^{m-1}) = \text{error} \tag{6.52}$$

The sum of the squares of these errors must be minimized.

The expression for the error must be differentiated with respect to each coefficient, c_1 to c_{m-1} and each differential equated to zero. The resulting set of simultaneous equations is then summed throughout the sets of data pairs and solved for the coefficients in the equation for the curve.

Multiple linear regression

If the independent variable, y, is linearly dependent upon more than one variable, x_1 to x_n, the residual, or error, equations for each data set are formed as before:

$$y = a_1 x_1 + \cdots + a_n x_n$$
$$y_1 - (a_1 x_1 + \cdots + a_n x_n)_1 = \text{error}_1$$
$$\vdots$$
$$y_m - (a_1 x_1 + \cdots + a_n x_n)_m = \text{error}_m$$

Step 1 Square the residual equations.

Step 2 Sum the equations for all data sets.

Step 3 Differentiate with respect to all coefficients.

Step 4 Solve the resulting set of simultaneous linear equations for the unknown coefficients.

Many standard computer packages are available for curve-fitting and multivariate regression.

6.10 CONTROLS

Figure 6.30 shows the basic elements of a simple closed-loop control system. The important parameter of the system to be controlled is measured using a suitable sensor, from which an output analogue voltage signal emanates. The controller device compares this signal with that expected from the specification. If the signal complies with the specification, no action is taken. If the signal deviates from the specification, an adjustment is made via the corrector, which brings the process variable into line with the specification. A simple example is a room thermostat.

System/control response lags

Because of the inertias of the system and the sensing instruments, the system takes time to respond to corrections requested by the comparator. For example, the thermal inertia of a mercury-in-steel thermometer delays the attainment of thermal equilibrium between the system being measured and the measuring device. If extra heat is delivered to the system, it takes time to heat up to the new steady-state condition.

Proportional control

In order to eliminate the hunting action which results from on/off control, the corrective signal can be made proportional to the deviation of the system variable from its desired set-point, according to the system/control response lags, determined by experimentation.

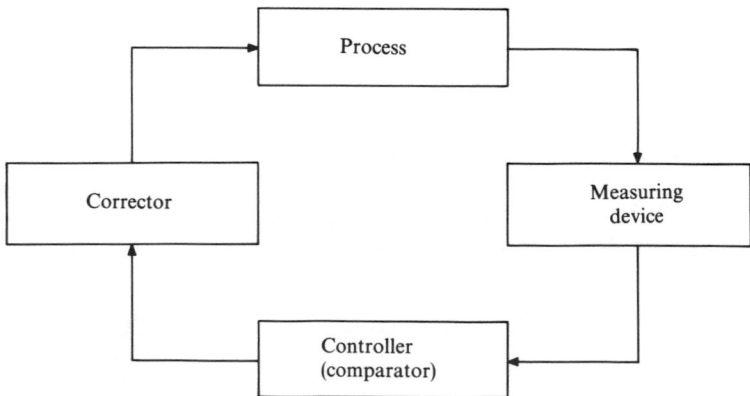

Figure 6.30 Closed-loop control system.

Computer-aided control systems

Computer-aided control requires a few extra components. Figure 6.31 shows the basic elements of a simple closed-loop computer-aided control system. The important parameter of the system to be controlled is measured using a suitable sensor, from which an output analogue voltage signal emanates. This must be amplified or reduced to match the range of voltages acceptable for input to the computer system (usually $-10\,V$ to $+10\,V$).

The analogue voltage signal must then be converted to digital information which the computer can understand. The analogue-to-digital converter (ADC) is an electronic device which accomplishes this. An 8-bit ADC converts voltages between $-10\,V$ and $+10\,V$ to binary bytes having integer values ranging from -127 to $+127$ (12.5 integers per V). For greater accuracy, a 12-bit ADC card will produce values from -2047 to $+2047$ (102.4 integers per V). Via suitable software, calculations and comparisons are made in the computer. When, the signal deviates from the specification, an adjustment digital signal is output. This must be converted back to a corrective analogue signal via a digital-to-analogue converter (DAC). The analogue signal controls a simple on/off switch or a power controller which actuates a system controller, which brings the process variable into line with the specification. Multiple

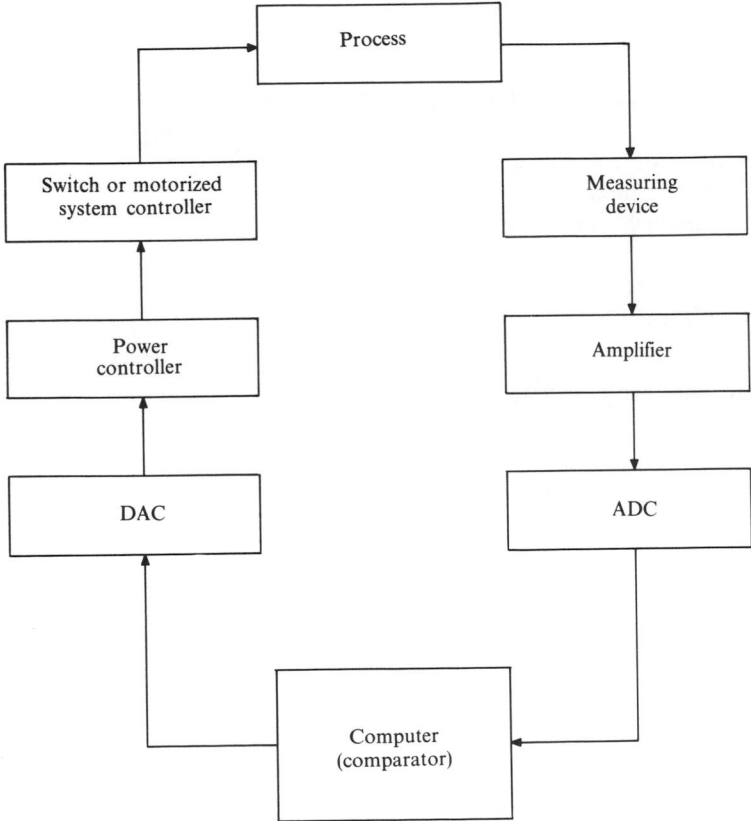

Figure 6.31 Closed-loop computerized control system.

variables can be monitored and controlled using multiple control circuits, made possible by introducing switching devices (multiplexers) to 'poll' each system variable to be controlled.

Energy management control systems

Many specialized instruments are available for measuring specialized compound parameters, such as the heat flowing through a pipe, steam or hot water flows. Heat meters measure both flowrate (usually with a turbine flow meter) and temperature (using a thermocouple probe, thermistor or resistance thermometer). Microchips analyse, record and aggregate heat flow data and can produce periodic reports.

Many proprietary *energy management control systems* are available to measure many parameters and to control hundreds of devices in a building or plant.

It is vital that a full and comprehensive energy audit be constructed in order to decide what should be measured and what should be controlled for maximum cost-effectiveness. The user, *not the supplier*, of the energy management control system should construct its specification.

Variables amenable to control include the following:

- Furnace/boiler control for maximum efficiency operation
- Room temperatures
- Humidities
- Time controls—system start-up and shut-down (optimum start/stop)
- Sequencing of multiple boilers
- Electrical maximum demand control via duty cycling and load shedding
- Air supply rates—time, temperatures and occupancy
- Controls according to occupancy—time and zone controls
- Lighting controls
- Fire and security matters
- Optimization of plant operation

Table 6.5 Measurement variables and associated control devices

Variables		Control devices
Temperature		Valves
Heat flux		Shutters
Thermal (e.g. solar) radiation		Pneumatic
Humidity		Pumps
Flue gas		Blower
Pressure		Dampers
Fluid velocities		Hydraulic
Fluid flow		Cut-outs
	Computer-Control System	
Displacement		Electrical circuits
Frequency		Switches
Electrical variables		Power
Fire, security		Alarms
Lighting levels		Displacement
Condition		Maintenance

A number of possible measurement variables and control devices are shown at Table 6.5.

The energy management control system can also provide energy management information for monitoring and targeting purposes and to produce reports and real-time cost data. An energy management checklist for controls is provided in Appendix 2.

REFERENCES

1. Jacob, M.: *Industrial Control Electronics*, Prentice-Hall, Englewood Cliffs, New Jersey, 1989.
2. Chesmond, C. J.: *Control System Technology*, Arnold, London, 1982.
3. Fanger, P. O.: *Thermal Comfort*, McGraw-Hill, New York, 1970.

FURTHER READING

Doebelin, E. O.: *Measurement Systems: Application and Design*, 4th edn. McGraw-Hill, New York, 1990.
Hertzfeld, C. M.: *Temperature—Its Measurement and Control in Science and Industry*, Volumes I, II and III, Reinhold, New York, 1962.
Morony, M. J.: *Facts from Figures*, Pelican, Harmondsworth, 1984.
Morris, A. S.: *Principles of Measurement and Instrumentation*, Prentice-Hall, Englewood Cliffs, New Jersey, 1988.
Topping, J.: *Errors of Observation and Their Treatment*, Science Paperbacks, Chapman Hall, London, 1972.

SEVEN
ECONOMICS AND FINANCE

7.1 INTRODUCTION

The energy manager should be concerned with initial designs and iterations leading to economically optimal systems and products, retrofit modifications to existing systems and the monitoring, control and co-ordination of operations and procedures to ensure that these activities fulfil their desired purposes with least use of energy and materials.

In Chapter 3, the use of the following symbols was recommended to track the energy and materials conversions through the manufacturing, use and disposal history of a product, or the resource flows through a built facility or service:

= energy/material storage
* energy release
∅ conversion of energy/materials
○ utilization of energy and materials
⌇ flow resistance
⌁ rejection to the environment

Each high-demand energy- or materials-consuming operation should be critically examined with a view towards greater energy efficiency and the minimization of pollution.

Design specifications, levels of performance and operating practices and procedures must be clearly defined, investigated and criticized with respect to the energy and materials expenditure involved and the pollution produced. The following must be ensured:

- Working lives are maximized
- The need for maintenance is minimized
- The total lifetime (capital cost plus running cost) is minimized
- Environmental pollution is minimized

Major centres for energy and materials waste and pollutant emission should be identified and investigated.

7.2 ECONOMICS

Economics is the study of how scarce resources should be apportioned among alternative ends. The basic ingredients of production are land, labour, capital, materials and energy supplies. Economic decisions try to obtain the best balance of these factors to achieve a desired result for least financial cost. As these decisions must be taken within a technical framework, *economic and technical analyses go hand-in-hand*.

All economic decisions are affected by two basic laws:

- The *benefits of scale* (the larger the system, the lower its unit cost)
- The law of *diminishing returns* (if successive increments of one factor are employed in conjunction with a fixed stock of all other factors, these increments will, after a point, yield successively smaller additions to the total return on the investment)

Economic optimizations are often performed parametrically in cost schedules, i.e. by varying each factor contributing to the total cost in turn, keeping the quantities of the remaining factors constant, until the extra advantage produced by the largest increment of the variable factor equals the cost of that increment.

As an example, consider the incremental application of thermal insulation to a wall, the optimal economic thickness results when the additional savings yielded by an additional incremental thickness of the insulant, $\Delta s/\Delta t$ equals the average cost per increment, C/t, where C is the total cost of the insulation of thickness, t.

Marginal cost = marginal revenue for maximum savings

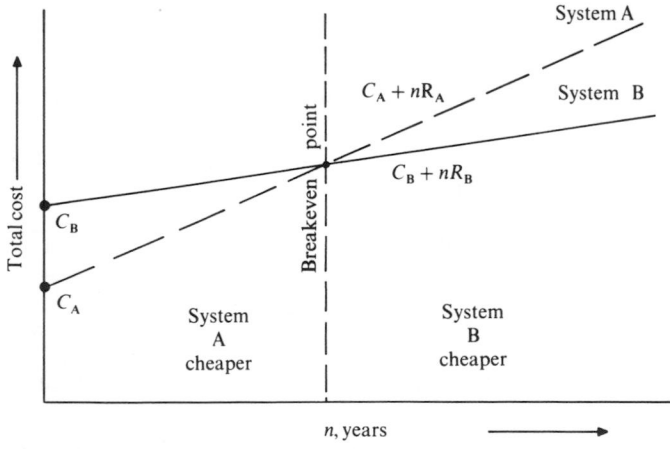

Figure 7.1 Breakeven point between two options.

Breakeven point between two alternative systems

If the capital cost of system A is C_A (£) and the running cost of system A is R_A (£ per year), and the capital cost of system B is C_B (£) and the running cost of system B is R_B (£ per year), the *breakeven period*, n years, is calculated from

$$C_A + nR_A = C_B + nR_B \tag{7.1}$$

i.e. when

$$n = \frac{C_B - C_A}{R_A - R_B} \tag{7.2}$$

This is illustrated in Fig. 7.1.

7.3 DISCOUNTED CASH FLOW

The present value concept

The *straight payback period* (years), calculated by dividing the capital cost of the investment by the savings resulting expected to result (i.e. in the first year of the project), takes no account of the following:

- The interest that the capital could earn if invested elsewhere
- The interest that the sums saved could earn
- The effect of inflation on the sums saved

A sum of money in the hand *now* is worth more than the same sum if received in one years time, because interest could be earned on it during the year. For example, if £100 is invested today at an annual interest rate of 10 per cent, it will be worth £110 in one year's time. Conversely, the *present value* of £110 received in one year's time is £100. The present value of £100 received in one year's time is £100/1.1 = £90.91.

A single investment

The accumulating worth of £100 invested for 25 years for various interest rates is given in Table 7.1.

   If the initial investment is C (£) and the annual interest rate is i (per cent), then the value, V (£) of the single investment after n years is given by

$$V = C(1 + i/100)n \tag{7.3}$$

For $C = 100$, $i = 20$ per cent, $n = 15$ years

$$V = 100(1 + 20/100)^{15} = £1541 \tag{7.4}$$

The factor, $(1 + i/100)^n$, is the *compound interest* earned by £1 over the period n (years), when the interest rate is i (per cent).

   The present value, PV, of a sum received in the future can be converted to its value *now* by the compound interest that £1 could earn, between now and then, at the expected interest

Table 7.1 A single investment of £100

End of year	Interest rate, %						
	5	10	15	20	25	30	35
1	105	110	115	120	125	130	135
2	110	121	132	144	156	169	182
3	116	133	152	173	195	220	246
4	122	146	175	207	244	286	332
5	128	161	201	249	305	371	448
6	134	177	231	299	381	483	605
7	141	195	266	358	477	627	817
8	148	214	306	430	596	816	1 103
9	155	236	352	516	745	1 060	1 489
10	163	259	405	619	931	1 379	2 011
11	171	285	465	743	1 164	1 792	2 714
12	180	314	535	892	1 455	2 330	3 664
13	189	345	615	1 070	1 819	3 029	4 947
14	198	380	708	1 284	2 274	3 937	6 678
15	208	418	814	1 541	2 842	5 119	9 016
16	218	459	936	1 849	3 553	6 654	12 171
17	229	505	1 076	2 219	4 441	8 650	16 431
18	241	556	1 238	2 662	5 551	11 246	22 182
19	253	612	1 423	3 195	6 939	14 619	29 946
20	265	673	1 637	3 834	8 674	19 005	40 427
21	279	740	1 882	4 601	10 842	24 706	54 577
22	293	814	2 164	5 521	13 553	32 118	73 679
23	307	895	2 489	6 625	16 941	41 754	99 466
24	323	985	2 863	7 950	21 176	54 280	134 280
25	339	1 083	3 292	9 540	26 470	70 564	181 278

rate. Thus, if S (£) is a saving received in year n, its present value is given by

$$PV = S/(1 + i/100)^n \qquad (7.5)$$

If $S = £100$, $i = 10$ per cent and $n = 2$

$$PV = 100/(1 + 10/100)^2 = £82.64$$

Table 7.2 lists present values for various savings occurring at various years in the future.

A saving of £100 received in 15 years time, when the expected interest rate is 15 per cent has a present value given by

$$PV = S/(1 + i/100)^n$$
$$= 100/(1 + 15/100)^{15} = £12.29$$

Annual investments

If £100 is invested in each of years 1 to 5, at an interest rate of 10 per cent, the accumulated capital sum is calculated as shown in Table 7.3.

Table 7.2 Present value for £100 saved in year n

Saving in year	Expected future interest rate, %						
	5.00	10.00	15.00	20.00	25.00	30.00	35.00
1	95.24	90.91	86.96	83.33	80.00	76.92	74.07
2	90.70	82.64	75.61	69.44	64.00	59.17	54.87
3	86.38	75.13	65.75	57.87	51.20	45.52	40.64
4	82.27	68.30	57.18	48.23	40.96	35.01	30.11
5	78.35	62.09	49.72	40.19	32.77	26.93	22.30
6	74.62	56.45	43.23	33.49	26.21	20.72	16.52
7	71.07	51.32	37.59	27.91	20.97	15.94	12.24
8	67.68	46.65	32.69	23.26	16.78	12.26	9.06
9	64.46	42.41	28.43	19.38	13.42	9.43	6.71
10	61.39	38.55	24.72	16.15	10.74	7.25	4.97
11	58.47	35.05	21.49	13.46	8.59	5.58	3.68
12	55.68	31.86	18.69	11.22	6.87	4.29	2.73
13	53.03	28.97	16.25	9.35	5.50	3.30	2.02
14	50.51	26.33	14.13	7.79	4.40	2.54	1.50
15	48.10	23.94	12.29	6.49	3.52	1.95	1.11
16	45.81	21.76	10.69	5.41	2.81	1.50	0.82
17	43.63	19.78	9.29	4.51	2.25	1.16	0.61
18	41.55	17.99	8.08	3.76	1.80	0.89	0.45
19	39.57	16.35	7.03	3.13	1.44	0.68	0.33
20	37.69	14.86	6.11	2.61	1.15	0.53	0.25
21	35.89	13.51	5.31	2.17	0.92	0.40	0.18
22	34.18	12.28	4.62	1.81	0.74	0.31	0.14
23	32.56	11.17	4.02	1.51	0.59	0.24	0.10
24	31.01	10.15	3.49	1.26	0.47	0.18	0.07
25	29.53	9.23	3.04	1.05	0.38	0.14	0.06

Table 7.3 Annual investment of £100 at 10% interest rate

Year	Investment in year				
	1	2	3	4	5
	Accumulating capital				
end of 1	110.00				
year 2	121.00	110.00			
3	133.10	121.00	110.00		
4	146.41	133.10	121.00	110.00	
5	161.05	146.41	133.10	121.00	110.00

The £100 invested at the start of year 1 is worth £100 $\times (1 + 10/100)^5 =$ £161 at the end of year 5.

The £100 invested at the start of year 2 is worth £100 $\times (1 + 10/100)^4 =$ £146 at the end of year 5.

The £100 invested at the start of year 3 is worth £100 $\times (1 + 10/100)^3 =$ £133 at the end of year 5.

The £100 invested at the start of year 4 is worth £100 $\times (1 + 10/100)^2 =$ £121 at the end of year 5.

The £100 invested at the start of year 5 is worth £100 $\times (1 + 10/100)^1 =$ £110 at the end of year 5.

The total amount invested is £500. The total worth of the investment at the end of year 5 is £671.56 and so the total interest earned over the 5-year period is £171.56.

Generalizing, if the annual sum invested is £C the expected annual interest rate is i per cent and the number of years is n years:

the £C invested at the start of year 1
is worth $£C(1 + i/100)^n$ at the end of year n

the £C invested at the start of year 2
is worth $£C(1 + i/100)^{n-1}$ at the end of year n

the £C invested at the start of year 3
is worth $£C(1 + i/100)^{n-2}$ at the end of year n

$$\vdots$$

the £C invested at the start of year n
is worth $£C(1 + i/100)^{n-(n-1)}$ at the end of year n

or

$$£C(1 + i/100)^1 \text{ at the end of year } n$$

The sum of the geometric series

$$1 + x + x^2 + x^3 + \cdots + x^{n-1} = \frac{x^n - 1}{x - 1} \tag{7.6}$$

thus if the number of terms, $n = 5$ and $x = 1.1$

$$1 + 1.1 + 1.21 + 1.331 + 1.4641 = 6.1051 = \frac{1.1^5 - 1}{1.1 - 1} = 6.1051$$

If $I = i/100$, the total worth, W, of n annual investments of £C at the end of year n is the sum of the series

$$C(1 + I)^1 + C(1 + I)^2 + C(1 + I)^3 + \cdots + C(1 + I)^n \tag{7.7}$$

or

$$W = C(1 + I)^1 + (1 + I)^2 + (1 + I)^3 + \cdots + (1 + I)^n) \tag{7.8}$$

Since

$$1 + x + x^2 + x^3 + \cdots + x^{n-1} = \frac{x^n - 1}{x - 1}$$

$$x + x^2 + x^3 + \cdots + x^n = \frac{x^n - 1}{x - 1} - 1 + x^n$$

$$= \frac{x^{n-1} - x}{x - 1}$$

$$= \frac{x(x^n - 1)}{x - 1} \tag{7.9}$$

In the series for total worth $x = 1 + I$, therefore

$$W = C\left(\frac{(1 + I)((1 + I)^n - 1)}{(1 + I) - 1}\right)$$

$$= C\left(\frac{1 + I}{I}((1 + I)^n - 1)\right) \tag{7.10}$$

For $C = £100$, $i = 10$ per cent ($I = 0.1$) and $n = 5$ years

$$W = 100\left(\frac{1.1}{0.1}((1.1)^5 - 1)\right)$$

$$= 671.56$$

Table 7.4 Accumulating capital

Year	Total invested	Value (end year)
1	100.00	110.00
2	200.00	231.00
3	300.00	364.10
4	400.00	510.51
5	500.00	671.56
6	600.00	848.72
7	700.00	1 043.59
8	800.00	1 257.95
9	900.00	1 493.74
10	1 000.00	1 753.12
11	1 100.00	2 038.43
12	1 200.00	2 352.27
13	1 300.00	2 697.50
14	1 400.00	3 077.25
15	1 500.00	3 494.97
16	1 600.00	3 954.47
17	1 700.00	4 459.92
18	1 800.00	5 015.91
19	1 900.00	5 627.50
20	2 000.00	6 300.25
21	2 100.00	7 040.27
22	2 200.00	7 854.30
23	2 300.00	8 749.73
24	2 400.00	9 734.71
25	2 500.00	10 818.18

Table 7.5 Present value

Year/ Saving in year	Accumulating present value				
	1	2	3	4	5
1	90.91				
2		82.60			
3			75.13		
4				68.30	
5					62.10

Table 7.4 lists the end worth of a £100 annual investment over n years at an annual interest rate of 10 per cent.

Present value If £100 is saved in each of years 1 to 5, at an interest rate of 10 per cent, the accumulated present value is as shown in Table 7.5.

The £100 saved in year 1 is worth $£100/(1 + 10/100)^1 = £90.91$ now.
The £100 saved in year 2 is worth $£100/(1 + 10/100)^2 = £82.60$ now.
The £100 saved in year 3 is worth $£100/(1 + 10/100)^3 = £75.13$ now.
The £100 saved in year 4 is worth $£100/(1 + 10/100)^4 = £68.30$ now.
The £100 saved in year 5 is worth $£100/(1 + 10/100)^4 = £62.10$ now.

The total amount saved is £500. The total present value of those future savings *now* is £379.04 and so that total discount on the savings over the 5-year period is £120.96.

Generalizing, if the annual sum saved is £C, the expected annual interest rate is i per cent, and the number of year is n years:

$$\text{the £}S \text{ saved in year } n \text{ is worth £}S/(1 + i/100)^n \text{ now}$$

$$\text{the £}S \text{ saved in year } n - 1 \text{ is worth £}S/(1 + i/100)^{n-1} \text{ now}$$

$$\vdots$$

$$\text{the £}S \text{ saved in year 1 is worth £}S/(1 + i/100)^1 \text{ now}$$

The total present value, PV, of n annual savings of £S is the sum of the series

$$S/((1 + I)^1) + S/((1 + I)^2) + S/((1 + I)^3) + \cdots + S/((1 + I)^n)$$

or

$$PV = S(1/(1 + I)^1 + 1/(1 + I)^2 + 1/(1 + I)^3 + \cdots + 1/(1 + I)^n) \qquad (7.11)$$

Since

$$x + x^2 + x^3 + \cdots + x^n = \frac{x(x^n - 1)}{x - 1}$$

and in the series for present value, $x = (1/(1 + I))$

$$PV = S\left(\frac{(1/(1 + I))((1/(1 + I))^n - 1)}{(1/(1 + I)) - 1}\right)$$

or

$$PV = S\left(\frac{1-(1/1+I)^n}{I}\right)$$ (7.12)

the *standard annuity formula.*

For example, if $S = £100$, $I = 0.1$ and $n = 5$ years:

$$PV = 100\left(\frac{1-(1/1.1)^5}{0.1}\right)$$

$$= £379.04$$

The net present value, NPV, of an investment, $£C$, which produces n annual savings of $£S$ for an annual interest rate of I, is given by

$$NPV = PV - C = S\left(\frac{1-(1+I)^n}{I}\right) - C$$ (7.13)

7.4 LOANS

Car dealer

The unscrupulous second-hand car salesman may quote an annual percentage interest rate of 10 per cent and then apply this rate to the total sum loaned at the start of the loan.

For example, if the amount of loan is £10 000.00, the number of years is 5, and the interest rate is 10.00 per cent ($I = 0.1$), then the value of the loan is $10\,000 \times (1 + 0.1 \times 5) = £15\,000.00$ and the annual repayment would be £3000.00. The repayment schedule would be as shown in Table 7.6.

Annual percentage rate (APR) loans

The present value expression

$$PV = \text{annuity factor} \times S$$ (7.14)

where annuity factor $= 1 - (1/1 + I)^n/I$, can be recast as follows:

amount of loan = annuity factor × annual repayment

Table 7.6 Repayment schedule

Year	Amount outstanding, £
1	15 000.00
2	12 000.00
3	9 000.00
4	6 000.00
5	3 000.00
6	0

and

$$\text{annual repayment} = \text{total repayment}/n$$

Therefore

$$\text{amount of loan} = \text{annuity factor} \times \text{total repayment}/n$$

or

$$\text{total repayment} = \text{amount of loan} \times n/\text{annuity factor} \qquad (7.15)$$

Table 7.7 lists annuity factors, Table 7.8 lists the ratios of the total repayments to amounts of loans and Table 7.9 lists the ratios of annual repayment to amounts of loans for various interest rates and loan periods.

Table 7.9 and Fig. 7.2 on page 308 indicate that there is little to be gained by taking out a loan or mortgage over a period greater than 10 years. For example, the annual repayment

Table 7.7 Annuity factors for various interest rates and loan periods

| Years | Interest rates I | | | | | | |
	0.05	0.10	0.15	0.20	0.25	0.30	0.35
1	0.95	0.91	0.87	0.83	0.80	0.77	0.74
2	1.86	1.74	1.63	1.53	1.44	1.36	1.29
3	2.72	2.49	2.28	2.11	1.95	1.82	1.70
4	3.55	3.17	2.85	2.59	2.36	2.17	2.00
5	4.33	3.79	3.35	2.99	2.69	2.44	2.22
6	5.08	4.36	3.78	3.33	2.95	2.64	2.39
7	5.79	4.87	4.16	3.60	3.16	2.80	2.51
8	6.46	5.33	4.49	3.84	3.33	2.92	2.60
9	7.11	5.76	4.77	4.03	3.46	3.02	2.67
10	7.72	6.14	5.02	4.19	3.57	3.09	2.72
11	8.31	6.50	5.23	4.33	3.66	3.15	2.75
12	8.86	6.81	5.42	4.44	3.73	3.19	2.78
13	9.39	7.10	5.58	4.53	3.78	3.22	2.80
14	9.90	7.37	5.72	4.61	3.82	3.25	2.81
15	10.38	7.61	5.85	4.68	3.86	3.27	2.83
16	10.84	7.82	5.95	4.73	3.89	3.28	2.83
17	11.27	8.02	6.05	4.77	3.91	3.29	2.84
18	11.69	8.20	6.13	4.81	3.93	3.30	2.84
19	12.09	8.36	6.20	4.84	3.94	3.31	2.85
20	12.46	8.51	6.26	4.87	3.95	3.32	2.85
21	12.82	8.65	6.31	4.89	3.96	3.32	2.85
22	13.16	8.77	6.36	4.91	3.97	3.32	2.85
23	13.49	8.88	6.40	4.92	3.98	3.33	2.85
24	13.80	8.98	6.43	4.94	3.98	3.33	2.86
25	14.09	9.08	6.46	4.95	3.98	3.33	2.86
26	14.38	9.16	6.49	4.96	3.99	3.33	2.86
27	14.64	9.24	6.51	4.96	3.99	3.33	2.86
28	14.90	9.31	6.53	4.97	3.99	3.33	2.86
29	15.14	9.37	6.55	4.97	3.99	3.33	2.86
30	15.37	9.43	6.57	4.98	4.00	3.33	2.86

Table 7.8 Total repayment/amount of loan

Years	Interest rates I						
	0.05	0.10	0.15	0.20	0.25	0.30	0.35
1	1.05	1.10	1.15	1.20	1.25	1.30	1.35
2	1.08	1.15	1.23	1.31	1.39	1.47	1.55
3	1.10	1.21	1.31	1.42	1.54	1.65	1.77
4	1.13	1.26	1.40	1.55	1.69	1.85	2.00
5	1.15	1.32	1.49	1.67	1.86	2.05	2.25
6	1.18	1.38	1.59	1.80	2.03	2.27	2.52
7	1.21	1.44	1.68	1.94	2.21	2.50	2.79
8	1.24	1.50	1.78	2.08	2.40	2.74	3.08
9	1.27	1.56	1.89	2.23	2.60	2.98	3.38
10	1.30	1.63	1.99	2.39	2.80	3.23	3.68
11	1.32	1.69	2.10	2.54	3.01	3.50	4.00
12	1.35	1.76	2.21	2.70	3.22	3.76	4.32
13	1.38	1.83	2.33	2.87	3.44	4.03	4.64
14	1.41	1.90	2.45	3.04	3.66	4.31	4.97
15	1.45	1.97	2.57	3.21	3.89	4.59	5.31
16	1.48	2.05	2.69	3.38	4.12	4.87	5.65
17	1.51	2.12	2.81	3.56	4.35	5.16	5.99
18	1.54	2.19	2.94	3.74	4.58	5.45	6.33
19	1.57	2.27	3.07	3.92	4.82	5.74	6.67
20	1.60	2.35	3.20	4.11	5.06	6.03	7.02
21	1.64	2.43	3.33	4.29	5.30	6.33	7.36
22	1.67	2.51	3.46	4.48	5.54	6.62	7.71
23	1.71	2.59	3.59	4.67	5.78	6.92	8.06
24	1.74	2.67	3.73	4.86	6.03	7.21	8.41
25	1.77	2.75	3.87	5.05	6.27	7.51	8.75
26	1.81	2.84	4.01	5.25	6.52	7.81	9.10
27	1.84	2.92	4.15	5.44	6.77	8.11	9.45
28	1.88	3.01	4.29	5.63	7.01	8.41	9.80
29	1.92	3.10	4.43	5.83	7.26	8.70	10.15
30	1.95	3.18	4.57	6.03	7.51	9.00	10.50
31	1.99	3.27	4.71	6.22	7.76	9.30	10.85
32	2.02	3.36	4.86	6.42	8.01	9.60	10.20
33	2.06	3.45	5.00	6.62	8.26	9.90	11.55

on a loan over 10 years at an interest rate of 15 per cent is 33 per cent more than that for the same loan over 25 years, but the total repayment over the 25 years period doubles.

Example 1

Amount of loan	10 000.00
Number of years	5
Interest rate, I	0.10
Annuity factor	3.79
Total repayment	13 189.87
Annual repayment	2 637.97

The repayment details are shown in Table 7.10.

Table 7.9 Annual repayment/amount of loan

	Interest rates I						
Years	0.05	0.10	0.15	0.20	0.25	0.30	0.35
1	1.05	1.10	1.15	1.20	1.25	1.30	1.35
2	0.54	0.58	0.62	0.65	0.69	0.73	0.78
3	0.37	0.40	0.44	0.47	0.51	0.55	0.59
4	0.28	0.32	0.35	0.39	0.42	0.46	0.50
5	0.23	0.26	0.30	0.33	0.37	0.41	0.45
6	0.20	0.23	0.26	0.30	0.34	0.38	0.42
7	0.17	0.21	0.24	0.28	0.32	0.36	0.40
8	0.15	0.19	0.22	0.26	0.30	0.34	0.38
9	0.14	0.17	0.21	0.25	0.29	0.33	0.38
10	0.13	0.16	0.20	0.24	0.28	0.32	0.37
11	0.12	0.15	0.19	0.23	0.27	0.32	0.36
12	0.11	0.15	0.18	0.23	0.27	0.31	0.36
13	0.11	0.14	0.18	0.22	0.26	0.31	0.36
14	0.10	0.14	0.17	0.22	0.26	0.31	0.36
15	0.10	0.13	0.17	0.21	0.26	0.31	0.35
16	0.09	0.13	0.17	0.21	0.26	0.30	0.35
17	0.09	0.12	0.17	0.21	0.26	0.30	0.35
18	0.09	0.12	0.16	0.21	0.25	0.30	0.35
19	0.08	0.12	0.16	0.21	0.25	0.30	0.35
20	0.08	0.12	0.16	0.21	0.25	0.30	0.35
21	0.08	0.12	0.16	0.20	0.25	0.30	0.35
22	0.08	0.11	0.16	0.20	0.25	0.30	0.35
23	0.07	0.11	0.16	0.20	0.25	0.30	0.35
24	0.07	0.11	0.16	0.20	0.25	0.30	0.35
25	0.07	0.11	0.15	0.20	0.25	0.30	0.35
26	0.07	0.11	0.15	0.20	0.25	0.30	0.35
27	0.07	0.11	0.15	0.20	0.25	0.30	0.35
28	0.07	0.11	0.15	0.20	0.25	0.30	0.35
29	0.07	0.11	0.15	0.20	0.25	0.30	0.35
30	0.07	0.11	0.15	0.20	0.25	0.30	0.35

Table 7.10 Repayment schedule, example 1

Year	Loan outstanding	Interest	Repayment
1	10 000.00	1 000.00	2 637.97
2	8 362.03	836.20	2 637.97
3	6 560.25	656.03	2 637.97
4	4 578.30	457.83	2 637.97
5	2 398.16	239.82	2 637.97
6	0.00		

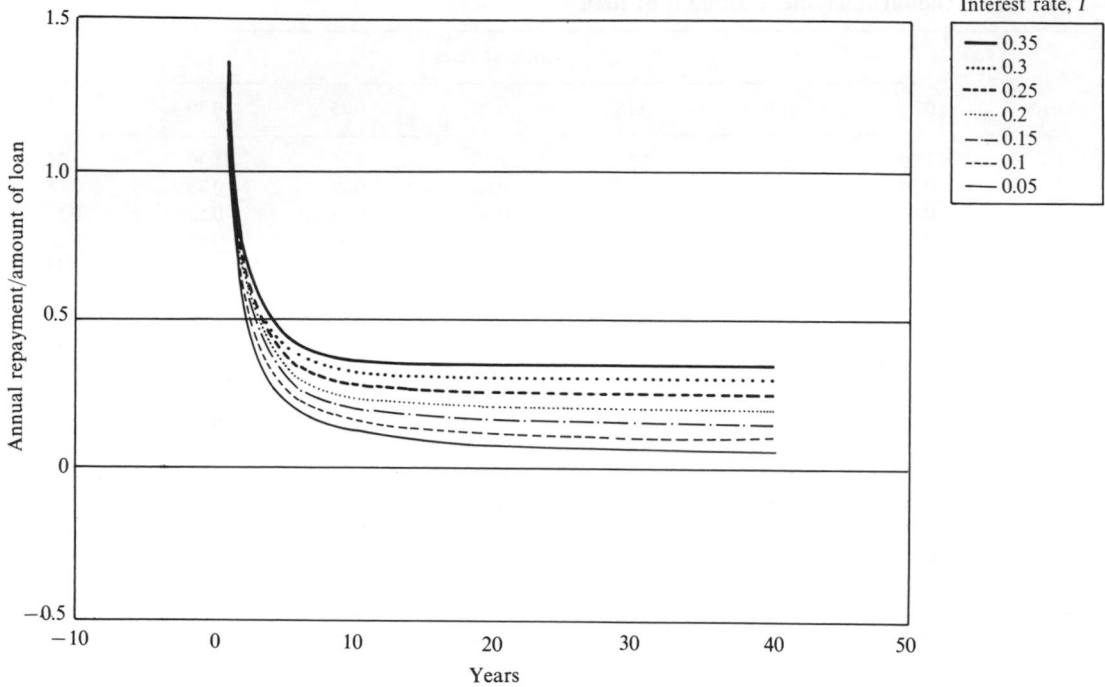

Figure 7.2 Variations of annual repayments with the amount of the loan.

Example 2 Credit card 1 This example shows that the loan from a credit card (28 to 30 per cent) could take 40 years to repay. Over this period, the total repayment would be 12 times the original loan!

Amount of loan	1 000.00
Number of years	40
Interest rate, I	0.30
Annuity factor	3.33
Total repayment	12 000.33
Annual repayment	300.01

The repayment details are shown in Table 7.11.

Example 3 Credit card 2 Most credit card companies insist on a minimum monthly payment of 5 per cent of the outstanding balance. The repayment schedule (Table 7.12) shows that, if this is all that is repaid per month the capital sum at outset will be paid off over a very long period of time.

The loan details are: £1000 at an annual interest rate, $I = 0.3$, repaying 5 per cent of the balance each month.

Table 7.11 Repayment schedule, example 2

Year	Loan outstanding	Interest	Repayment
1	1 000.00	300.00	300.01
2	999.99	300.00	300.01
3	999.98	299.99	300.01
4	999.97	299.99	300.01
5	999.95	299.98	300.01
6	999.92	299.98	300.01
7	999.89	299.97	300.01
8	999.85	299.96	300.01
9	999.80	299.94	300.01
10	999.73	299.92	300.01
11	999.65	299.89	300.01
12	999.53	299.86	300.01
13	999.38	299.81	300.01
14	999.19	299.76	300.01
15	998.94	299.68	300.01
16	998.61	299.58	300.01
17	998.19	299.46	300.01
18	997.63	299.29	300.01
19	996.91	299.07	300.01
20	995.98	298.79	300.01
21	994.77	298.43	300.01
22	993.19	297.96	300.01
23	991.14	297.34	300.01
24	988.47	296.54	300.01
25	985.00	295.50	300.01
26	980.49	294.15	300.01
27	974.63	292.39	300.01
28	967.01	290.10	300.01
29	957.10	287.13	300.01
30	944.23	283.27	300.01

7.5 INVESTMENTS

It has been shown that the present value of the total annual savings (£S) arising from an investment of £C at the outset is given by

$$PV = S\left(\frac{1-(1/1+I)^n}{I}\right) \tag{7.16}$$

Loans Versus investments

Suppose that a £100 000 mortage is taken out over 25 years at an average interest rate of 12.5 per cent.

Amount	100 000.00
Number of years	25
Interest rate, I	0.125

Table 7.12 Repayment schedule, example 3

Month	Loan outstanding	Interest	Repayment	Total repaid
1	1 000.00	25.00	50.00	50.00
2	975.00	24.38	48.75	98.75
3	950.63	23.77	47.53	146.28
4	926.86	23.17	46.34	192.62
5	903.69	22.59	45.18	237.81
6	881.10	22.03	44.05	281.86
7	859.07	21.48	42.95	324.82
8	837.59	20.94	41.88	366.70
9	816.65	20.42	40.83	407.53
10	796.24	19.91	39.81	447.34
11	776.33	19.41	38.82	486.16
12	756.92	18.92	37.85	524.00
13	738.00	18.45	36.90	560.90
14	719.55	17.99	35.98	596.88
15	701.56	17.54	35.08	631.96
16	684.02	17.10	34.20	666.16
17	666.92	16.67	33.35	699.51
18	650.25	16.26	32.51	732.02
19	633.99	15.85	31.70	763.72
20	618.14	15.45	30.91	794.62
21	602.69	15.07	30.13	824.76
22	587.62	14.69	29.38	854.14
23	572.93	14.32	28.65	882.79
24	558.61	13.97	27.93	910.72
25	544.64	13.62	27.23	937.95
26	531.03	13.28	26.55	964.50
27	517.75	12.94	25.89	990.39
28	504.81	12.62	25.24	1 015.63
29	492.19	12.30	24.61	1 040.24
30	479.88	12.00	23.99	1 064.23
31	467.88	11.70	23.39	1 087.63
32	456.19	11.40	22.81	1 110.43
33	444.78	11.12	22.24	1 132.67
34	433.66	10.84	21.68	1 154.36

Annuity factor   7.72
Total repayment   323 834
Annual repayment   12 953.4

If £12 953.4 were invested each year for 25 years at an average interest rate of 12.5 per cent, the capital sum accrued would be

$$W = 12\,953.4 \left(\frac{1.125}{0.125} ((1.125)^{25} - 1) \right)$$

$$= £209\,8754$$

Net present value

The net present value, NPV, of an investment is the total present value of the future savings minus the capital cost of the investment, i.e.

$$NPV = PV - C \tag{7.17}$$

The *payback period*, PBP, for the investment is the number of years for which the NPV is zero, i.e. PBP is when $NPV = 0$, or $PV = C$.
Now

$$S\left(\frac{1 - (1/1 + I)^n}{I}\right) = C$$

or

$$1 - (1/1 + I)^n = CI/S$$

from which

$$PBP, n = \frac{\ln(1 - CI/S)}{\ln(1/1 + I)} \tag{7.18}$$

Thus, for $C = £1000$, $S = £500$ and $I = 0.1$:

$$PBP, n = \frac{\ln(1 - 1000 \times 0.1/500)}{\ln(1/1.1)}$$

$$= 2.34 \text{ years.}$$

Note that the straight payback period is $1000/500 = 2$ years.

Check:

$$NPV = S\left(\frac{1 - (1/1 + I)^n}{I}\right) - C$$

$$NPV = 500\left(\frac{1 - (1/1.1)2.34}{0.1}\right) - 1000 = 0$$

Rate of return

The *rate of return*, ROR, on an investment is the value of the interest rate I which makes the NPV zero over the life of the project, i.e. I for which $NPV = PV - C = 0$, or $PV = C$. Now

$$S\left(\frac{1 - (1/1 + I)^n}{I}\right) = C$$

or

$$1 - (1/1 + I)^n = CI/S \tag{7.19}$$

Because this equation is transcendental in I, it must be solved by iteration. Consider, for example, $C = £1000$, $S = £500$ and, $n = 3$ years:

$$I = (1 - (1/1 + I)^n)S/C$$
$$I = 0.5(1 - (1/1 + I)^3)$$

Iterating:

$$I = 0.1243$$
0.1482
0.1697
0.1875
0.2014
0.2117
0.2189
0.2239
0.2272
0.2295
0.2309
0.2319
0.2325
0.2329
0.2332
0.2334
0.2335
0.2336
0.2336
0.2336
0.2337
0.2337
0.2337
0.2337

Resulting in a true rate of return on the capital invested of 23.37 per cent. Note that the straight rate of return $SROR$ is $500/1000 = 0.5$ ($i = 50$ per cent).

Check:

$$NPV = S\left(\frac{1 - (1/1 + I)^n}{I}\right) - C$$

$$NPV = 500\left(\frac{1 - (1/1.2337)^3}{0.2337}\right) - 1000 = 0$$

From the equation for the true rate of return, ROR, i.e. when

$$I = (1 - (1/1 + I)^n)S/C$$

it can be seen that the value of ROR depends upon the life of the project, n as well as the savings per year, S, and the capital cost of the project, C. When the life is infinity,

$I = S/C = SROR$, the straight rate of return. Otherwise ROR is less than the $SROR$ as the following example demonstrates.

For $C = £1000$ and $S = £500$

$$ROR = I, \text{ when } I = (1 - (1/1 + I)^n)S/C$$

n	$ROR = I$
3	0.2337
4	0.3490
5	0.4014
6	0.4451
7	0.4656
8	0.4780
9	0.4858
10	0.4907
$\vdots$	$\vdots$
infinity	0.5

Modification for resource conservation

If the value of the resource saved, e.g. energy, water or materials, is expected to inflate annually at r per cent, setting $R = r/100$, the value of the resource saved in successive years will be as follows:

Year 1 $S(1 + R)$
Year 2 $S(1 + R)^2$
Year 3 $S(1 + R)^3$, etc.

so that the modified present value of these savings is

$$PV' = S(((1 + R)/(1 + I))^1 + ((1 + R)/(1 + I))^2 + ((1 + R)/(1 + I))^3 + \cdots + ((1 + R)/(1 + I))^n)$$
(7.20)

which can be summed by

$$PV' = S \frac{1 + R}{I - R}\left(1 - \left(\frac{1 + R}{1 + I}\right)^n\right)$$
(7.21)

When $R = 0$, this equation reduces to the annuity formula

$$PV = S \frac{1}{I}\left(1 - \left(\frac{1}{1 + I}\right)^n\right)$$

We can also denote

$$\text{Factor}' = \frac{1 + R}{I - R}\left(1 - \left(\frac{1 + R}{1 + I}\right)^n\right)$$

Table 7.13 lists UK energy prices for the period 1970–1988 as well as the index of personal disposable income.[1]

Table 7.13 UK energy prices for the period 1970–1988 with the index of personal disposable income

Year	RFL	WFGE	PDI
1970	100	100	100
1971	110	106	107
1972	118	115	127
1973	118	127	149
1974	141	142	172
1975	220	192	214
1976	263	224	244
1977	314	267	270
1978	325	294	316
1979	353	326	395
1980	441	415	450
1981	536	440	494
1982	610	470	535
1983	656	502	575
1984	676	552	619
1985	704	577	634
1986	714	503	653
1987	721	494	672
1988	—	—	—
1989	—	—	—
1990	—	—	—

RFL—Retail prices of fuel and light.
WFGE—Wholesale prices of fuel, gas and electricity.
PDI—Personal disposable income.

Table 7.14 shows that inflation rates for the retail prices of fuel and light and for the wholesale prices of fuel, gas and electricity were 12 per cent and 10 per cent, respectively, over the period chosen. The values have been projected beyond 1987.

Table 7.15 shows that the retail prices of fuel and light and the wholesale prices of fuel, gas and electricity did not alter substantially in real terms (i.e. when normalized by personal disposable income) over the period, retail prices being relatively lowest in 1973 and 1979 and relatively highest in 1977 and 1982, with wholesale prices continuously falling to a low in 1987. This situation does not help to promote the desire for energy conservation.

Prevailing UK interest rates over the period are listed in Table 7.16, together with average house (property) prices for interest's sake.

It may be deduced that the average (geometric mean) interest rate and house price inflation over the period were 11 and 13 per cent, respectively.

Figure 7.3, obtained from Table 7.16 indicates that, apart from three peaks in the rate of increase of property prices, the advantage of investing in property *appears to decline* over the period studied.

These analyses have shown that, in the UK, the energy price inflation rate over the period studied approximately equalled prevailing interest rates. If it can be assumed that this trend will continue over the life of the energy conservation investment project, then the two compound interest factors in the modified present value equation cancel out. The present value of the energy saved in each of all future years in the life of the project is then equal to the annual sum expected at the outset.

Table 7.14 Inflation rate of UK energy prices for the period 1970–1988

	RFL	1.12	WFGE	1.10
1970	100	100.00	100	100.00
1971	110	112.33	106	109.86
1972	118	126.18	115	120.69
1973	118	141.74	127	132.59
1974	141	159.21	142	145.67
1975	220	178.85	192	160.03
1976	263	200.90	224	175.81
1977	314	225.67	267	193.14
1978	325	253.49	294	212.19
1979	353	284.75	326	233.11
1980	441	319.86	415	256.09
1981	536	359.30	440	281.34
1982	610	403.60	470	309.08
1983	656	453.36	502	339.56
1984	676	509.26	552	373.04
1985	704	572.05	577	409.82
1986	714	642.59	503	450.23
1987	721	721.82	494	494.62
1988	—	810.82	—	543.39
1989	—	910.79	—	596.97
1990	—	1023.09	—	655.83

RFL—Retail prices of fuel and light.
WFGE—Wholesale prices of fuel, gas and electricity.

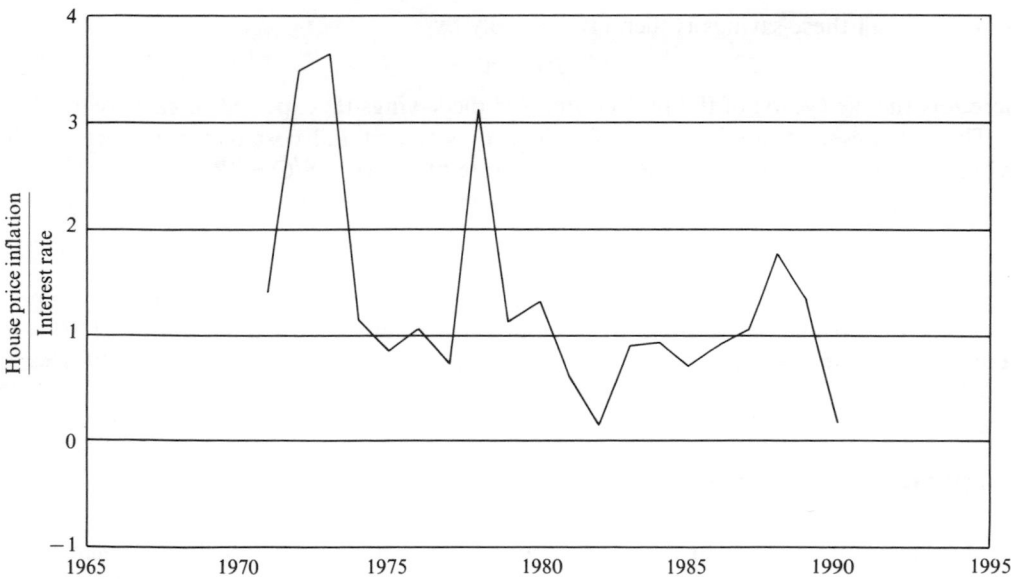

Figure 7.3 Historical ratios of property price increases/interest rates.

Table 7.15 Normalized UK energy prices for the period 1970–1988

Year	RFL/PDI	WFGE/PDI
1970	1.00	1.00
1071	1.03	0.99
1972	0.93	0.91
1973	0.79	0.85
1974	0.82	0.83
1975	1.03	0.90
1976	1.08	0.92
1977	1.16	0.99
1978	1.03	0.93
1979	0.89	0.83
1980	0.98	0.92
1981	1.09	0.89
1982	1.14	0.88
1983	1.14	0.87
1984	1.09	0.89
1985	1.11	0.91
1986	1.09	0.77
1987	1.07	0.74
1988	—	—
1989	—	—
1990	—	—

RFL—Retail prices of fuel and light.
WFGE—Wholesale prices of fuel, gas and electricity.
PDI—Personal disposable income.

The value of all these savings is then given simply by

$$PV = nS \tag{7.22}$$

where n is the life (years) of the project and S is the savings (£) expected in each year.

The net present value is $C - nS$ (£) where C is the capital cost of the project (£), the payback period (years) is $C/S = SPBP$ and the rate of return $= S/C = SROR$.

The *total lifetime cost*, TLC, of an energy/materials-consuming system is given by

$$TLC = C + \sum \Delta C + n(R - \sum S) \tag{7.23}$$

where C(£) is the basic capital cost of the system (e.g. a building) and ΔC is the extra capital cost of each additional (or retrofit) investment which results is an annual saving of S(£). To maximize TLC for any particular system, all investments which yield a positive NPV should be included.

7.6 OPTION IDENTIFICATION AND ANALYSES

The various retrofit options for the system may be formulated using critical assessment charts and alternatives trees.

Table 7.16 Prevailing UK interest rates, together with average house (property) prices

Year	IR	W	HP	HPI	HPW	HPW/W	HPI/IR
1970	8.00	1.00	4 917.00		1.00	1.00	1.41
1971	8.00	1.08	5 472.00	11.29	1.11	1.03	3.49
1972	8.50	1.17	7 093.00	29.62	1.44	1.23	3.64
1973	10.20	1.29	9 728.00	37.15	1.98	1.53	1.14
1974	11.00	1.43	10 949.00	12.55	2.23	1.55	0.85
1975	11.00	1.59	11 969.00	9.32	2.43	1.53	1.05
1976	11.40	1.77	13 406.00	12.01	2.73	1.54	0.72
1977	10.40	1.96	14 405.00	7.45	2.93	1.50	3.13
1978	10.00	2.15	18 917.00	31.32	3.85	1.79	1.11
1979	13.80	2.45	21 802.00	15.25	4.43	1.81	1.31
1980	14.00	2.79	25 811.00	18.39	5.25	1.88	0.60
1981	14.00	3.18	27 990.00	8.44	5.69	1.79	0.14
1982	11.80	3.56	28 456.00	1.66	5.79	1.63	0.90
1983	11.30	3.96	31 345.00	10.15	6.37	1.61	0.93
1984	11.50	4.42	34 701.00	10.71	7.06	1.60	0.71
1985	13.30	5.00	37 969.00	9.42	7.72	1.54	0.90
1986	11.80	5.59	41 991.00	10.59	8.54	1.53	1.05
1987	10.80	6.20	46 766.00	11.37	9.51	1.53	1.79
1988	11.40	6.91	56 297.00	20.38	11.45	1.66	1.35
1989	14.00	7.87	66 948.00	18.92	13.62	1.73	0.17
1990	15.40	9.08	68 660.00	2.56	13.96	1.54	

IR—UK interest rates.
W—Worth of 1 unit invested in 1970.
HP—Average house prices.
HPI—House price inflation.
HPW—Worth of 1 unit invested in property in 1970.

Alternative energy source options

Internal sundry gains

- with distribution system
  - air
  - water
  - with or without ducting
- without distribution system

Passive solar gains

- with passive control
- with active control
- uncontrolled

Active solar energy collection

- flat plate
  - naturally convective

with storage
diurnal
other
without storage
– forced convective
with storage
diurnal
other
without storage

- concentrating

Use of reject heat

Increased utilization efficiency

Reduce friction and drag

- lubricate
- streamline

Enhance distribution

- increase transfer functions
  - heat
  - mass
  - solids, liquids, gases

Controls

- active
- passive

Options to suppress energy rejection

Insulate

- walls
- windows
- roof
- doors
- base
- pipes
- processes

Isolate

- repair leaks
- eliminate random occurrences
- make secure

Reclaim

- energy
  – types
- materials
  –types

7.7 OPTIMIZATION

Optimizing insulation thickness

When the level of insulation to be applied to a flat surface is arbitrary, the *SPBP* or *SROR* may be optimized by computing the optimal thickness of insulation.

Let

U_o = current (uninsulated) overall heat transfer coefficient for the section, $W\,m^{-2}\,K$

ΔT = mean annual temperature difference over the section, °C

n = is the number of heating hours per year, hours

δ = thickness of insulant to be applied, m

k = thermal conductivity of the insulant, $W\,m^{-1}\,K^{-1}$

A = area of the section, m^2

c_f = fixed capital cost for installing the insulant, $£\,m^{-2}$

c_i = unit capital cost of the for insulant, $£\,m^{-3}$

c_q = unit cost for heat, $£\,kWh^{-1}$.

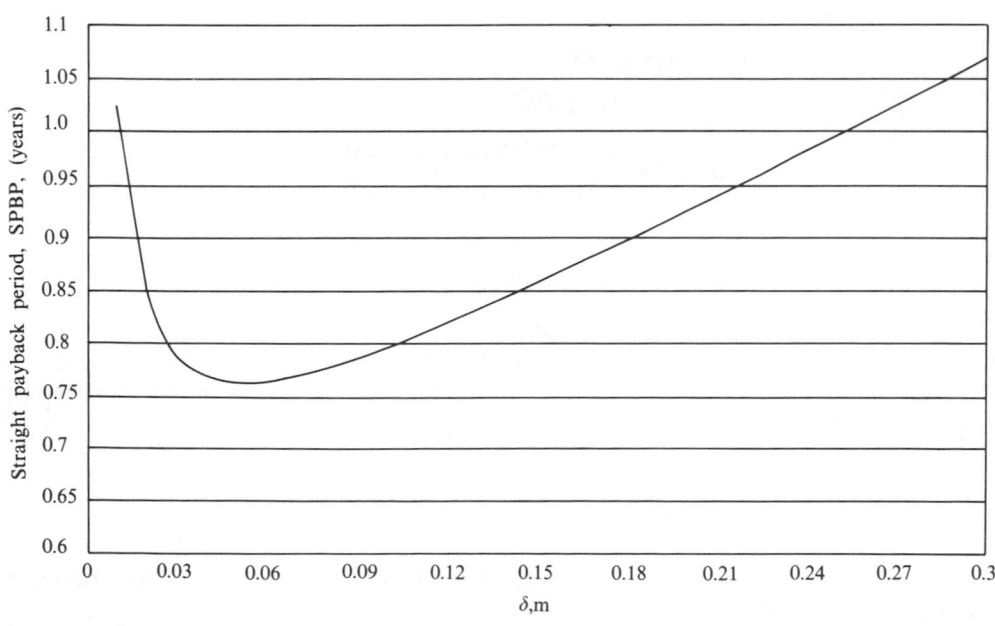

Figure 7.4 Optimal insulation thickness.

If a thickness, δ, of the insulant is applied, the modified U-value for the section becomes

$$U = kU_o/(k + \delta U_o) \qquad (7.24)$$

The heat saved per year (kWh) is then given by

$$Q = (U_o - U)A\Delta Tn/1000$$

$$= \delta U_o^2 A\Delta Tn/(k + \delta U_o)1000 \qquad (7.25)$$

The capital cost for installing the insulant is

$$C_i = (c_f + c_i\delta)A \qquad (7.26)$$

The value of the heat saved is $c_q Q$, or

$$c_q \delta U_o^2 A\Delta Tn/(k + \delta U_o)1000 \qquad (7.27)$$

and so the *SPBP* is then $C_i/c_q Q$, or

$$(c_f + c_i\delta)(k + \delta U_o)1000/c_q \delta U_o^2 \Delta Tn \qquad (7.28)$$

which is the expression to be minimized by varying δ.

Note that the SPBP is independent of the area, A, to be insulated.

With the following values:

$$U_o = 6.0 \, \text{W m}^{-2}\,\text{K}^{-1}$$

$$\Delta T = 10.8 \, °\text{C}$$

$$n = 6448 \, \text{hours per year}$$

$$k = 0.04 \, \text{W m}^{-1}\,\text{K}^{-1}$$

$$c_f = 10 \, £\,\text{m}^{-2}$$

$$c_i = 25 \, £\,\text{m}^{-3}$$

$$c_q = 0.04 \, £\,\text{kWh}^{-1}$$

$$SPBP = \frac{(10 + 25\delta)(0.04 + 6\delta)1000}{0.04 \times 36 \times 10.8 \times 6448\delta}$$

$$= \frac{(10 + 25\delta)(0.04 + 6\delta)1000}{100.28\delta}$$

Figure 7.4 shows the relationship between *SPBP* and δ, where it can be seen that *SPBP* minimizes at 0.763 years (9 months) for an insulation thickness of 0.05 m (5 cm).

A similar optimization can be conducted to optimize the thickness of insulation to be applied to pipework.

7.8 CONFLICT CORRECTION

Consider the schematic diagram for the energy flows through a heated built environment shown in Fig. 7.5. The energy, inputs to the system are as follows:

- E, Electricity
- S, Solar energy
- F, Fossil fuels
- M, Metabolic heat gains from people
- P, Fuels for miscellaneous fired processes

The electrical energy may be used directly for space heating or in appliances, from which some of the reject heat may become sundry gain (SG) and the rest directly rejected (DR) to

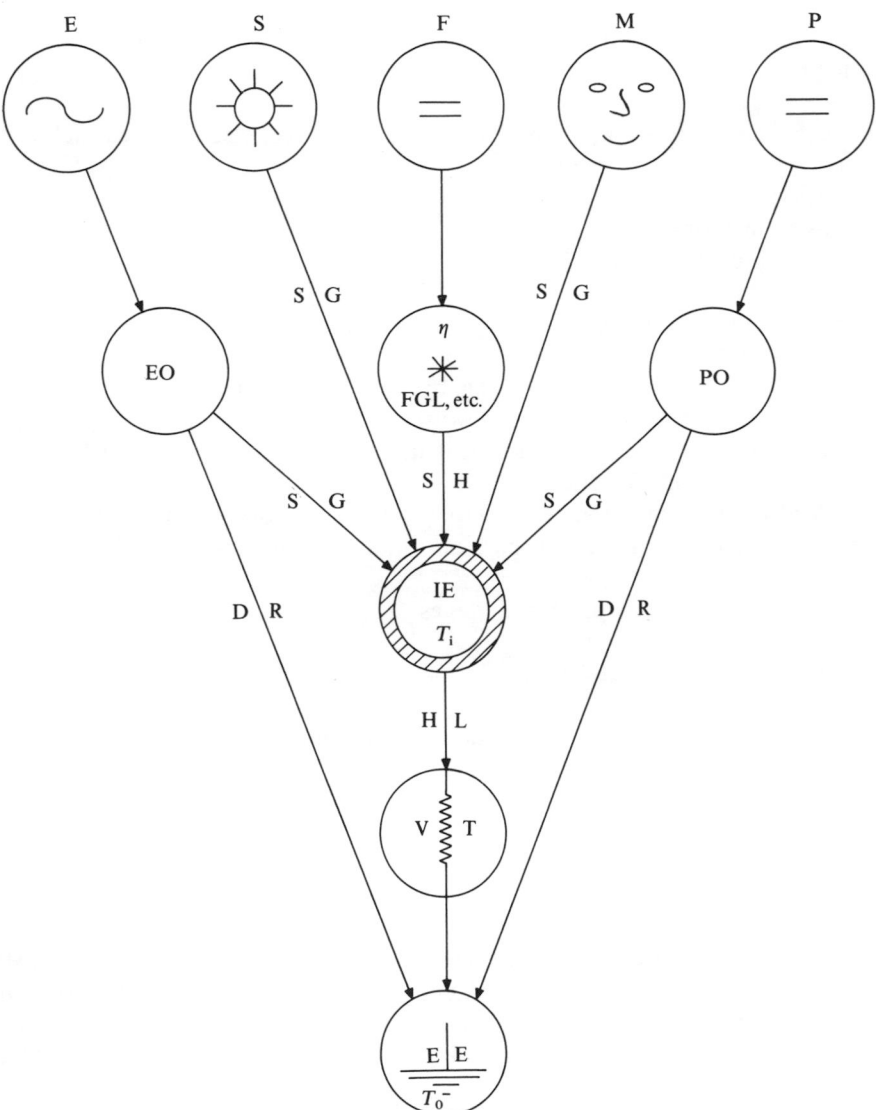

Figure 7.5 Energy audit—option conflict correction chart.

the external environment (EE). In the diagram, electrical heating has been subsumed into the sundry gain.

The heat from the fuel energy supplied for fired appliances (P) becomes similarly part sundry gain (i.e. casing heat losses) and part directly rejected.

Heat gains from solar energy and people are direct sundry gains to the internal environment, IE.

The space heating fuel is combusted to produce space heat with an efficiency, η, which takes account of stack flue gas loss (FGL), furnace/boiler, and distribution losses, and then delivered to the internal environment, IE. If the space heat delivered, SH, is increased by ΔSH, then the heating fuel supplied to the combustion process must be increased by ΔSH/η.

The total heat delivered to the internal environment at temperature T_i is then the sum of the following items:

- The sundry gains from electrical operations (EO) (including direct electrical heating)
- Solar heat gains
- The space heat delivered, SH
- Sundry heat gains from people
- The sundry gains from processes

This is rejected (HL) to the external environment at temperature T_o via fabric heat transmission, T, and the heat in exhaust ventilating air, V.

A reduction in sundry gains from electricity (e.g. by insulating electrical appliances), solar energy (variable), people (e.g. nights, annual shut-downs, variable), or processes (e.g. by insulating process appliances), requires that the space heat delivered be increased accordingly to maintain the internal temperature, T_i, constant.

Reductions in heat losses from the internal to the external environment may be accomplished by reducing heat losses via fabric transmission using the following methods:

- Insulating, and hence reducing component U-values
- Partitioning and zoning
- Decreasing the internal air temperature, T_i

or by reducing heat losses via ventilation using the following methods:

- Reducing the number of fresh air changes
- Introducing heat recovery from exhaust ventilating air
- Partitioning and zoning
- Decreasing the internal air temperature, T_i

In addition, heat may be recovered from the direct rejects from electrical and fossil-fired plant and processes to provide space heating. A reduction in the heating load, HL, results in a corresponding reduction of the space heat delivered, SH. If the space heat delivered, SH, is decreased by ΔSH, then the heating fuel supplied to the combustion process is decreased by ΔSH/η.

Consider the basic system shown in Fig. 7.6, where the fossil fuel to heating energy conversion efficiency is 75 per cent, and 1000 units of electrical energy are supplied to the

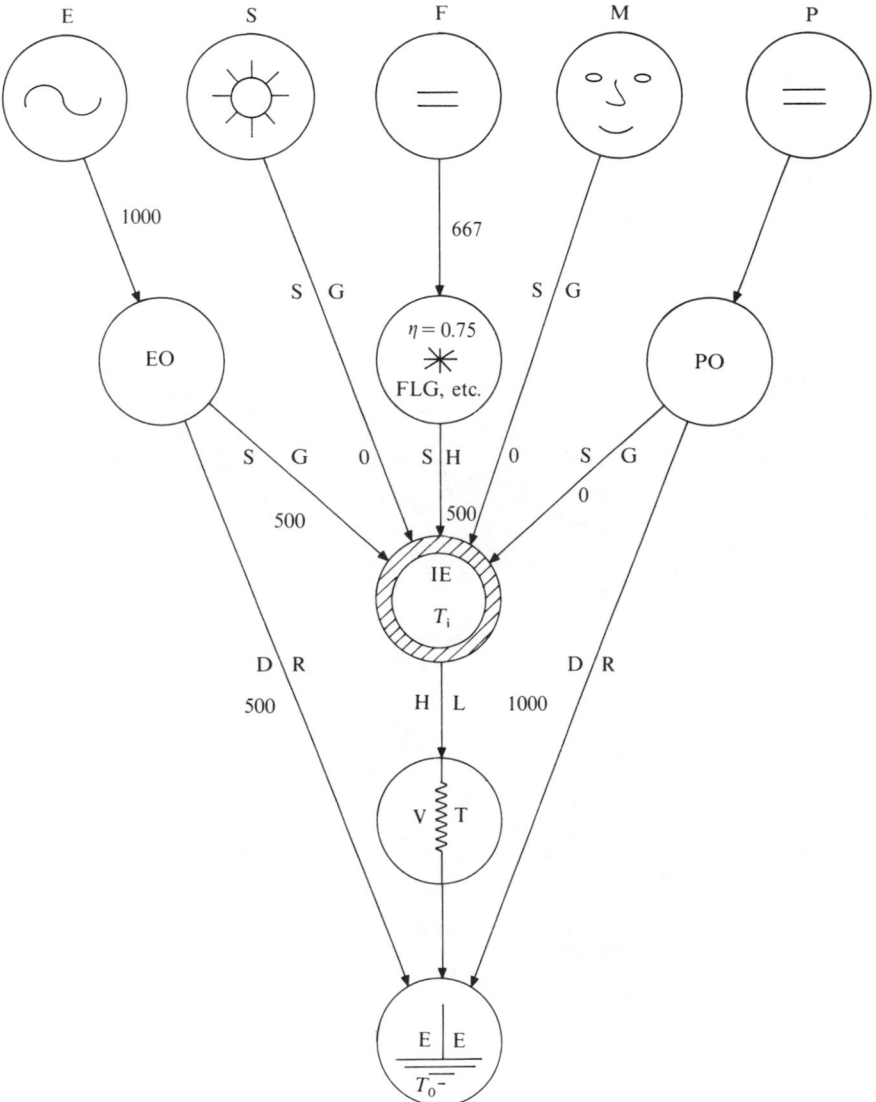

Figure 7.6 Energy audit—conflict correction, basic system.

electrical operations, resulting in a sundry heat gain to the internal environment of 500 units. The heat load on the internal environment to offset fabric transmission and ventilation losses is 1000 units. Discounting this load by the electrical sundry gains, the space heat delivered is 500 units, for which 667 (500/0.75) units of fossil fuel is required to be combusted. The total energy input in 1000 (electricity) + 667 (fossil fuel) = 1667 units.

The total energy output is 500 (electrical direct reject) + 1000 (ventilation and transmission) + 167 (fossil fuel conversion losses) = 1667 units.

There are two options for financial savings:

- Option A   Improve conversion efficiency from 75 to 80 per cent.
- Option B   Reduce the heating load from 1000 units to 800 units.

Figure 7.7 shows the effect of applying option A to the system. Again, 1000 units of electrical energy are supplied to the electrical operations, resulting in a sundry heat gain to the internal environment of 500 units. The heat load on the internal environment to offset

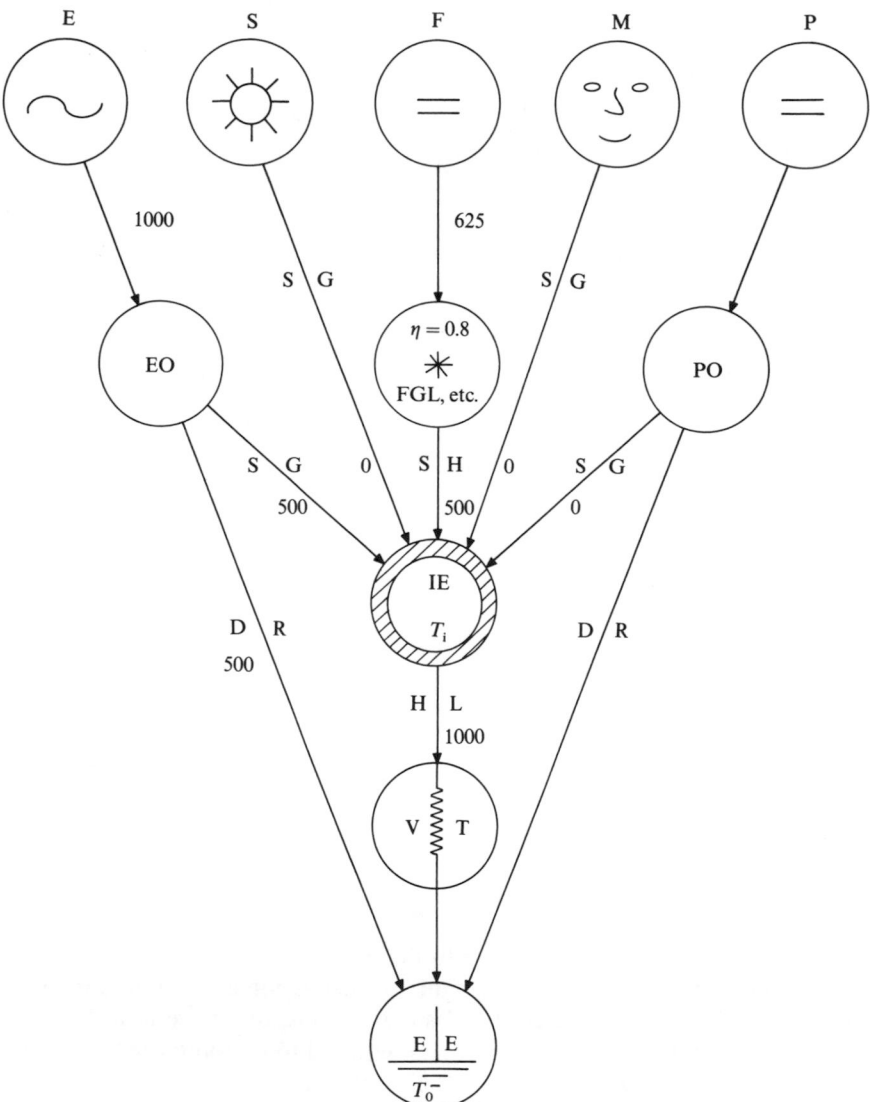

Figure 7.7 Energy audit—conflict correction, basic system + option A.

fabric transmission and ventilation losses is 1000 units. Discounting this load by the electrical sundry gains, the space heat delivered is 500 units, for which 625 (500/0.8) units of fossil fuel is required to be combusted. The total energy input is 1000 (electricity) + 625 (fossil fuel) = 1625 units.

The total energy output is 500 (electrical direct reject) + 1000 (ventilation and transmission) + 125 (fossil fuel conversion losses) = 1625 units. The total saving arising from the implementation of option A is then 1667 − 1625 = 42 units.

Figure 7.8 shows the effect of applying option B to the system. Again, 1000 units of

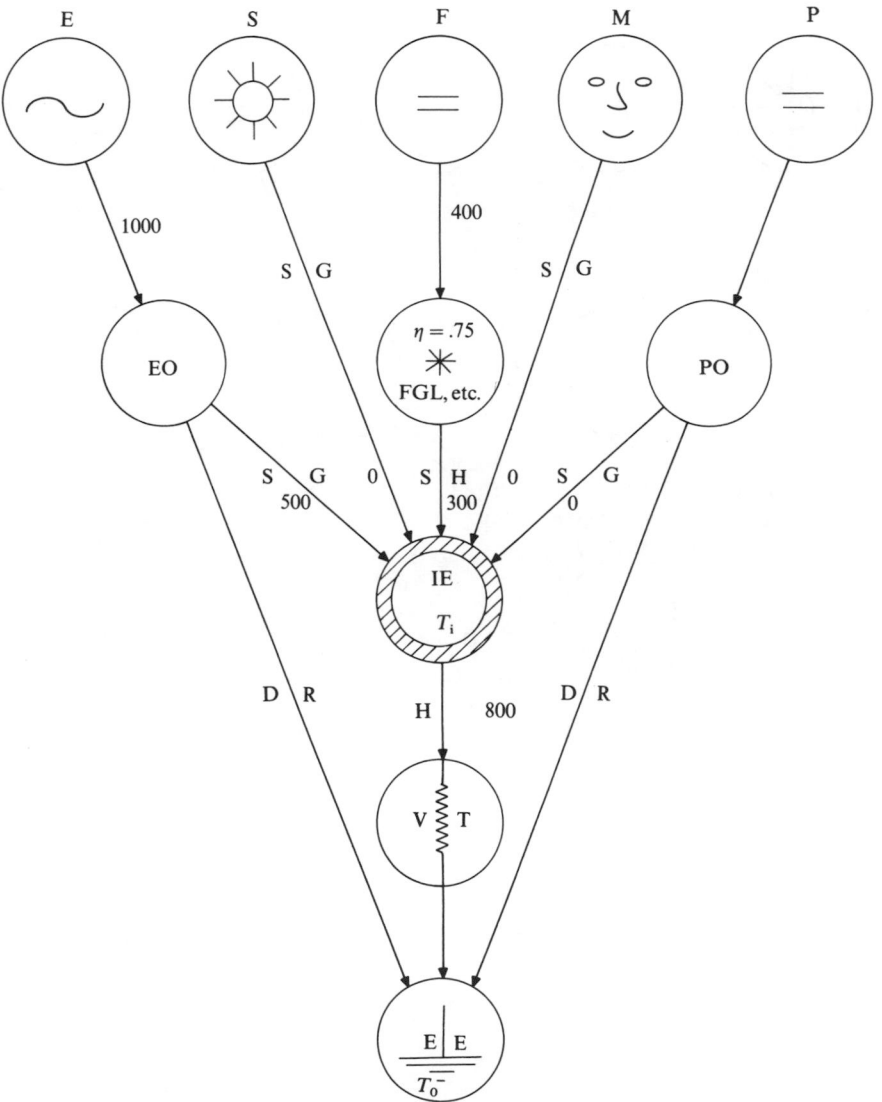

Figure 7.8 Energy audit—conflict correction, basic system + option B.

electrical energy are supplied to the electrical operations, resulting in a sundry heat gain to the internal environment of 500 units. The heat load on the internal environment to offset fabric transmission and ventilation losses is 800 units. Discounting this load by the electrical sundry gains, the space heat delivered is 300 units, for which 400 (300/0.75) units of fossil fuel is required to be combusted. The total energy input is 1000 (electricity) + 400 (fossil fuel) = 1400 units.

The total energy output is 500 (electrical direct reject) + 800 (ventilation and transmission) + 100 (fossil fuel conversion losses) = 1400 units. The total saving arising from the implementation of Option B is then 1667 − 1400 = 267 units.

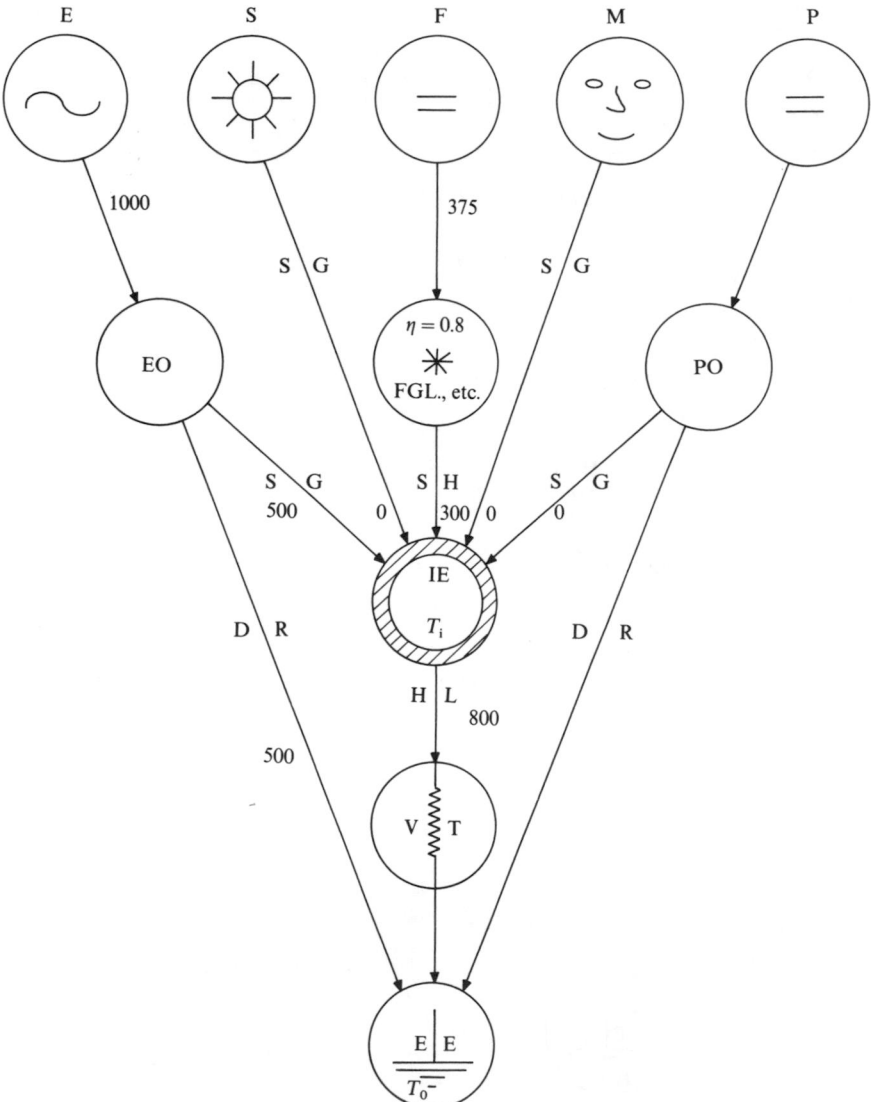

Figure 7.9 Energy audit—conflict correction, basic system + options A and B.

Figure 7.9 shows the result of applying both options A and B to the system. Again, 1000 units of electrical energy are supplied to the electrical operations, resulting in a sundry heat gain to the internal environment of 500 units. The heat load on the internal environment to offset fabric transmission and ventilation losses is 800 units. Discounting this load by the electrical sundry gains, the space heat delivered is 300 units, for which 400 (300/0.75) units of fossil fuel is required to be combusted. The total energy input is 1000 (electricity) + 375 (fossil fuel) = 1375 units.

The total energy output is 500 (electrical direct reject) + 800 (ventilation and transmission) + 75 (fossil fuel conversion losses) = 1375 units. The total saving arising from the implementation of both options A and B is then 1667 − 1375 = 292 units.

If the savings arising from applying options A and B are added, it might be wrongly inferred that total energy savings of 42 + 267 = 309 units would result. The 'lost' savings (309 − 292 = 17 units) arise from the *conflict of the two options*.

Another example of conflicting options concerns the individual effects of installing double glazing, blinds and curtains and the overall result of installing all three options simultaneously. The law of diminishing returns applies. Sequential energy-conserving investment exercises should select the best cost-effective option to be applied to the base case, implement this and then reappraise all other remaining options, thus correcting for conflict. An optimal sequence of investments will then result, which will form the basis of a prediction of financial expenditures through the life of the project as the target schedule for project management and appraisal.

7.9 CONSTRUCTING THE OPTIMAL TARGET INVESTMENT SCHEDULE

In Chapter 2, the summary of investment opportunities for the site audited ranked in order of cost-effectiveness as shown in Table 7.17, and the simple investment plan (Table 7.18) was produced which resulted in the projected savings shown in Table 7.19.

Some of these options are however in conflict and the investment schedule must be corrected as follows: on average, oil cost is £0.0333 per kWh and gas cost is £0.0136 per kWh, so that oil + gas cost is £0.0188 per kWh or, £18 800 per 10^6 kWh.

Dayrate electricity cost is £0.0683 per kWh and nightrate electricity cost is £0.0376 per kWh, so that all electricity + MD charges cost is £0.0776 per kWh or £77 600 per 10^6 kWh.

Table 7.17 Summary of investment opportunities

Retrofit measure	Annual savings, £	Capital cost, £	Straight rate of return, %
A. Reduce electrical supply capacity	3 000	0	infinity
B. Inside air temperature reduction	37 120	1 000	3 712
C. Abolish electrical heating	12 571	2 000	629
D. Boiler control	5 340	2 000	267
E. Use of compressor cooling air	8 447	10 000	85
F. Roof insulation	18 000	40 000	45
Totals	84 478	55 000	154

Table 7.18 Investment plan

Year	Retrofit measure	Capital cost, £	Savings year on year, £	Accumulated savings, £
1	A	0	3 000	3 000
2	B and C	3 000	52 691	52 691
3	D, E and F	52 000	84 478	85 169

Table 7.19 Projected savings over three years

Initial energy bill	£244 000
Gross invested	£55 000
Net invested	£ zero
Final energy bill	£160 000
Residual capital	£85 169

Table 7.20 Marginal investment opportunities

Retrofit measure	Annual savings, £	Capital cost, £	Straight rate of return, %
A. Reduce electrical supply capacity	3 000	0	infinity
+ B. Inside air temperature reduction	37 000	1 000	3 700
+ C. Abolish electrical heating	13 000	2 000	650
+ D. Boiler control	6 000	2 000	300
+ E. Use of compressor cooling air	8 000	10 000	80
+ F. Roof insulation	17 000	40 000	43
Totals	84 000	55 000	153

Table 7.21 Revised investment plan

Year	Retrofit measure	Capital cost, £	Savings year on year, £	Accumulated savings, £
1	A	0	3 000	3 000
2	B and C	3 000	53 000	53 000
3	D, E and F	52 000	84 000	85 000

Table 7.22 Revised projected savings

Initial energy bill (inc. MD charges)	£277 000
Gross invested	£55 000
Net invested	£ zero
Final energy bill	£196 000
Residual capital	£85 000

Implementation of Option A results in no conflict and £3000 will be saved each year thereafter for zero investment.

Figure 7.10 then shows the simplified conflict correction chart for the base case, having an overall annual energy bill of £277 000 with 5.57×10^6 kWh per year flowing through the building. This base case state-point is indicated in Fig. 7.11. The state-points arising from applying either of options, C, D, E and F to the conflict correction chart are also indicated. The option which results in the best *SROR* ($= 3760$ per cent) is option B (Fig. 7.12), which is therefore chosen for adoption, altering the audit balance to the new state point indicated (B).

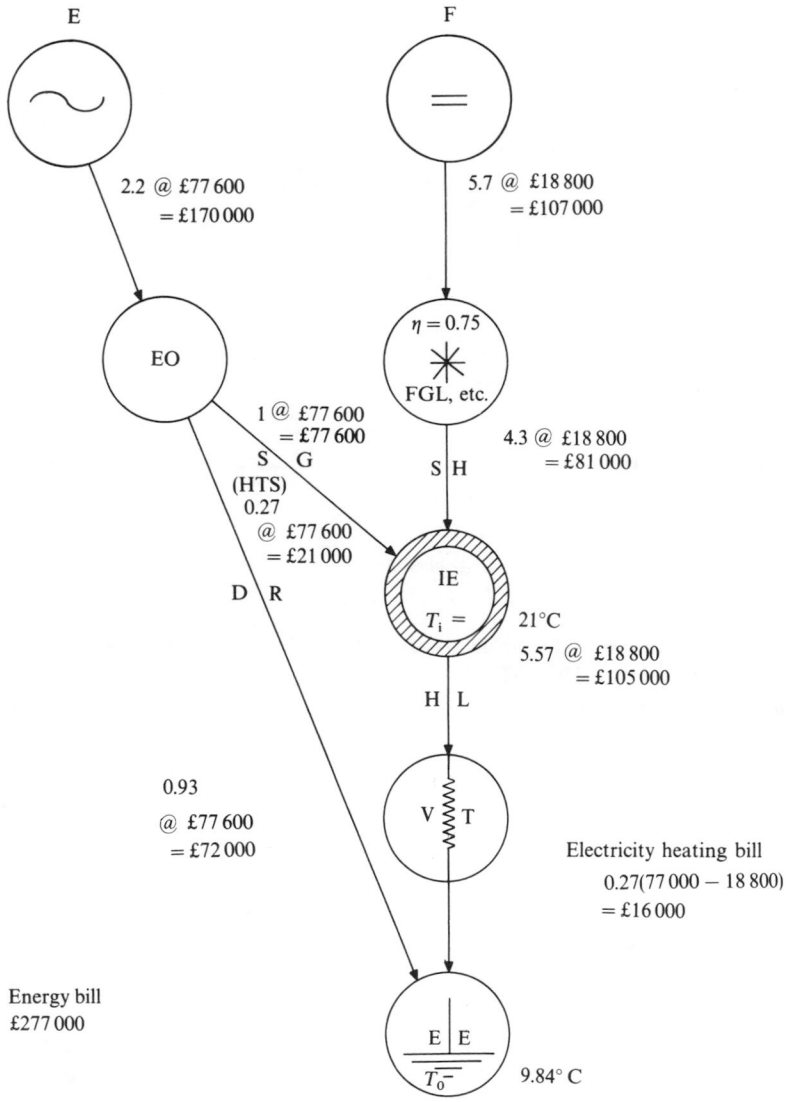

Figure 7.10 Energy audit—Bitusa Industries, conflict correction, base case.

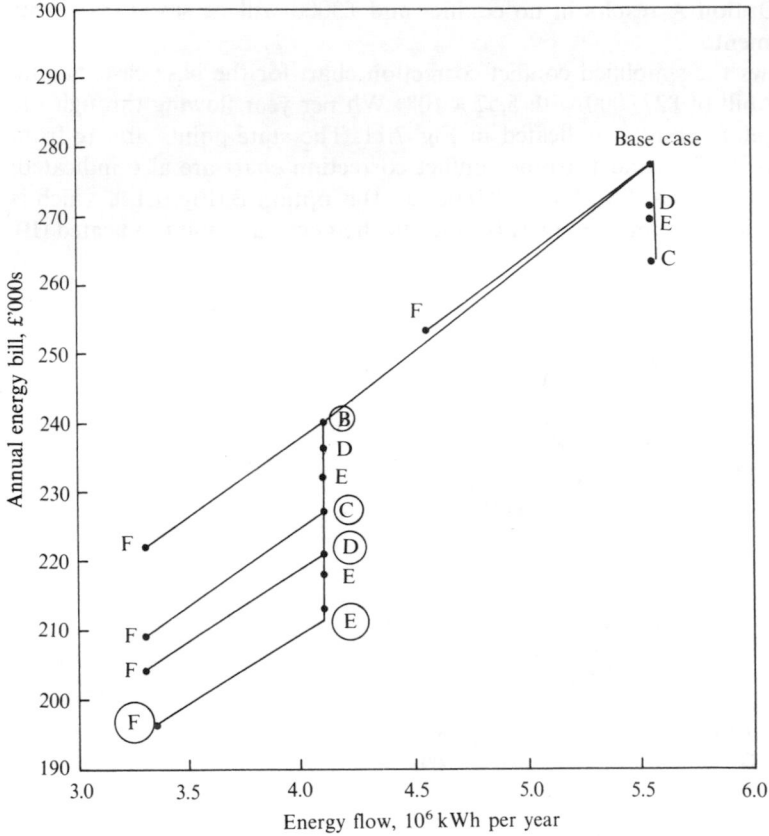

Figure 7.11 Energy audit—Bitusa Industries, optimal investment schedule.

The procedure is repeated with the remaining options to produce the optimal conflict-corrected investment schedule (B) to (C) to (D) to (E) to (F).

This results in the amended investment portfolio of *marginal* investment opportunities as shown in Table 7.20.

The corrected investment plan (Table 7.21) may be produced, together with revised projected savings, Table 7.22.

7.10 PROJECT MANAGEMENT, MONITORING AGAINST THE TARGET FINANCIAL SCHEDULE

This monthly schedule for the three-year plan is shown in Table 7.23, the projected savings each month have been corrected for degree-day, DD, differences, the total annual degree days being 2067. In reality the savings should be corrected continuously for actual degree-days per month, fuel price changes, altering capital costs and changes in the levels of activity.

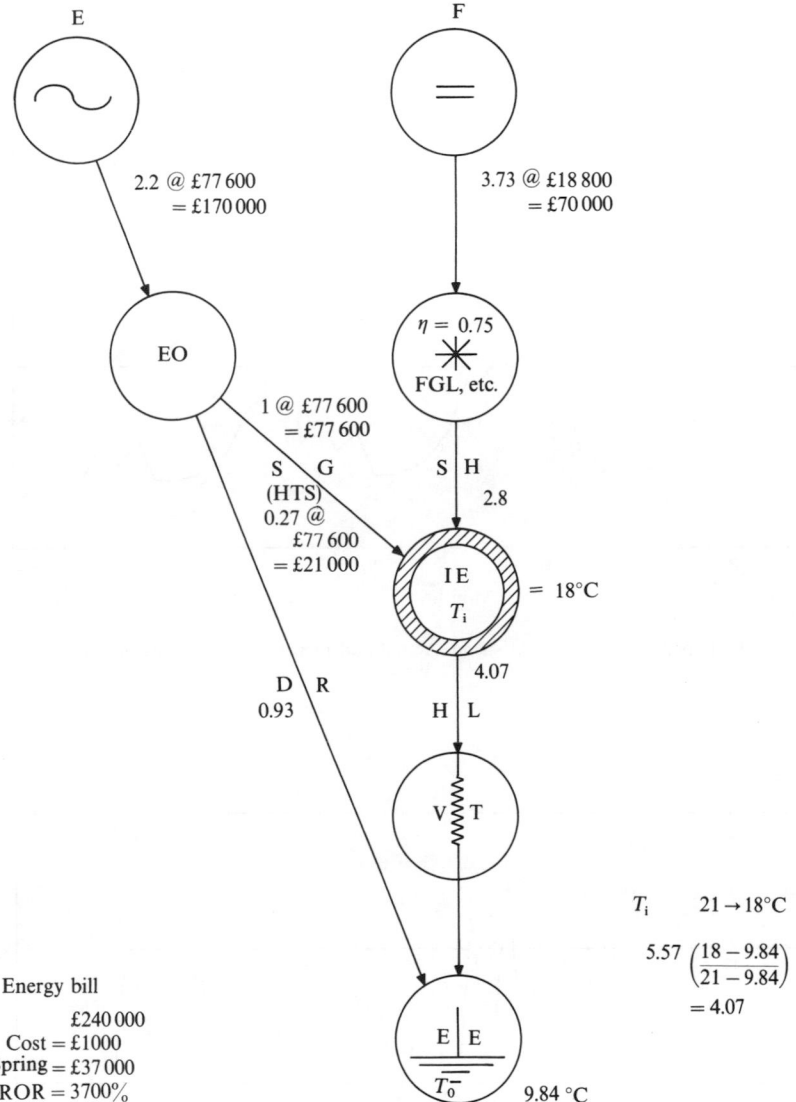

E

F

2.2 @ £77 600
= £170 000

3.73 @ £18 800
= £70 000

EO

$\eta = 0.75$

FGL, etc.

1 @ £77 600
= £77 600

S   G
(HTS)
0.27 @
£77 600
= £21 000

S H

2.8

I E

T_i = 18°C

4.07

D   R
0.93

H   L

V   T

T_i 21 → 18°C

$$5.57 \left(\frac{18 - 9.84}{21 - 9.84} \right)$$
= 4.07

Energy bill
£240 000
Cost = £1000
Spring = £37 000
SROR = 3700%

E   E

T_0^- 9.84 °C

Figure 7.12 Energy audit—Bitusa Industries, conflict correction, base case + option B.

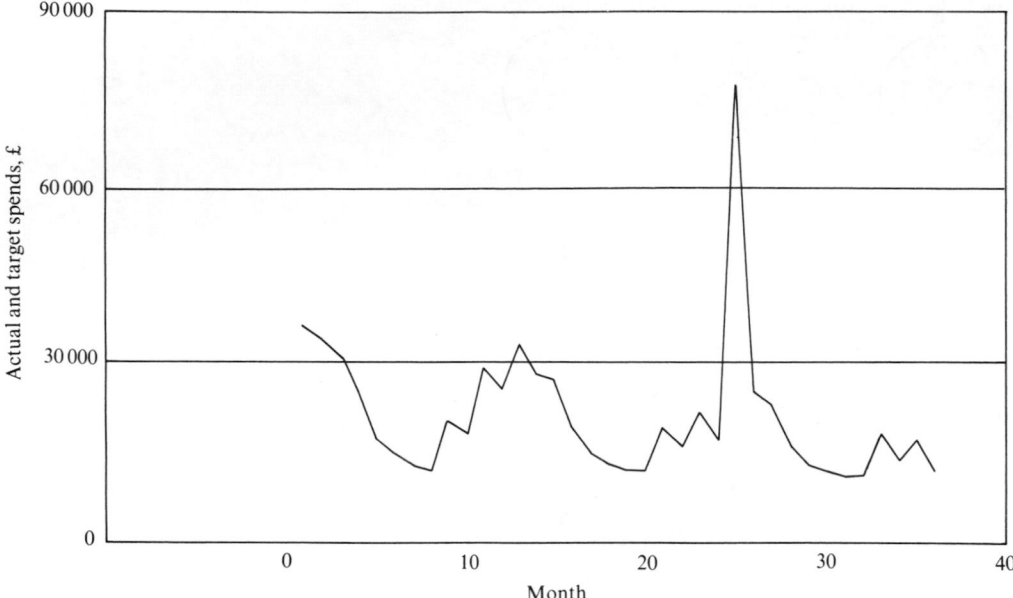

Figure 7.13 Comparison of actual spend with the target spend. Actual and target are coincident on this scale.

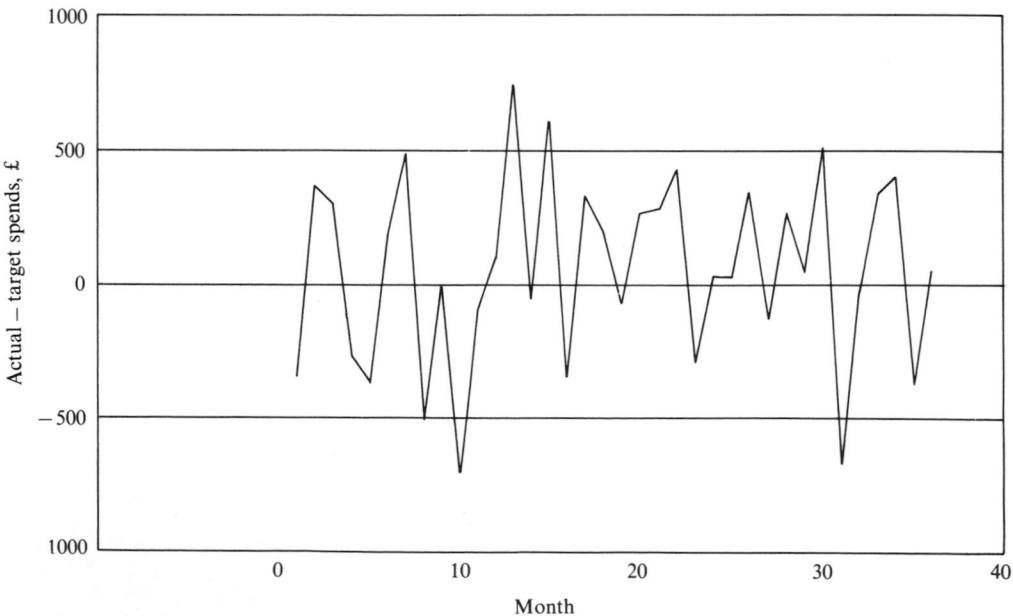

Figure 7.14 Deviations from the target spend.

Table 7.23 Monthly schedule for 3-year investment plan

Month	DD	B 91–92 costs	C Capital costs	Savings per month
1	286	36 595	0	250
2	227	33 885	0	250
3	217	31 955	0	250
4	241	25 525	0	250
5	115	17 625	0	250
6	88	15 066	0	250
7	27	12 766	0	250
8	40	12 760	0	250
9	64	20 263	0	250
10	132	18 966	0	250
11	295	29 350	0	250
12	335	25 560	0	250
13	286	36 595	3 000	7 333
14	227	33 885	0	5 820
15	217	31 955	0	5 564
16	241	25 525	0	6 179
17	115	17 625	0	2 948
18	88	15 066	0	2 256
19	27	12 766	0	6 927
20	40	12 760	0	1 025
21	64	20 263	0	1 641
22	132	18 966	0	3 384
23	295	29 350	0	7 564
24	335	25 560	0	8 589
25	286	36 595	52 000	11 622
26	227	33 885	0	9 224
27	217	31 955	0	8 818
28	241	25 525	0	9 793
29	115	17 625	0	4 673
30	88	15 066	0	3 576
31	27	12 766	0	1 097
32	40	12 760	0	1 625
33	64	20 263	0	2 600
34	132	18 966	0	5 364
35	295	29 350	0	11 988
36	335	25 560	0	13 613

Comparison of actual and target spends

Monthly target and actual spends on fuel, electricity and capital investments are shown in Fig. 7.13 and Table 7.24. The monthly deviations from the target are given in Fig. 7.14.

The cumulative sum (cu.sum) of the deviations from the target schedule (Fig. 7.15) shows some relaxation over years 2 and 3, although the comparisons of actual monthly spend with those appertaining in the base year, shown in Table 7.25 and Fig. 7.16, show that the project followed its projected schedule on course to hit the final target as predicted.

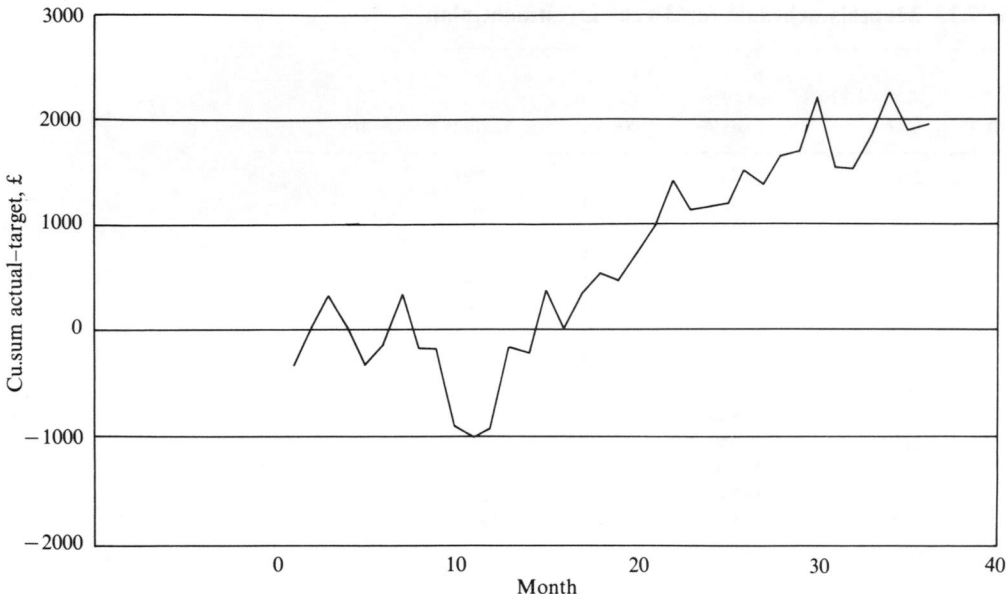

Figure 7.15 Cumulative sums of the deviations from the target spend.

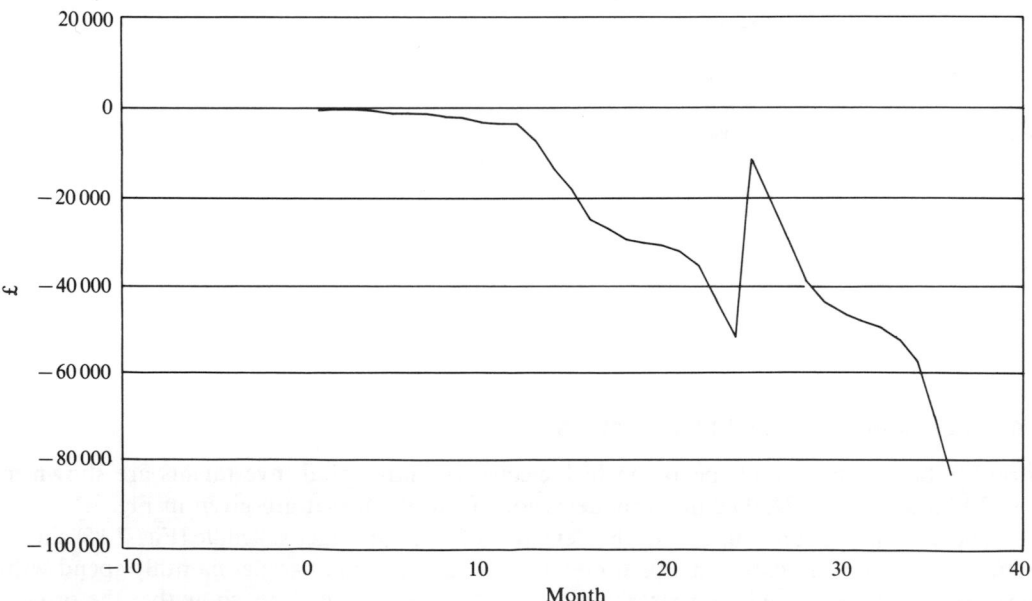

Figure 7.16 Accumulated savings on the base year (£ negative indicates positive saving)

Table 7.24 Target and actual spends

Month	T Target	A Actual	A − T	Cu.sum (A − T)
1	36 345	36 000	− 345	− 345
2	33 635	34 000	365	20
3	31 705	32 000	295	315
4	25 275	25 000	− 275	40
5	17 375	17 000	− 375	− 335
6	14 816	15 000	184	− 151
7	12 516	13 000	484	333
8	12 510	12 000	− 510	− 177
9	20 013	20 000	− 13	− 190
10	18 716	18 000	− 716	− 906
11	29 100	29 000	− 100	− 1006
12	25 310	25 400	90	− 916
13	32 261	33 000	738	− 177
14	28 064	28 000	− 64	− 242
15	26 390	27 000	609	366
16	19 345	19 000	− 345	21
17	14 676	15 000	323	345
18	12 809	13 000	190	535
19	12 073	12 000	− 73	461
20	11 734	12 000	265	727
21	18 621	18 900	278	1005
22	15 581	16 000	418	1424
23	21 785	21 500	− 285	1138
24	16 970	17 000	29	1168
25	76 972	77 000	27	1195
26	24 660	25 000	339	1535
27	23 136	23 000	− 136	1399
28	15 731	16 000	268	1668
29	12 951	13 000	48	1716
30	11 489	12 000	510	2266
31	11 668	11 000	− 668	1557
32	11 134	11 110	− 24	1533
33	17 662	18 000	337	1871
34	13 601	14 000	398	2269
35	17 361	17 000	− 361	1908
36	11 946	12 000	53	1962

Table 7.25 Comparison of actual spend and accumulated savings (shown negative) on base year

Month	$A-(B+C)$ Diff	Cu.sum $(A-(B+C))$
1	−595	−595
2	115	−480
3	45	−435
4	−525	−960
5	−626	−1585
6	−66	−1651
7	234	−1417
8	−760	−2177
9	−263	−2440
10	−966	−3406
11	−350	−3756
12	−160	−3916
13	−3595	−7511
14	−5885	−13396
15	−4955	−18351
16	−6525	−24876
17	−2625	−27501
18	−2066	−29567
19	−766	−30333
20	−760	−31093
21	−1363	−32456
22	−2966	−35422
23	−7850	−43272
24	−8560	−51832
25	40405	−11427
26	−8885	−20312
27	−8955	−29267
28	−9525	−38792
29	−4625	−43417
30	−3066	−46483
31	−1766	−48249
32	−1650	−49899
33	−2263	−52162
34	−4966	−57128
35	−12350	−69478
36	−13560	−83038

REFERENCE

1. *UK Digest of Statistics*, U.K. Statistical Office, 1990.

FURTHER READING

Helcke, G.: *The Energy Saving Guide*, Pergamon Press, Oxford, 1981.

CHAPTER
EIGHT

MINIAUDIT—USE OF COMPUTATIONAL AIDS

8.1 ENERGY MANAGEMENT INFORMATION SYSTEMS

When confronted with a collection of sites to audit, the energy manager must first institute an *energy management information system* which contains data on fuels and electricity consumed and monthly degree-days (DD) for all sites.

Monitoring and targeting of multiple sites

The energy management information system is prepared using the following facilities/activities:

1. Database
2. Analyses
3. Construct ratios $kWh\,m^{-2}\,DD^{-1}$, $GJ\,m^{-3}\,DD^{-1}$
4. Make comparisons between sites
5. Develop yardsticks and targets to monitor

The pro-forma sheets, Tables 8.1 and 8.2, are intended to be used for comparing several audits and for comparing individual buildings within a single audit.

8.2 THE THIRTY-NINE STEPS FOR ENERGY MANAGEMENT

There are 39 steps to follow to optimize energy management investments at any individual site:

1. Identify site and scope of audit.
2. Obtain fuel and electricity bills for a recent representative year.
3. Obtain degree-days.

(*Continued on page 340*)

Table 8.1 Energy audit: cross-audit comparators

Date Period of analysis Audit no. Description Auditors	
Number of buildings Total floor area (m^2) Total enclosed volume (m^3) Total exposed area for heat flow (m^2) Number of boilers Combustion/boiler efficiencies (%) Distribution losses (%) Total fuel used (kWh per year) Total fuel bill (£) Total electricity used (kWh per year) Total electricity bill (£/annum) Degrees-days per year, DD (over occ.hrs.) Mean annual outside air temperature (°C) Mean internal temperature (and range) (°C) Mean ΔT (°C) No. of people Occupied hours Sundry gains (kWh per year) ● electricity ● other fuels ● people Direct reject (kWh per year) ● electricity ● other fuels Domestic hot water (kWh per year) All energy $kWh\,m^{-2}$ DD(over occ.hrs.) $kWh\,m^{-3}$ DD(over occ.hrs.) kWh per capita DD(over occ.hrs.) Electricity $kWh\,m^{-2}$ DD(over occ.hrs.) $kWh\,m^{-3}$ DD(over occ.hrs.) kWh per capita DD(over occ.hrs.) Fuel $kWh\,m^{-2}$ DD(over occ.hrs.) $kWh\,m^{-3}$ DD(over occ.hrs.) kWh per capita DD(over occ.hrs.) Heating energy $kWh\,m^{-2}$ DD(over occ.hrs.) $kWh\,m^{-3}$ DD(over occ.hrs.) kWh per capita DD(over occ.hrs.)	

Table 8.1 (*Continued*)

Total space heat delivered $kWh\,m^{-2}\,DD$(over occ.hrs.) $kWh\,m^{-3}\,DD$(over occ.hrs.) kWh per capita DD(over occ.hrs.) Mean overall U-value $(W\,m^{-2}\,K^{-1})$ Mean number of air changes per hour, n Fabric transmission losses (kWh per year) Ventilation losses (kWh per year) T/V (fabric transmission/vent. losses) $G\,((T+V)\,m^{-3}\Delta T^{-1})$ Cooling requirements (kWh per year) Compressed air (kWh per year) All other uses (kWh per year) Environmental factors	
Obvious wastages	
Maintenance and repair requirements	
Retrofit recommendations	

Table 8.2 Energy audit: building comparators

Audit no. Building number Description Floor area (m^2) Enclosed volume (m^3) Exposed area for heat flow (m^3) Combustion/boiler efficiency/s (%) Distribution losses (%) Fuel used (kWh per year) Electricity used (kWh per year) Mean internal temperature (and range) (°C) Mean ΔT (°C) No. of people Occupied hours Sundry gains (kWh per year) • Electricity • other fuels • People Direct reject (kWh per year) • electricity • other fuels Domestic hot water (kWh per year)	

4. Identify thermal or other changes that have occurred over or since the year under examination.
5. Examine electrical supply capacity, maximum demand variations and all tariffs.
6. Feed data to spreadsheet.
7. Convert quantities to common units.
8. Convert degree-days to mean monthly temperatures during operating hours.
9. Plot kVA versus degree-days.
10. Plot kVA versus mean monthly temperatures during operating hours.
11. Correlate with linear regression and quantify scatter.
12. Identify electricity use for heating.
13. Plot kWh of electricity versus degree-days.
14. Quantify electricity use for heating.
15. Plot kWh of electricity versus monthly temperatures during operating hours.
16. Correlates with linear regression and quantify scatter.
17. Check electricity use for heating.
18. Obtain annual mean furnace/boiler efficiency to discount fuel supplied to obtain heating energy.
19. Plot kWh of heating energy versus degree-days.
20. Plot kWh of heating energy (including electricity use for heating) versus monthly temperatures during operating hours.
21. Correlate with linear regression and quantify scatter.
22. Obtain annual mean inside air temperature.
23. Obtain base temperatures.
24. Quantify sundry gains and direct rejects of energy.
25. Obtain equation relating overall U-value to number of air changes, n.
26. External site survey—areas and volumes and capita—UA values.
27. Deduce n.
28. Construct output analysis—ventilation and fabric transmission losses.
29. Investigate other effluents and stack losses.
30. Internal surveys if process plant and equipment significant.
31. Construct throughput analysis.
32. Construct energy audit balance sheet.
33. Identify options for energy conservation and money-saving options.
34. Analysis of options—straight rate of return.
35. Produce ranked investment portfolios.
36. Identify option conflicts.
37. Re-rank investment portfolios.
38. Establish 5-year investment plan.
39. Set up monitoring and targeting and project management systems.

8.3 MINIAUDIT

The following miniaudit illustrates the procedure and demonstrates how the use of computers and software aids can speed the analysis.

Raw data

The following details are available for Townsville Town Hall:

Required inside air temperature 20 °C
Heating: 5 days/week
Start: 0800 hours
Stop: 1800 hours

Therefore, 2600 hr/Year
217 hr/month

Step 1   Identify site and scope of audit
Step 2   Obtain fuel and electricity bills for a recent representative year.
Step 3   Obtain degree-days.
Step 4   Identify thermal or other changes that have occurred over or since the year under examination.

Table 8.3 Gas supplied

Month	Therms	Degree-days
Jan	14 043	441
Feb	9 780	318
Mar	12 513	345
Apr	7 054	132
May	5 123	132
Jun	3 236	30
Jul	2 764	20
Aug	2 625	43
Sep	4 318	102
Oct	4 999	211
Nov	7 654	234
Dec	10 961	298

Table 8.4 Heating energy delivered

Month	Therms	Degree-days
Jan	11 234	441
Feb	7 824	318
Mar	10 010	345
Apr	5 643	132
May	4 098	132
Jun	2 589	30
Jul	2 211	20
Aug	2 100	43
Sep	3 454	102
Oct	3 999	211
Nov	6 123	234
Dec	8 769	298

Step 5   Examine electrical supply capacity, maximum demand variations and all tariffs.

(Step 5 is not applicable here.)

Gas is supplied as shown in Table 8.3, and the gas boiler has an efficiency of 80 per cent so that the heating energy delivered is as shown in Table 8.4.

Construction of software aids

'Temprise' is a software routine to synthesize daily outside air temperature variations and to chop the data to obtain mean monthly outside air temperatures over the heating period. It has been primed with mean monthly temperatures for the latitude 51.7 °N in the UK. Daily temperature swings are estimated by fitting a sinewave across the months of the year connecting the summer and winter design points' diurnal temperature swings. Representative 24-hour diurnal temperatures for each of the 12 months are provided in a scrolling speedometer display. Sample screendumps are also provided.

'Regress' is a program for linear regression, primed with two simple examples, each using four data pairs.

Both these programs are listed at the end of this chapter.

The energy manager is then armed with an IBM PS/2 personal computer containing the following facilities:

- A wordprocessor (Wordstar)
- A spreadsheet (Supercalc 5)
- The basic programming language
- The basic programs 'temprise' and 'regress'
- A document preparation package (Latex)

Step 6   Feed data to spreadsheet.
Step 7   Convert quantities to common units.

The therms versus degree-day data is fed to the spreadsheet, where it is converted to kWh versus mean monthly temperatures as shown in Table 8.5. The calculations assume a conversion factor 29.3 kWh per therm.

Table 8.5 kWh versus mean monthly temperatures

Month	Therms	Degree-days	Days per month	kWh	Temp. °C
Jan	11 234	441	31	329 156.2	1.274 194
Feb	7 824	318	28	229 243.2	4.142 857
Mar	10 010	345	31	293 293.0	4.370 968
Apr	5 643	132	30	165 339.9	11.1
May	4 098	132	31	120 071.4	11.24 194
Jun	2 589	30	30	75 857.7	14.5
Jul	2 211	20	31	64 782.3	14.85 484
Aug	2 100	43	31	61 530.0	14.11 290
Sep	3 454	102	30	101 202.2	12.1
Oct	3 999	211	31	117 170.7	8.693 548
Nov	6 123	234	30	179 403.9	7.7
Dec	8 769	298	31	256 931.7	5.887 097

Table 8.5 is then modified by the document preparation package to produce the following output:

Townsville Town Hall
   Required inside air temperature 20°C
Heating: 5 days/week
Start: 0800 hours
Stop: 1800 hours

   Therefore, 2600 hr/annum, 217 hr/month

Month	Therms	Degree-days	Days/month	kWh	Temp. °C
Jan	11 234	441	31	329 156.2	1.274 194
Feb	7 824	318	28	229 243.2	4.142 857
Mar	10 010	345	31	293 293.0	4.370 968
Apr	5 643	132	30	165 339.9	11.1
May	4 098	132	31	120 071.4	11.241 94
Jun	2 589	30	30	75 857.7	14.5
Jul	2 211	20	31	64 782.3	14.854 84
Aug	2 100	43	31	61 530.0	14.112 90
Sep	3 454	102	30	101 202.2	12.1
Oct	3 999	211	31	117 170.7	8.693 548
Nov	6 123	234	30	179 403.9	7.7
Dec	8 769	298	31	256 931.7	5.887 097

Step 8   Convert degree-days to mean monthly temperatures during operating hours.
Step 9   Plot kVA versus degree-days.
Step 10   Plot kVA versus mean monthly temperatures during operating hours.

(Steps 9 and 10 are not applicable here.)

The mean monthly temperatures are then modified by the 'temprise' routine. The procedure used was to extract the temperature column from the spreadsheet by outputting its range to a csv (comma separated variables) file, temp. csv:

   1.274 194
   4.142 857
   4.370 968
  11.1
  11.241 94
  14.5
  14.854 84
  14.112 90
  12.1
   8.693 548
  7.7
  5.887 097

and inputting this file to 'temprise'.

```
340 REM MEAN MONTHLY TEMPERATURES
350 MMT(1)= 1.274194
360 MMT(2)= 4.142857
370 MMT(3)= 4.370968
380 MMT(4)= 11.1
390 MMT(5)= 11.24194
400 MMT(6)= 14.5
410 MMT(7)= 14.85484
420 MMT(8)= 14.1129
430 MMT(9)= 12.1
440 MMT(10)=8.693548
450 MMT(11)=7.7
460 MMT(12)=5.887097
```

This replacement was conducted by saving 'temprise' as an ASCII file and then overlying the new temperature data using the wordprocessor. The following subroutine was then added

```
905 OPEN"tempchop"FOR OUTPUT AS #1
906 FOR MON=1 TO 12:PRINT#1,TT(MON):NEXT MON
907 CLOSE #1
910 COLOR 2:SCREEN 0:RUN 20
```

to output the 'chopped' csv file

```
5.547507
8.301605
8.216008
14.51616
14.22885
17.1722
17.41111
16.78305
15.08337
12.10561
11.54149
10.04379
```

to be imported into the document preparation package to produce

Month	kWh	Temp. °C
Jan	329 156.2	5.547 507
Feb	229 243.2	8.301 605
Mar	293 293.0	8.216 008
Apr	165 339.9	14.516 16
May	120 071.4	14.228 85
Jun	75 857.7	17.172 2
Jul	64 782.3	17.411 1
Aug	61 530.0	16.783 05
Sep	101 202.2	15.083 37
Oct	117 170.7	12.105 61
Nov	179 403.9	11.541 49
Dec	256 931.7	10.043 79

Step 11   Correlate with linear regression and quantity scatter.

Step 12   Identify electricity use for heating.

Step 13   Plot kWh of electricity versus degree-days.

Step 14   Quantify electricity use for heating.

Step 15   Plot kWh of electricity versus monthly temperatures during operating hours.

Step 16   Correlate with linear regression and quantify scatter.

Step 17   Check electricity use for heating.

Step 18   Obtain annual mean furnace/boiler efficiency to discount fuel supplied to obtain heating energy.

Step 19   Plot kWh of heating energy versus degree-days.

Step 20   Plot kWh of heating energy (including electricity use for heating) versus monthly temperatures during operating hours.

Step 21   Correlate with linear regression and quantify scatter.

Step 22   Obtain annual mean inside air temperature.

(Steps 12 to 17 are not applicable here. Steps 18 and 22 have been taken earlier.)

The monthly kWh and mean temperatures over the heating hours are then extracted and exported from the spreadsheet to produce the file, kWhtemp.csv:

```
12
5.547507        329156.2
8.301605        229243.2
8.216008        293293
14.51616        165339.9
14.22885        120071.4
17.1722         75857.71
17.41111        64782.3
16.78305        61530
15.08337        101202.2
12.10561        117170.7
11.54149        179403.9
10.04379        256931.7
```

which is then fed to 'regress', modified as follows:

```
90 REM ********************************************************************
100 REM * INPUT DATA                                                     *
110 REM *      nn   x, y pairs                                           *
120 REM ********************************************************************
180 GOSUB 1000

570 GOSUB 2000:GOSUB 620:END

1000 OPEN "kWHTEMP.CSV" FOR INPUT AS #1
1010 INPUT#1,NN:PRINT NN
1015 DIM X(NN),Y(NN), XD(NN),YD(NN)
1020 FOR N = 1 TO NN:INPUT#1, X(N), Y(N):PRINT X(N),Y(N):NEXT N
1030 CLOSE #1
1040 GOSUB 620:CLS:GOSUB 580:RETURN

2000 OPEN "equation.csv" FOR OUTPUT AS #2
2010 PRINT#2, A,B,SEYX,R
2020 CLOSE #2
2030 RETURN
```

Step 23   Obtain base temperature.
Step 24   Quantify sundry gains and direct rejects of energy.
Step 25   Obtain equation relating overall *U*-value to number of air changes, *n*.

The output file, equation.csv:

a	b	SEyx	r
−22295.18	446621.5	27476.24	−.950183

contains the coefficients of the linear equation, the standard error and the coefficient of correlation, which are then imported to the spreadsheet to produce the correlated 'energy signature'.

−22295.2
446621.5
27476.24
−.950183

Month	kWh	kWhpred	Tempchop C
Jan	329 156.2	322 938.8	5.54
Feb	229 243.2	261 535.7	8.30
Mar	293 293.0	263 444.1	8.21
Apr	165 339.9	122 981.1	14.51
May	120 071.4	129 386.7	14.22
Jun	75 857.7	63 764.2	17.17
Jul	64 782.3	58 437.6	17.41
Aug	61 530.0	72 440.3	16.78
Sep	101 202.2	110 335.1	15.08
Oct	117 170.7	176 724.7	12.10
Nov	179 403.9	189 301.9	11.54
Dec	256 931.7	222 693.4	10.04
Total	1 993 982.0	1 993 984.0	12.57
Average	166 165.2		Tmean

The 'energy signature' for the site then appears as follows:

Temp. C	kWhpred	+2SEyx	−2SExyt
0	446 621.5	501 574.0	391 669.0
1	424 326.3	479 278.8	369 373.8
2	402 031.1	456 983.6	347 078.7
3	379 736.0	434 688.4	324 783.5
4	357 440.8	412 393.3	302 488.3
5	335 145.6	390 098.1	280 193.1
6	312 850.4	367 802.9	257 897.9
7	290 555.2	345 507.7	235 602.8
8	268 260.1	323 212.5	213 307.6

9	245 964.9	300 917.4	191 012.4
10	223 669.7	278 622.2	168 717.2
11	201 374.5	256 327.0	146 422.0
12	179 079.3	234 031.8	124 126.9
13	156 784.2	211 736.6	101 831.7
14	134 489.0	189 441.5	79 536.5
15	112 193.8	167 146.3	57 241.3
16	89 898.6	144 851.1	34 946.1
17	67 603.4	122 555.9	12 650.9
18	45 308.2	100 260.7	−9 644.2
19	23 013.1	77 965.7	−31 939.4
20	717.9	55 670.4	−54 234.6
21	−21 577.3	33 375.2	−76 529.8
22	−43 872.5	11 080.0	−98 824.9
23	−66 167.6	−112 15.2	−121 120.0

and this is shown in Fig. 8.1, together with the original data (which is seen to cluster within two standard errors either side of the mean line), the mean annual inside air temperature, the mean annual outside air temperature during the heating periods, and the annual mean monthly predicted kWh.

It is seen that the intercept of the energy characteristic on the y-axis, i.e. when no heating is required, corresponds to the annual mean inside air temperature, and so it is concluded that, for this site, sundry gains are insignificant.

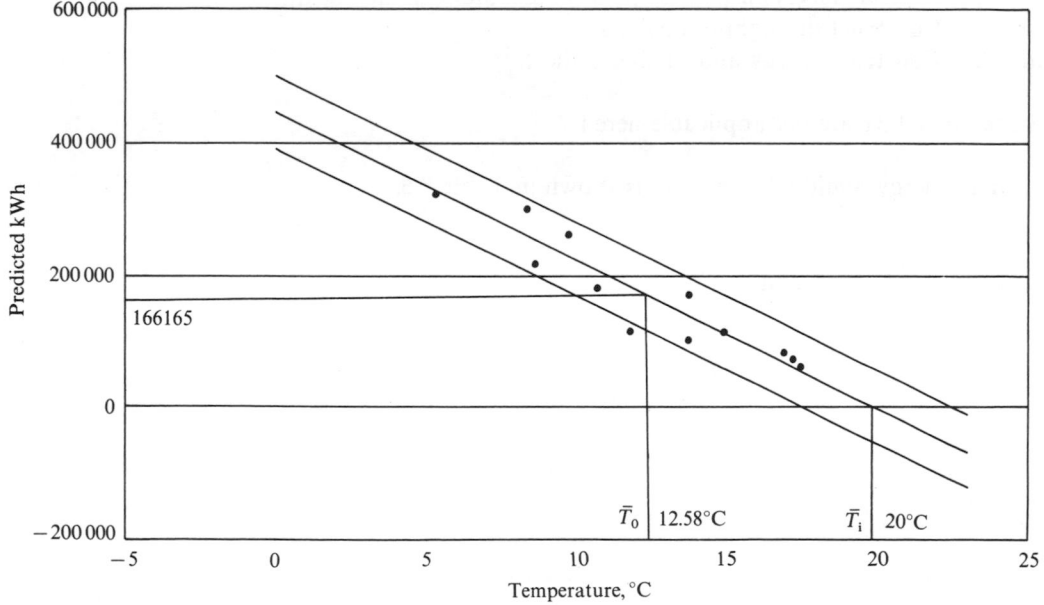

Figure 8.1 Energy signature—Townsville Town Hall.

The annual mean monthly heating energy is thus given by:

$$kWh = -22\,295.2to + 446\,621.5$$
$$= -22\,295.2to + 22\,331.08ti$$
$$166165.2 = 22394(ti\text{-}to)$$

Step 26   External site survey—areas and volumes and capita—UA values.
Step 27   Deduce n.

Therefore $0.217(UA + nV/3) = 22394$
where $0.217 =$ hours/month/1000(W/kW)

$$n = 1.94$$
$$A = 15\,000\,m^2$$
$$U = 1.5\,W\,m^{-2}\,K^{-1}$$
$$V = 125\,000\,m^3$$
$$UA = 22\,500\,W\,K^{-1}$$
$$nV/3 = 80\,698\,m^3\,hr^{-1}$$
$$sum = 103\,198.2$$
$$mean(ti\text{-}to) = 7.42$$
$$kWh/month = 166\,180.7$$
$$kWh/annum = 1\,994\,169.0$$

Step 28   Construct output analysis—ventilation and fabric transmission losses.
Step 29   Investigate other effluents and stack losses.
Step 30   Internal surveys if process plant and equipment significant.
Step 31   Construct throughput analysis.
Step 32   Construct energy audit balance sheet.

(Steps 30 and 31 are not applicable here.)

The energy audit balance sheet is shown in Table 8.6.

Table 8.6 Annual miniaudit

Parameter	kWh	Cost at £0.0136 per kWh, £
Fuel delivered	2 492 711	33 901
Sundry gains	0	0
Total fuel input	2 492 711	33 901
Stack losses	498 092	6 774
Fabric transmission losses	434 738	5 912
Ventilation losses	1 559 221	21 205
Total fuel output	2 492 051	33 891

Table 8.7 Investment options

Option	Installed capital cost, £	Annual savings, £	Unit *SPBP*, (years)
A. Temperature reduction and control	1 000	8000	0.13
B. Local extracts and reduced general air changes	10 000	6000	1.66
C. Recover heat from stack	9 000	2000	4.50
D. Recover heat from exhaust air	35 000	7000	5.00
E. Improved insulation	11 000	2000	5.50
F. Substitute condensing boiler	25 000	3000	8.33

Step 33   Identify options for energy conservation and money-saving options
Step 34   Analysis of options—straight rate of return

The analysis of options is shown in Table 8.7.

All that then remains are the final steps:

Step 35   Produce ranked investment portfolio.
Step 36   Identifying option conflicts.
Step 37   Re-rank investment portfolios.
Step 38   Establish 5-year investment plan.
Step 39   Set up monitoring and targeting and project management systems.

Extracts from this analysis are then fed to the document preparation package to produce the report:

8.4 TOWNSVILLE TOWN HALL ENERGY AUDIT

Input data

Required inside air temperature 20 °C
Heating: 5 days/week
Start: 0800 hours
Stop: 1800 hours
Therefore, 2600 hr per year, 217 hr per month

Report

The report is shown in Tables 8.8 to 8.12.

Table 8.8 Energy, degree-days and temperatures

Month	Therms	Degree-days	Days per month	kWh	Temp. °C
January	11 234	441	31	329 156.2	1.27
February	7 824	318	28	229 243.2	4.14
March	10 010	345	31	293 293.0	4.37
April	5 643	132	30	165 339.9	11.1
May	4 098	132	31	120 071.4	11.24
June	2 589	30	30	75 857.7	14.5
July	2 211	20	31	64 782.3	14.85
August	2 100	43	31	61 530.0	14.0
September	3 454	102	30	101 202.2	12.1
October	3 999	211	31	117 170.7	8.69
November	6 123	234	30	179 403.9	7.7
December	8 769	298	31	256 931.7	5.88

Table 8.9 Energy and mean monthly temperatures over occupied periods. (The report would also contain the diagram shown in Fig. 8.1)

Month	kWh	Temp. °C
January	329 156.2	5.547 507
February	229 243.2	8.301 605
March	293 293.0	8.216 008
April	165 339.9	14.516 16
May	120 071.4	14.228 85
June	75 857.7	17.172 2
July	64 782.3	17.411 11
August	61 530.0	16.783 05
September	101 202.2	15.083 37
October	117 170.7	12.105 61
November	179 403.9	11.541 49
December	256 931.7	10.043 79

Table 8.10 Summary data

Number of air changes, n	1.94/hr
Exposed area, A	15 000 m^2
Overall U-value, U	1.5 W m^{-3} K^{-1}
Internal volume, V	125 000 m^3
UA	22 500 W K^{-1}
$nV/3$	80 698 m^3 per hr
Mean ΔT	7.42 °C
Heating kWh per year	1 994 169.0

Table 8.11 Annual miniaudit

Parameter	kWh	Cost at 0.0136 per kWh
Fuel delivered	2 492 711	33 901
Sundry gains	0	0
Total fuel input	2 492 711	33 901
Stack losses	498 092	6 774
Fabric transmission losses	434 738	5 912
Ventilation losses	1 559 221	21 205
Total fuel output	2 492 051	33 891

Table 8.12 Investment options

Option	Installed capital cost, £	Annual savings, £	Unit SPBP, (years)
A. Temperature reduction and control	1 000	8000	0.13
B. Local extracts and reduced general air changes	10 000	6000	1.66
C. Recover heat from stack	9 000	2000	4.50
D. Recover heat from exhaust air	35 000	7000	5.00
E. Improved insulation	11 000	2000	5.50
F. Substitute condensing boiler	25 000	3000	8.33

8.5 FURTHER SOFTWARE ROUTINES

Discounted cash flow

The BASIC program 'DCF' runs on IBM pcs and compatible machines. It allows the appraisals of loans and investments in the following terms:

Loans

- Monthly repayment
- Number of repayments
- Amount of loan
- Interest charged

Investments

- Lump sum to capital
- Monthly saver
- Net present value

- Rate of return
- Payback period

Monthly Schedules are also constructed.

Optimum insulation thickness wall/pipe

The BASIC program 'OPTIN' runs on IBM pcs and compatible machines. It estimates the optimal thickness of insulation to be applied to a wall or a pipe in the following terms:

- Straight payback period
- Straight rate of return
- Net present value
- True rate of return
- True payback period

No responsibility can be accepted for malfunction of any computer software developed from these listings by the reader.

SOFTWARE LISTINGS

This is the batch file go.bat which enables the suite of software to start up on the command 'go'.

```
@echo off
basic go
```

This is the master menu go.bas

```
10 CLS:COLOR 11:PRINT "SOFTWARE ROUTINES"
20 COLOR 30:LOCATE 4,20:PRINT "1 TEMPRISE"
30 LOCATE 6,20:PRINT "2 REGRESS"
40 LOCATE 8,20:PRINT "3 DCF"
50 LOCATE 10,20:PRINT "4 OPTIN"
60 LOCATE 12,20:PRINT "5 END"
70 COLOR 7:LOCATE 16,20:INPUT "SELECT ROUTINE (1,2,3,4 OR 5)", RT
80 IF RT < > 1 AND RT < > 2 AND RT < > 3 AND RT < > 4 AND RT < > 5 THEN 70
90 IF RT = 5 THEN 1000
100 ON RT GOSUB 200,300,400,500
200 LOAD"TEMPRISE.BAS",R
210 RETURN
300 LOAD"REGRESS.BAS",R
310 RETURN
400 LOAD"DCF.BAS",R
410 RETURN
500 LOAD"OPTIN.BAS",R
510 RETURN
1000 CLS:SYSTEM
```

```
20 REM   ****************************************************************
30 REM   *                                                              *
40 REM   *                      TEMPRISE                                *
50 REM   *                                                              *
60 REM   *         A routine to synthesize outside daily                *
70 REM   *                temperature variations                       *
80 REM   *       and to chop data to obtain monthly mean temperatures   *
90 REM   *              for heating periods less than 24 hours          *
100 REM  *                                                              *
110 REM  ****************************************************************
120 DIM MMT(12),T(12),S(12), TT(12)
130 REM  ****************************************************************
140 REM  *               Synthesize air temperatures C                 *
150 REM  *                                                              *
160 REM  * INPUT DATA                                                   *
170 REM  *            mon mean monthly temperatures                     *
180 REM  *             sd summer design point temperature               *
190 REM  *             wd winter design point temperature               *
200 REM  *                                                              *
210 REM  * Primed values for UK 51.7 N latitude                         *
220 REM  ****************************************************************
230 GOSUB 880
240 PRINT"                                                   "
250 LOCATE 5,20:COLOR 5:PRINT"TEMPRISE        Paul W O'Callaghan"
260 LOCATE 7,6:PRINT "Temperatures"
270 LOCATE 8,6:COLOR 6:PRINT"Month      Mean          Swing        High
Low        Chopped"
280 SD=22:REM summer design point temperature
290 WD=-3:REM winter design point temperature
300 REM MEAN MONTHLY TEMPERATURES
310 MMT(1)=4!
320 MMT(2)=4.5
330 MMT(3)=7!
340 MMT(4)=9!
350 MMT(5)=13!
360 MMT(6)=16!
370 MMT(7)=19!
380 MMT(8)=17!
390 MMT(9)=14!
400 MMT(10)=10!
410 MMT(11)=6.5
420 MMT(12)= 5
430 PRINT "Jan"
440 PRINT "Feb"
450 PRINT "Mar"
460 PRINT "Apr"
470 PRINT "May"
480 PRINT "Jun"
490 PRINT "Jul"
500 PRINT "Aug"
510 PRINT "Sep"
520 PRINT "Oct"
530 PRINT "Nov"
540 PRINT "Dec"
550 COLOR 9
560 SS = SD - MMT(7): REM july swing
570 SW = MMT(1) - WD: REM january swing
580 MS= (SS + SW)/2:REM mean annual swing
590 AMPS= ABS((SS - SW)/2):REM annual swing
600 FOR MON= 1 TO 12
610 S(MON) = MS + AMPS*SIN(2*3.14*((2+MON)/12))
620 LOCATE 8+MON,5:PRINT MON,MMT(MON),S(MON),MMT(MON)+S(MON),MMT(MON)-S(MON)
630 NEXT MON
640 GOSUB 930
650 REM  ****************************************************************
```

```
660 REM *                      Diurnal Swings                                    *
670 REM *********************************************************************
680 COLOR 11
690 FOR MON = 1 TO 12
700 FOR TH =1 TO 24
710 TEMP = MMT(MON)+S(MON)*SIN(2*3.14*(TH-9)/24)
720 LOCATE 21,6:COLOR 6:PRINT"Month        hour            Mean Temp       Swing
    Temp"
730 COLOR 14:LOCATE 22,5:PRINT MON,TH,MMT(MON),S(MON),TEMP
740 NEXT TH:NEXT MON
750 REM *********************************************************************
760 REM *              Chop for Time on/time off                            *
770 REM *********************************************************************
780 COLOR 15
790 GOSUB 930
800 LOCATE 23,6:INPUT"Time on (1 - 24)";ONT
810 LOCATE 23,6:INPUT"Time off(1 - 24)";OFT
820 FOR MON= 1 TO 12
830 GOSUB 980
840 TT(MON) = TEMPSUM/N:LOCATE 8+MON,67:PRINT TT(MON)
850 NEXT MON
860 GOSUB 930
870 COLOR 2:SCREEN 0:RUN 20
880 REM *********************************************************************
890 REM *                  Screen set-up routine                           *
900 REM *********************************************************************
910 COLOR 2:SCREEN 0
920 RETURN
930 REM *********************************************************************
940 REM *                     Paging routine                               *
950 REM *********************************************************************
960 A$=INKEY$:IF A$=""THEN 960
970 RETURN
980 TEMPSUM=0:N=0
990 FOR TH =1 TO 24
1000 IF TH<ONT OR TH>OFT THEN 1040
1010 TEMP = MMT(MON)+S(MON)*SIN(2*3.14*(TH-9)/24)
1020 REM LOCATE 23,5:PRINT TD;TH;TT(MON);S(MON);TEMP
1030 TEMPSUM=TEMPSUM + TEMP:N=N+1
1040 NEXT TH
1050 RETURN
```

```
10 CLS
20 REM *********************************************************************
30 REM                                                                     *
40 REM                          REGRESS                                    *
50 REM                                                                     *
60 REM               A routine for linear regression                       *
70 REM                                                                     *
80 REM *********************************************************************
90 REM *********************************************************************
100 REM * INPUT DATA                                                       *
110 REM *          x, y pairs                                              *
120 REM *********************************************************************
130 GOSUB 580
140 LOCATE 1,20:COLOR 5:PRINT"REGRESS          Paul W O'Callaghan"
150 LOCATE 5,5:COLOR 2:PRINT"number of data pairs"
160 NN=4:LOCATE 5,70:COLOR 3:PRINT"4"
```

```
170 LOCATE 5,68:INPUT NX:IF NX<>0 THEN NN=NX
180 FOR N = 1 TO NN:PRINT N;:INPUT"x,y";X(N),Y(N):NEXT N
190 REM linear regression
200 XS=0:YS=0:XYS=0:XXS=0:YYS=0
210 FOR N = 1 TO NN
220 XS=XS+X(N):YS=YS+Y(N)
230 XYS=XYS+X(N)*Y(N):XXS=XXS+X(N)*X(N):YYS=YYS+Y(N)*Y(N)
240 NEXT N
250 A = (NN*XYS- XS*YS)/(NN*XXS - XS*XS)
260 B = (YS*XXS-XS*XYS)/(NN*XXS-XS*XS)
270 LOCATE 10,30:COLOR 4:PRINT "y = ";A;"x + ";B:COLOR 3
280 REM mean values
290 XM = XS/NN:YM=YS/NN
300 REM sums of deviations
310 XD = 0:YD = 0:XXD=0:YYD=0:XYD=0
320 FOR N = 1 TO NN
330 XD(N)=X(N)-XM:YD(N)=Y(N)-YM
340 XD=XD+XD(N):YD=YD+YD(N)
350 XXD= XXD+XD(N)*XD(N):YYD=YYD+YD(N)*YD(N):XYD=XYD+XD(N)*YD(N)
360 NEXT N
370 XD=XD/NN:YD=YD/NN:XXD=XXD/NN:YYD=YYD/NN:XYD=XYD/NN
380 REM Variances
390 VARX=XXD:VARY=YYD:SIGMAX = SQR(VARX):SIGMAY=SQR(VARY)
400 PRINT"Variance of x-values ";VARX
410 PRINT"Variance of y-values ";VARY
420 PRINT
430 PRINT"Standard Deviation of x-values ";SIGMAX
440 PRINT"Standard Deviation of y-values ";SIGMAY
450 REM Covariance
460 COVXY = XYD
470 PRINT"Covariance of x and y ";COVXY
480 PRINT
490 R = COVXY/SQR(VARX*VARY)
500 PRINT"Product Moment Correlation Coefficient ";R
510 PRINT
520 REM standard errors
530 SEYX = SIGMAY*SQR(1-R*R)
540 SEXY = SIGMAX*SQR(1-R*R):COLOR 4
550 PRINT"Standard Error of the Estimate of y from x " ; SEYX
560 PRINT"Standard Error of the Estimate of x from y " ; SEXY
570 GOSUB 620:END
580 REM ************* ***************************************************
590 REM *                  Screen set-up routine                        *
600 REM ********************************************************************
610 COLOR 2:SCREEN 0:RETURN
620 REM ********************************************************************
630 REM *                  Paging routine                               *
640 REM ********************************************************************
650 A$=INKEY$:IF A$=""THEN 650
660 RETURN
```

```
10 CLS
20 REM ********************************************************************
30 REM *                         DCF                                     *
40 REM *                                                                 *
50 REM *            Appraisals of Loans and Investments                  *
60 REM *                                                                 *
70 REM ********************************************************************
80 REM *                                                n                *
90 REM *          PRESENT VALUE = S(1+R) (1 - [(1+R)/(1+I)] )            *
100 REM *                         -------                                *
110 REM *                          (I-R)                                 *
120 REM *                                                                *
130 REM *         where S = annual savings                               *
140 REM *                R = R = r/100,                                  *
150 REM *                    where r% is the value                       *
160 REM *                         that the resource saved               *
170 REM *                    e.g. energy, water or materials,            *
180 REM *                    is expected to inflate per annum            *
190 REM *                                                                *
200 REM *                                                                *
210 REM *                                                                *
220 REM *             I = i/100,                                         *
230 REM *                 where i% is the interest that might be         *
240 REM *                 earned by investing the money elsewhere        *
250 REM *                                                                *
260 REM *             n (years) is the life of the project              *
270 REM *                     or payback period                         *
280 REM *                                                                *
290 REM *      When C = the Capital Installed Cost of the Project        *
300 REM *                                                                *
310 REM *      NET PRESENT VALUE = PV - C                                *
320 REM *                                                                *
330 REM ********************************************************************
340 GOSUB 1900
350 LOCATE 1,20:COLOR 5:PRINT"DCF            Paul W O'Callaghan"
360 REM ********************************************************************
370 REM * MASTER MENU                                                    *
380 REM********************************************************************
390 LOCATE 3,10:COLOR 12:PRINT"Loans"
400 LOCATE 4,10:COLOR 11:PRINT"Investments"
410 COLOR 2
420 LOCATE 6,10:INPUT"l, i or q(quit)";A$
430 IF A$<>"l" AND A$<>"i" AND A$<>"q" THEN 420
440 IF A$ = "q"THEN CLS:LOAD"go",R
450 IF A$ = "l" THEN GOSUB 480
460 IF A$ = "i" THEN GOSUB 900
470 RUN
480 REM loans
490 COLOR 12:LOCATE 3,40:PRINT"Loans"
500 COLOR 3:LOCATE 4,40:PRINT"Monthly Repayment"
510 LOCATE 5,40:PRINT"Number of Monthly Repayments"
520 LOCATE 6,40:PRINT"Amount of Loan"
```

```
530 LOCATE 7,40:PRINT"Interest Charged"
540 COLOR 2:LOCATE 10,40:INPUT"m, n, a, i or q(quit)";B$
550 IF B$<>"m" AND B$<>"n" AND B$<>"a" AND B$<>"i" AND B$<>"q" THEN 540
560 IF B$ = "q" THEN 620
570 IF B$ = "m" THEN GOSUB 630
580 IF B$ = "n" THEN GOSUB 690
590 IF B$ = "a" THEN GOSUB 750
600 IF B$ = "i" THEN GOSUB 810
610 GOTO 490
620 RETURN
630 REM monthly repayment
640 CLS:COLOR 5:PRINT"Loans: Monthly Repayment":FLAG = 1
650 GOSUB 1990:GOSUB 2110
660 S=PV/FACTOR
670 GOSUB 2140:GOSUB 2330
680 CLS:RETURN
690 REM number of monthly repayments
700 CLS:COLOR 5:PRINT"Loans: Number of monthly repayments":FLAG = 2
710 GOSUB 1990:GOSUB 2110
720 N = (LOG(1/(1-IM*PV/S)))/(LOG(1+IM)):N=INT(N+.5)
730 GOSUB 2140:GOSUB 2330
740 CLS:RETURN
750 REM amount of loan
760 CLS:COLOR 5:PRINT"Loans: Amount of Loan":FLAG = 3
770 GOSUB 1990:GOSUB 2110
780 PV=S*FACTOR
790 GOSUB 2140:GOSUB 2330
800 CLS:RETURN
810 REM interest charged
820 CLS:COLOR 5:PRINT"Loans: Interest Charged":FLAG = 4
830 GOSUB 1990
840 FOR I= 100 TO 0 STEP -.1
845 XX=1+I/100:YY=XX^(1/12):IM=YY-1
846 GOSUB 2110
850 DIFF = PV-S*FACTOR:IF DIFF<=0 THEN 880
860 NEXT I
870 XX=1+IM:YY=XX^(12):ZZ=YY-1:I=ZZ*100
880 GOSUB 2140:GOSUB 2330
890 CLS:RETURN
900 REM investments
910 COLOR 3:LOCATE 3,40:PRINT"Investments"
920 COLOR 11:LOCATE 4,40:PRINT"Lump Sum to Capital"
930 LOCATE 5,40:PRINT"Monthly Saver"
940 LOCATE 6,40:PRINT"Net Present Value"
950 LOCATE 7,40:PRINT"Rate of Return"
960 LOCATE 8,40:PRINT"Payback Period"
970 COLOR 10:LOCATE 10,40:INPUT"l,m,n,r,p,q(quit) ",B$
980 IF B$<>"l" AND B$<>"m" AND B$<> "n" AND B$<>"r" AND B$<>"p" AND B$<>"q"
THEN 970
990 IF B$ = "q" THEN 1060
1000 IF B$ = "l" THEN GOSUB 1070
1010 IF B$ = "m" THEN GOSUB 1170
```

```
1020 IF B$ = "n" THEN GOSUB 1280
1030 IF B$ = "r" THEN GOSUB 1450
1040 IF B$ = "p" THEN GOSUB 1600
1050 CLS:GOTO 900
1060 RETURN
1070 REM lump sum to capital
1080 CLS:COLOR 13:PRINT"Investments:Lump Sum to Capital"
1090 PRINT
1100 INPUT"Capital Sum Invested ", C
1110 INPUT"Number of Years ", NY:N =12*NY
1120 INPUT"Annual Interest Rate ",I
1125 XX=1+I/100:YY=XX^(1/12):IM=YY-1
1130 V = C*(1+IM)^N
1140 COLOR 15:PRINT"Capital Sum Accumulated",V
1150 GOSUB 2380
1160 GOSUB 1940:RETURN
1170 REM mothly saver
1180 CLS:COLOR 13:PRINT"Investments:Monthly Saver"
1190 PRINT
1200 INPUT"Monthly Sum Invested ", S
1210 INPUT"Number of Years ", NY:N =12*NY
1220 INPUT"Annual Interest Rate ",I
1225 XX=1+I/100:YY=XX^(1/12):IM=YY-1
1230 V = S*((1+IM)/IM) * ((1+IM)^N-1)
1240 COLOR 15:PRINT"Capital Sum Accumulated",V
1250 COLOR 15:PRINT"Total Saved", S*N
1260 GOSUB 2530
1270 GOSUB 1940:RETURN
1280 REM net present value
1290 CLS:COLOR 13:PRINT"Net Present Value"
1300 PRINT
1310 INPUT"Annual Sum Saved                      ", S
1320 INPUT"Number of Years                      ", N
1330 INPUT"Annual Interest Rate (%)             ",I:I=I/100
1340 INPUT"Annual Commodity Price Inflation Rate (%) ",R:R=R/100
1350 INPUT"Capital Cost of the Project          ",CC
1360 PV = S*((1+R)/(I-R))*(1-((1+R)/(1+I))^N):NPV = PV-CC
1370 PRINT
1380 COLOR 15:PRINT"Total Savings               ", S*N
1390 PRINT"Present Value                   ", PV
1400 PRINT"Net Present Value               ",NPV
1410 PRINT"Straight Payback Period         ",CC/S
1420 PRINT"Straight Rate of Return         ",(S/CC)*100 "%"
1430 PRINT"Undiscounted Profit at End of Project   ", S*N-CC
1440 GOSUB 1940:RETURN
1450 REM rate of return
1460 CLS:COLOR 13:PRINT"Rate of Return"
1470 PRINT
1480 INPUT"Annual Sum Saved                     ", S
1490 INPUT"Project Life (years)                 ", N
1500 INPUT"Annual Commodity Price Inflation Rate (%) ",R
1510 INPUT"Capital Cost of the Project          ",CC
```

```
1520 FOR I = 100 TO 0 STEP -.1
1530 NPV = CC - S*((1+R)/(I-R))*(1-((1+R)/(1+I))^N):IF NPV <=0 THEN 1550
1540 NEXT I
1550 PRINT:COLOR 15:PRINT"Annual Rate of Return              ",
I
1560 PRINT"Straight Payback Period              ",CC/S
1570 PRINT"Straight Rate of Return              ",(S/CC)*100 "%"
1580 PRINT"Undiscounted Profit at End of Project      ", S*N-CC
1590 GOSUB 1940:RETURN
1600 REM payback period
1610 CLS:COLOR 13:PRINT"Payback Period"
1620 PRINT
1630 INPUT"Annual Sum Saved              ", S
1640 INPUT"Annual Interest Rate (%)              ",I
1650 INPUT"Annual Commodity Price Inflation Rate (%) ",R
1660 INPUT"Capital Cost of the Project              ",CC
1670 FOR N = 1 TO 1000 STEP .01
1680 NPV = CC - S*((1+R)/(I-R))*(1-((1+R)/(1+I))^N):IF NPV <=0 THEN 1710
1690 NEXT N
1700 COLOR 15
1710 COLOR 15:PRINT"Payback Period              ", N*12 "months "N
"years"
1720 PRINT"Straight Payback Period              ",CC/S
1730 PRINT"Straight Rate of Return              ",(S/CC)*100 "%"
1740 GOSUB 1940:RETURN
1750 REM input routine
1760 CLS:PRINT
1770 LOCATE 20,20:INPUT"Amount of Loan";C$:C=VAL(C$)
1780 IF C$="?" THEN 1800
1790 IF C=0 THEN 1770
1800 LOCATE 21,20:INPUT"Period of the Loan(years)";N$:N=VAL(N$)
1810 IF N$="?" THEN 1830
1820 IF N=0 THEN 1800
1830 LOCATE 22,20:INPUT"Annual Repayment"; AR$:AR=VAL(AR$)
1840 IF AR$="?" THEN 1860
1850 IF AR=0 THEN 1830
1860 PRINT C$,N$,AR$, C,N,AR
1870 END
1880 RETURN
1890 REM input routine for investments
1900 REM ****************************************************************
1910 REM *                Screen set-up routine                       *
1920 REM ****************************************************************
1930 COLOR 2:SCREEN 0:RETURN
1940 REM ****************************************************************
1950 REM *                Paging routine                              *
1960 REM ****************************************************************
1970 A$=INKEY$:IF A$=""THEN 1970
1980 RETURN
1990 REM input routine
2000 PRINT
2010 IF FLAG = 1 THEN 2030
```

```
2020 INPUT"Monthly Repayment";S
2030 IF FLAG = 2 THEN 2050
2040 INPUT"Number of Years to Repay";NY:N=12*NY
2050 IF FLAG = 3 THEN 2070
2060 INPUT"Amount of Loan";PV
2070 IF FLAG = 4 THEN 2100
2080 INPUT"Annual Interest Rate (%)",I
2090 XX=1+I/100:YY=XX^(1/12):IM=YY-1
2100 FLAG= 0 :RETURN
2110 REM FACTOR
2120 FACTOR = (1/IM)*(1-(1/(1+IM)^N))
2130 RETURN
2140 COLOR 15
2150 PRINT"Amount of Loan"TAB(25)PV
2160 PRINT"Number of Monthly Repayments"TAB(25)N
2170 PRINT"Number of years to repay"TAB(25)N/12
2180 PRINT"Annual Interest Rate (%)"TAB(25)I
2190 PRINT"Monthly Repayment"TAB(25)S
2200 PRINT"Annual Repayment"TAB(25)S*12
2210 PRINT"Total Repayment"TAB(25)N*S
2220 RETURN
2230 REM repayment schedule
2240 CLS:COLOR 13:PRINT"Repayment Schedule":COLOR 14
2250 PRINT "Month"TAB(20)"Amount Outstanding"TAB(40)"Amount Paid"
2260 AMO = PV
2270 FOR I = 1 TO N
2280 AMO = AMO*(1+IM)-S:AMP=I*S
2290 PRINT I TAB(20) AMO TAB(40) AMP
2300 IF I/12=INT(I/12) THEN COLOR 13:PRINT"Year";INT(I/12):LOCATE 3,1:COLOR
14:GOSUB 1940
2310 NEXT I
2320 GOSUB 1940:CLS:RETURN
2330 PRINT
2340 LOCATE 14,1:INPUT"Repayment Schedule (y or n)"; YN$
2350 IF YN$<>"n" AND YN$<>"y" THEN 2340
2360 IF YN$="y" THEN GOSUB 2230
2370 RETURN
2380 PRINT
2390 LOCATE 14,1:INPUT"Monthly Schedule (y or n)"; YN$
2400 IF YN$<>"n" AND YN$<>"y" THEN 2390
2410 IF YN$="y" THEN GOSUB 2430
2420 RETURN
2430 REM monthly schedule
2440 CLS:COLOR 13:PRINT"Monthly Schedule":COLOR 14
2450 PRINT "Month"TAB(20)"Amount Accumulated"
2460 AMO = C
2470 FOR I = 1 TO N
2480 AMO = AMO*(1+IM)
2490 PRINT I TAB(20) AMO
2500 IF I/12=INT(I/12) THEN COLOR 13:PRINT"Year";INT(I/12):LOCATE 3,1:COLOR
14:GOSUB 1940
2510 NEXT I
```

```
2520 GOSUB 1940:CLS:RETURN
2530 PRINT
2540 LOCATE 14,1:INPUT"Monthly Schedule (y or n)"; YN$
2550 IF YN$<>"n" AND YN$<>"y" THEN 2540
2560 IF YN$="y" THEN GOSUB 2580
2570 RETURN
2580 REM monthly schedule
2590 CLS:COLOR 13:PRINT"Monthly Schedule":COLOR 14
2600 PRINT "Month"TAB(20)"Amount Accumulated"TAB(40)"Amount Saved"
2610 AMO = 0
2620 FOR I = 1 TO N
2630 AMO = AMO*(1+IM)+S
2640 PRINT I TAB(20) AMO TAB(40) I*S
2650 IF I/12=INT(I/12) THEN COLOR 13:PRINT"Year";INT(I/12):LOCATE 3,1:COLOR
14:GOSUB 1940
2660 NEXT I
2670 GOSUB 1940:CLS:RETURN
```

```
10 CLS
20 REM *********************************************************************
30 REM *                        OPTIN                                      *
40 REM *                                                                   *
50 REM *           Optimising Insulation Thickness                        *
60 REM *                                                                   *
70 REM *********************************************************************
80 REM *                                                                   *
90 REM *         U    Initial U-value for a Wall  W/m2K                    *
100 REM *        DT   Annual Mean Temperature Difference                   *
110 REM *             Across the Insulation K                              *
120 REM *        NH   Number of Heating Hours per Annum                    *
130 REM *        CQ   Cost of Heating Energy          Cost per kWh         *
140 REM *                                                                   *
150 REM *        K    Thermal Conductivity of the Insulant W/mK            *
160 REM *        CV   Cost of Insulant per Unit Volume    Cost per m3      *
170 REM *        F    Fixed Cost for Installing Insulant                   *
180 REM *                                                                   *
190 REM *        R    Annual Energy Cost Inflation Rate %                  *
200 REM *        I    Annual Interest Rate %                               *
210 REM *                                                                   *
220 REM *********************************************************************
230 GOSUB 810
240 LOCATE 1,20:COLOR 5:PRINT"OPTIN         Paul W O'Callaghan"
250 REM *********************************************************************
260 REM * MASTER MENU                                                      *
270 REM*********************************************************************
280 LOCATE 3,10:COLOR 12:PRINT"Wall"
290 LOCATE 4,10:COLOR 11:PRINT"Pipe"
300 COLOR 11
310 LOCATE 6,10:INPUT"w, p or q(quit)";A$
```

```
320 IF A$<>"w" AND A$<>"p" AND A$<>"q" THEN 310
330 IF A$ = "q"THEN CLS:END
340 IF A$ = "w" THEN GOSUB 370
350 IF A$ = "p" THEN GOSUB 1600
360 RUN
370 REM wall
380 CLS:COLOR 12:PRINT"Wall":COLOR 11
390 INPUT"Initial U-value, W/m2K                                    ",U
400 INPUT"Annual Mean Temperature Difference Across the Insulation K
",DT
410 INPUT"Number of Heating Hours per Annum                       ",NH
420 INPUT"Cost of Heating Energy per kWh                          ",CQ
430 INPUT"Thermal Conductivity of the Insulant W/mK               ",K
440 INPUT"Fixed Cost for Installing Insulant         Cost/m2      ",F
450 INPUT"Variable Cost of Installing Insulant       Cost/m3      ",CV
460 COLOR 12:PRINT"Optimising Objective"
470 PRINT"A Straight Payback Period/Straight Rate of Return"
480 PRINT"B Net Present Value"
490 PRINT"C Rate of Return"
500 PRINT"D Payback Period"
510 COLOR 11:INPUT "a,b,c,d, or q(quit) ",A$
520 IF A$<>"a" AND A$<>"b" AND A$<>"c" AND A$<>"d" AND A$<>"q" THEN 510
530 IF A$ = "a" THEN GOSUB 590
540 IF A$ = "b" THEN GOSUB 890
550 IF A$ = "c" THEN GOSUB 1080
560 IF A$ = "d" THEN GOSUB 1350
570 IF A$ = "q" THEN CLS:END
580 RUN
590 REM Straight Payback Period
600 COLOR 12:LOCATE 9,1:PRINT "Straight Payback Period"
610 WAS = 9.999999E+37
620 FOR DELTA=.001 TO 1 STEP .001:GOSUB 1520
630 SPBP = CC/S:REM if SPBP>WAS THEN 820
640 WAS=SPBP
650 NEXT DELTA
660 COLOR 14
670 PRINT"Minimum Straight Payback Period = "SPBP" years"
680 PRINT"Maximum Straight Rate of Return = "(1/SPBP)*100" %"
690 PRINT"when Insulation Thickness      = " 1000*DELTA "mm"
700 COLOR 15:PRINT"Cost of Insulation /m2         = " CC
710 PRINT"New U=-Value (W/m2K)           = " U1 "mm"
720 PRINT"Heat Saved per Annum m2        = " Q-Q1" kWh "
730 PRINT"Money Saved per Annum m2       = "CQ*(Q-Q1)
740 GOSUB 840
750 RUN
760 SPBP = CC/S:IF SPBP>WAS THEN 660
770 WAS=SPBP
780 NEXT DELTA
790 COLOR 14
800 PRINT"Minimum Straight Payback Period = "SPBP" years"
810 REM setup
```

```
820 REM ****************************************************************
830 COLOR 2:SCREEN 0:RETURN
840 REM ****************************************************************
850 REM *                    Paging routine                          *
860 REM ****************************************************************
870 A$=INKEY$:IF A$=""THEN 870
880 RETURN
890 REM Net Present Value
900 CLS:COLOR 12:PRINT "Net Present Value"
910 INPUT"Annual Energy Inflation Rate %              ",R:R=R/100
920 INPUT"Annual Interest Rate %                      ",I:I=I/100
930 INPUT"Project Life (years)                        ",NY
940 WAS = 0
950 FOR DELTA=.001 TO 1 STEP .001:GOSUB 1520
960 PV = S*((1+R)/(I-R))*(1-((1+R)/(1+I))^NY):NPV = PV-CC:IF NPV<WAS
THEN 1000
970 WAS=NPV
980 NEXT DELTA
990 RETURN
1000 COLOR 14
1010 PRINT"Maximum Net Present Value     = "NPV
1020 PRINT"when Insulation Thickness     = " 1000*DELTA "mm"
1030 COLOR 15:PRINT"Cost of Insulation /m2       = " CC
1040 PRINT"New U-Value               = " U1 "W/m2K"
1050 PRINT"Heat Saved per Annum m2       = " Q-Q1" kWh "
1060 PRINT"Money Saved per Annum m2      = "CQ*(Q-Q1)
1070 GOSUB 840
1080 REM Rate of return
1090 PRINT"Rate of Return"
1100 INPUT"Annual Energy Inflation Rate %              ",R:R=R/100
1110 INPUT"Project Life (years)                        ",NY
1120 FOR DELTA=.01 TO 1 STEP .01:GOSUB 1520
1130 FOR I=.001 TO 3 STEP .001
1140 PV = S*((1+R)/(I-R))*(1-((1+R)/(1+I))^NY):NPV = PV-CC
1150 CL=CL+1:IF CL>15 THEN CL=1
1160 COLOR CL
1170 LOCATE 18,1:PRINT "
1180 LOCATE 18,1:PRINT I:COLOR 13
1190 PRINT "DELTA ";1000*(DELTA-.01)"mm","Maximum Rate of Return "100*WAS"
%"
1200 IF NPV<=0 THEN 1220
1210 NEXT I
1220 IF I<=WAS THEN 1280
1230 WAS = I
1240 NEXT DELTA
1250 LOCATE 18,1:PRINT "
1260 LOCATE 19,1:PRINT "
1270 LOCATE 18,1
1280 COLOR 14:PRINT"Maximum Rate of Return        = "I*100"%"
1290 PRINT"when Insulation Thickness     = " 1000*DELTA "mm"
1300 COLOR 15:PRINT"Cost of Insulation /m2       = " CC
1310 PRINT"New U-Value               = " U1 "W/m2K"
1320 PRINT"Heat Saved per Annum m2       = " Q-Q1" kWh "
```

```
1330 PRINT"Money Saved per Annum m2          = "CQ*(Q-Q1)
1340 GOSUB 840
1350 REM Payback Period
1360 PRINT"Payback Period"
1370 INPUT"Annual Interest Rate %                        ",I:I=I/100
1380 INPUT"Annual Energy Inflation Rate %                ",R:R=R/100
1390 WAS = 9.999999E+37
1400 FOR DELTA=.001 TO 1 STEP .001:GOSUB 1520
1410 NY=(LOG(1-(CC*(I-R))/(S*(1+R)))/LOG((1+R)/(1+I)))
1420 IF NY<WAS THEN 1440
1430 GOTO 1450
1440 NEXT DELTA
1450 PRINT"Minimum Payback Period "NY" years"
1460 PRINT"when Insulation Thickness    = " 1000*DELTA "mm"
1470 COLOR 15:PRINT"Cost of Insulation /m2        = " CC
1480 PRINT"New U-Value                  = " U1 "W/m2K"
1490 PRINT"Heat Saved per Annum m2      = " Q-Q1" kWh "
1500 PRINT"Money Saved per Annum m2     = "CQ*(Q-Q1)
1510 GOSUB 840
1520 REM Wall Savings
1530 Q=U*DT*NH/1000:REM kWh per annum
1540 U1=K*U/(K+DELTA*U):REM new U-value
1550 Q1=U1*DT*NH/1000:REM new kWh/annum
1560 S=CQ*(Q-Q1):REM savings
1570 CC=DELTA*CV+F
1580 RETURN
1590 CLS:FOR I = 0 TO 15:COLOR I:PRINT I:NEXT I:END
1600 REM pipe
1610 CLS:COLOR 12:PRINT"PIPE":COLOR 12
1620 INPUT"Pipe Outer Diameter mm                        ";DI
1630 DI = DI/1000:RI=DI/2
1640 INPUT"Initial Insulation thickness mm               ";TI
1650 TI = TI/1000:RO=RI+TI
1660 INPUT"Annual Mean Temperature Difference Across the Insulation K
";DT
1670 INPUT"Number of Heating Hours per Annum             ";NH
1680 INPUT"Cost of Heating Energy per kWh                ";CQ
1690 INPUT"Thermal Conductivity of the Insulant W/mK     ";K
1700 INPUT"Fixed Cost for Installing Insulant    Cost/m length  ";F
1710 INPUT"Variable Cost of Installing Insulant  Cost/m3        ";CV
1720 GOSUB 2510
1730 COLOR 13:PRINT"Optimising Objective"
1740 PRINT"A Straight Payback Period/Straight Rate of Return"
1750 PRINT"B Net Present Value"
1760 PRINT"C Rate of Return"
1770 PRINT"D Payback Period"
1780 COLOR 11:INPUT "a,b,c,d, or q(quit) ",A$
1790 IF A$<>"a" AND A$<>"b" AND A$<>"c" AND A$<>"d" AND A$<>"q" THEN
510
1800 IF A$ = "a" THEN GOSUB 1860
1810 IF A$ = "b" THEN GOSUB 2030
1820 IF A$ = "c" THEN GOSUB 2240
```

```
1830 IF A$ = "d" THEN GOSUB 2550
1840 IF A$ = "q" THEN CLS:END
1850 RUN
1860 REM Straight Payback Period
1870 CLS:COLOR 12:PRINT "Straight Payback Period"
1880 WAS = 9.999999E+37
1890 FOR R=.001 TO 1 STEP .001:GOSUB 2740
1900 SPBP = CC/S:LOCATE 3,3:PRINT R,SPBP,WAS:REM if SPBP>WAS THEN 2710
1910 LOCATE 3,3:PRINT"
1920 WAS=SPBP
1930 NEXT R
1940 COLOR 14
1950 PRINT"Minimum Straight Payback Period = "SPBP" years"
1960 PRINT"Maximum Straight Rate of Return = "(1/SPBP)*100" %"
1970 PRINT"when Insulation Thickness       = " 1000*R" mm"
1980 COLOR 15:PRINT"Cost of Insulation /m       = " CC
1990 PRINT"Heat Saved per Annum m2         = " Q-Q1" kWh "
2000 PRINT"Money Saved per Annum m2        = "CQ*(Q-Q1)
2010 GOSUB 840
2020 RUN
2030 REM Net Present Value
2040 CLS:COLOR 12:PRINT "Net Present Value"
2050 INPUT"Annual Energy Inflation Rate %                    ",RE:RE=RE/100
2060 INPUT"Annual Interest Rate %                            ",I:I=I/100
2070 INPUT"Project Life (years)                              ",NY
2080 GOSUB 2510
2090 WAS = -1E-37
2100 FOR R=.001 TO 1 STEP .001:GOSUB 2740
2110 PV = S*((1+RE)/(I-RE))*(1-((1+RE)/(1+I))^NY):NPV = PV-CC
2120 IF NPV<WAS AND NPV>0 THEN 2160
2130 WAS=NPV
2140 NEXT R
2150 RETURN
2160 COLOR 14
2170 PRINT"Maximum Net Present Value      = "NPV
2180 PRINT"when Insulation Thickness      = " R*1000 " mm"
2190 COLOR 15:PRINT"Cost of Insulation /m         = " CC
2200 PRINT"Heat Saved per Annum m2        = " Q-Q1" kWh "
2210 PRINT"Money Saved per Annum m2       = "CQ*(Q-Q1)
2220 GOSUB 840
2230 RUN
2240 REM Rate of return
2250 CLS:PRINT"Rate of Return"
2260 INPUT"Annual Energy Inflation Rate %                    ",RE:RE=RE/100
2270 INPUT"Project Life (years)                              ",NY
2280 GOSUB 2510
2290 FOR R=.01 TO 1 STEP .01:GOSUB 2740
2300 FOR I=.001 TO 3 STEP .001
2310 PV = S*((1+RE)/(I-RE))*(1-((1+RE)/(1+I))^NY):NPV = PV-CC
2320 CL=CL+1:IF CL>15 THEN CL=1
2330 COLOR CL
2340 LOCATE 18,1:PRINT "
```

```
2350 LOCATE 18,1:PRINT I:COLOR 13
2360 PRINT "r ";1000*(R-.01)"mm","Maximum Rate of Return "100*WAS" %"
2370 IF NPV<=0 THEN 2390
2380 NEXT I
2390 IF I<=WAS THEN 2450
2400 WAS = I
2410 NEXT R
2420 LOCATE 18,1:PRINT "
2430 LOCATE 19,1:PRINT "
2440 LOCATE 20,1:PRINT "
2450 LOCATE 18,1
2460 COLOR 14:PRINT"Maximum Rate of Return          = "I*100"%"
2470 PRINT"when Insulation Thickness       = " 1000*R " mm"
2480 COLOR 15:PRINT"Cost of Insulation /m           = " CC
2490 GOSUB 840
2500 RUN
2510 REM Initial Pipe Loss
2520 RR1=LOG(RO/RI)/(2*3.14*K*1):REM per metre length
2530 Q=DT*NH/(RR1*1000):REM kWh per annum
2540 RETURN
2550 REM Payback Period
2560 CLS:PRINT"Payback Period"
2570 INPUT"Annual Interest Rate %                      ",I:I=I/100
2580 INPUT"Annual Energy Inflation Rate %              ",RE:RE=RE/100
2590 GOSUB 2510
2600 WAS = 9.999999E+37
2610 FOR R=.001 TO 1 STEP .001:GOSUB 2740
2620 NY=(LOG(1-(CC*(I-RE))/(S*(1+RE)))/LOG((1+RE)/(1+I)))
2630 IF NY<WAS THEN 2650
2640 GOTO 2660
2650 NEXT R
2660 PRINT"Minimum Payback Period "NY" years"
2670 PRINT"when Insulation Thickness       = " 1000*R " mm"
2680 COLOR 15:PRINT"Cost of Insulation /m           = " CC
2690 PRINT"Heat Saved per Annum m2         = " Q-Q1" kWh "
2700 PRINT"Money Saved per Annum m2        = "CQ*(Q-Q1)
2710 GOSUB 840
2720 RUN
2730 RETURN
2740 REM Pipe Savings
2750 RR2=LOG((RO+R)/RI)/(2*3.14*K*1)
2760 Q1=DT*NH/(RR2*1000):REM kWh per annum
2770 S=CQ*(Q-Q1):REM savings
2780 V=3.14*((RO+R)^2-RO^2)
2790 CC=V*CV+F
2800 RETURN
```

NINE

THE PAST IS OUR FUTURE

9.1 ENERGY AND MATERIALS

Chapter 1 reviewed world fossil fuel reserves and current energy consumptions in various types and geographical sectors. It was shown that, although the lives of the fossil fuels are limited, new finds still tend to increase these lives. The global greenhouse effect was discussed and no firm conclusion as to whether human activities are leading to significant global climate change could be deduced from the data available.

Nevertheless, the needs for ecologically-safe clean local environments make pollution prevention desirable. As most environmental pollution stems, directly or indirectly, from fossil fuel combustion, energy conservation is seen as a priority area for attention. Furthermore, energy savings result in financial savings and also result in reduced environmental pollution.

Chapter 2 developed a systematic procedure for energy management. This involves statistical analyses of monthly fuel and electricity bills leading to the production of an energy audit and report, the identification of cost-effective investment opportunities and an optimal project plan, in which cash savings are reinvested over the project life to produce further, greater, financial returns. Herein lies the key to *profitable* environmental conservation.

Consumerism

In the UK, USA and Europe, well over 25 per cent of all energy and materials are consumed in road and rail building, bridge construction, maintenance, vehicle manufacture, garages, vehicle repairs, petrol and diesel fuelds, and the direct use of electricity.

More suburbs lead to more traffic, more roads, declining public transport, more traffic, distributed manufacturing sites, more traffic, more roads, more suburbs...! Sixty per cent of all air pollution emanates from motor vehicles.

Chapter 3 considered the energy and environmental costs of manufacturing systems and products. Energy is used and pollution produced in the exploration, extraction, supply, conversion and utilization of fossil fuels and raw materials, in transportation, and in the manufacture, advertising, marketing, distribution, retailing and utilization and disposal of products. Conservation-conscious planning and design can reduce energy and materials consumption and concomitant environmental pollution in each of these activities.

Environmental auditing should adopt a 'cradle-to-grave' approach. Environmental ratings and comparison among products should consider the energy and raw materials requirements and the air, water and soil pollution produced by a product in production, distribution, utilization and ultimate disposal to the environment. The environmental impact of a product should begin at the design stage, commencing with the question: *Is this product really necessary?* 'Green' consumerism, acting locally in response to perceived global problems, will reject items considered to be unnecessary or undesirable.

SustainAbility on CEEFAX lists the following ten 'unnecessary' items to avoid purchasing on the basis that if the item is not bought it will not be produced:

- Bottled water and purifiers
- Disposable razors
- Kitchen rolls
- Disposable cameras
- Hairsprays
- Loo blocks
- Air fresheners
- Disposable plates, cups and cutlery
- Individually-wrapped sweets, etc.
- Aerosols

SustainAbility consumer rules include:

- Using mains electricity—not expensive and polluting batteries
- Using gas not electricity
- Showering—not bathing (Jacussi, Sauna, Solarium?)
- Walking, cycling and using public transport—not cars
- Avoiding junk mail and free newspapers

Toxic materials, unnecessary packaging, disposable products, unnecessary products, dangerous products, irreparable items, energy-intensive products and systems, and environmentally polluting items (in manufacture, use or disposal) should be avoided.

Some of the various items entering these categories include clingfilm, aluminium foil, food processors, electric carving knifes, electric can openers, toasted sandwich makers, coffee machines, cosmetics, shower gels, bath foams and oils, hair gels and sprays, home perm kits, plastic furniture, foamed upholstery, dishwashers, detergents and washing powders, bleaches, dry cleaner fluids, cleaning liquids, fly and wasp killers, air fresheners, toilet blocks, descalers, toilet cleaners, bath foams, power tools, batteries, electrical lawn mowers, shredders, strimmers, etc., synthetic fertilizers, artificial insecticides, weed controllers, insect repellents, etc., etc.—the list is endless!

'Throwaway' society

Over 15 per cent by weight and 25 per cent by value of all products purchased in the UK are designed to be thrown away almost immediately! Twenty-five per cent of the plastic produced and 50 per cent of the paper produced is used for packaging, newspapers and plastic cups and cutlery.

Each year, the UK consumer produces 23 million tonnes of domestic rubbish, 100 million tonnes of industrial waste, 130 million tonnes of mineral spoil, 200 million tonnes of agricultural waste and 30 million tonnes of sewage. It costs £650 million to collect, transport and dispose of these discarded materials.

Sixty per cent of domestic refuse might be recoverable, worth £2 billion/per year. This would involve waste sorting and collection, shredding organic materials to produce compost, recycling paper, glass, cans and waste oils. Metals, toxic chemicals, paint, sump oil, organic materials, bottles and paper should thus be separated out prior to disposal.

It is greatly more cost-effective to reuse a product rather than to recycle its constituent materials. Products should be:

Built to last
Built simple
Built modular

Almost every discarded material can be reclaimed, reused, recycled or combusted to produce heat, work, fuels, chemicals or materials. There exist no technical barriers to waste recovery, but the recovery of energy and materials requires the expenditure of energy and materials. Thus recovery projects, albeit stimulated by concerns for environmental conservation, are motivated by economics and social factors.

9.2 ENERGY AND ENVIRONMENTAL CONSERVATION

Rules for the efficient conservation of energy and materials and to avoid environmental pollution are listed throughout this book and Appendix 2 contains a comprehensive checklist for energy managers.

Chapter 4 detailed the fundamental concepts involved: thermophysical transport properties; density, thermal conductivity, specific heat, heat, expansion, work, viscosity, thermodynamics and exergy, temperature, energy conservation, work, efficiency, fluid flow, heat transfer, conduction, convection and radiation, psychrometry, air conditioning, mass transfer and solar energy.

Chapter 5 reviewed the technologies of energy use: fuels and combustion, furnaces and boilers, insulated pipework systems, building heat balance, comfort and climate, heat gains to buildings, heat losses from buildings, energy conservation in buildings, thermal insulation, waste heat recovery, thermal storage, combined heat and power, heat pumps and refrigerators.

Chapter 6 dealt with instrumentation and measurement for the monitoring and control of temperatures, heat fluxes, thermal radiation, psychrometric variables, fluid velocities and flow rates, pressures in fluids. Data collection, computer interfacing, measurement errors, data analyses and presentations were also covered.

Chapter 7 handled economics and financial matters, discounted cash flow, the present value concept for loans and investments, payback periods and rates of return on investment capital, option identification and analyses, investment optimization and conflict correction, the construction of target investment schedules and project management.

A miniaudit, illustrating the use of computational software aids, was constructed in Chapter 8, in which the *thirty-nine steps* for energy management were reviewed.

9.3 BEST INVESTMENTS

The following lists of optimal energy-saving investments have been extracted from over 100 energy audits:

Summary of low cost options (0 to 1 year payback)

Accounts, records, and maintenance

- Management and organization
- Accountability and responsibility
- Good housekeeping
- Good maintenance
- Repair of leaks
- Metering and records
- Records and accessibility of data
- Education of occupants
- Staff awareness campaigns

Temperatures

- Temperature reductions
- Zonal temperature specifications

Heating

- Adjustment of space heating controls
- Heating period reductions
- Zoning space heating
- Shutting down heat to non-occupied rooms and areas
- Use of occupancy sensors
- Use and adjustment of auto-off switches
- Locking frost protection thermostats
- Use and settings of time clocks—reset after power cuts
- Use of thermostats
- Rationalization of heated space
- Installation of reflective foil behind radiators
- Fitting night set-back controls

Electrical heating

- Elimination of electrical heating
- Use of pay-as-you-use meters

Ventilation

- Re-setting ventilation plant time switches
- Installing ventilation plant time switches
- Reduced fan speeds
- Using occupation sensors as controllers for ventilation plant

Boilers

- Modification of stand-by boiler policy
- Use of the more efficient boiler as the lead boiler
- Improved boiler control monitoring
- Reduced steam boiler pressure at night

Hot water

- Reductions in hot water temperatures
- Water heating on night tariffs
- Use of timeclocks for water heaters
- Use of flow restrictors on domestic hot water taps

Insulation and weatherstripping

- Insulation of boiler plant
- Insulation of distribution systems
- Insulation behind radiant panels
- Weatherstripping doors and windows
- Loft insulation
- Fitting automatic door closures

Lighting

- Rearranged light switching
- Elimination of unnecessary lighting
- Illuminance survey to cut unnecessary lighting
- Reducing illumination levels
- Use of lighting controls based upon occupancy sensors
- Use of lighting controls based upon master switching
- Use of task lighting
- Use of lighting switch-off controls using timers
- Decreased lighting periods in common areas

- Use of low-energy lamps
- Loss of glazed areas

Refrigeration systems

- Relocating refrigeration equipment away from heat sources
- Timeswitches for chiller plant

Air compressors

- Ducting outside air to compressors
- Installing maximum demand meters at compressor houses

Kitchens

- Use of time switches
- Minimization of oven pre-heating periods and for the hot storage of food
- Switching of extract or fans when not required
- Maximizing dish washing loads
- Use of auto-off switches for hot water tanks

Controls

- Turning off equipment when not in use
- Use of thermostatic controls
- Optimizing thermostat positions

Electricity

- Changing to night tariffs

Steam

- Flash steam recovery

Summary of medium cost options (1 to 2 year payback)

Accounts, records and maintenance

- Demolishing certain high energy demand low-occupancy buildings

Ventilation

- Use of occupancy sensors for mechanical ventilation control
- Destratification in full height areas
- Installing two-speed fans in air-conditioning plant

Heating

- Use of improved radiator valves
- Use of zonal heating controls
- Moving heaters to ground level

Boilers

- Use of automatic boiler control
- Use of compensators for boilers

Water

- All aspects of water conservation

Hot water

- Installation of showers
- Replacing steam by water heater for domestic hot water
- Installing spray taps for hand washing facilities

Insulation and weatherstripping

- Insulation of hot oil storage tanks
- Insulation of piping, ductwork, flanges and fittings
- Insulation of distribution pipework
- Trench sealing
- Loft insulation
- Insulation of ceilings
- Insulation of calorifiers
- Weatherstripping doors and windows
- Use of automatic door closures
- Insulation of water heaters and storage tanks
- Use of skirts to prevent floor heat losses

Lighting

- Use of low-energy bulbs
- Use of fluorescent lights

Controls

- Optimal heating controls
- Use of thermostatic control
- Use of time switches
- Installation of optimum start/stop controls

Electrical maximum demand

- Improving electrical power factors
- Use of adequate metering

Summary of high cost options (2–3 year payback)

Ventilation

- Heat recovery from extract air
- Recovery of heat/cold from ventilating air

Heating

- Substitution of distributed gas heating

Boilers

- Use of compensators on boilers
- Flue gas heat recovery
- Decentralizing boiler plant

Hot water

- Heat recovery from waste hot water

Insulation and weatherstripping

- Weatherstripping windows and doors
- Roof insulation
- Bulding vestibules
- Installing suspended ceilings
- Cavity wall insulation

Thermal storage

- Introduction of thermal storage
- Installation of storage heaters to replace electrical heaters

Controls

- Use of optimal start/stop controls
- Zonal heating/ventilation control
- Installation of building energy management systems

Combined heat and power

• Use of combined heat and power

Electricity

• Power factor correction

9.4 SUSTAINED REDUCTIONS IN ENERGY USE

In Chapter 2, it was demonstrated that, in order to achieve a renewable energy economy without incurring great social strife, world energy consumption will have to be reduced annually by 1 per cent indefinitely.

In the short term, and in order to conform to the Toronto Protocol, the widespread application of sensible and systematic energy and environmental management procedures and practices, as outlined in this book, will allow the necessary reductions to be made in the short term.

In the medium term, consumer attitudes *vis-à-vis* consumerism versus conservationism are beginning to swing to favour environmental protection.

In the long term, social changes with respect to transportation systems and the work ethic will ensure renewable resource economies, where nations, having very much reduced demands for energy and material resources, are sustained by renewable energy technologies: biomass, solar, wind, waves, tides and hydroelectric power.

FURTHER READING

Allaby, M., *The Survival Handbook*, Pan, London, 1975.
Button, J. (Friends of the Earth), *How to be Green*, Century Hutchinson, London, 1989.
Cipolla, C. M., *The Economic History of World Population*, Pelican, Penguin, 1962.
Elkington, J. and Hailes, J., *The Green Consumer Guide*, Gollancz, London, 1989.
Elkington, J. and Hailes, J., *The Green Consumer's Supermarket Shopping Guide*, Gollancz, London, 1989.
O'Callaghan, P. W., *Design and Management for Energy Conservation*, Pergamon Press, Oxford, 1981.

DATABANK

The energy management data contained in this appendix are presented primarily in tabular form (Tables A1 to A1.40). The tables are generally self-explanatory, but where necessary equations and/or additional data are given in the text.

INTERNATIONAL SYSTEM OF UNITS

Throughout this text SI units (Système International d'Unités) have been used. This system was recommended by the International Organization for Standardization in 1960 and has since been adopted by many countries. The SI system is superior to others in current use

Table A1.1 Basic and supplementary SI units

Quantity	Name of unit	Symbol
Basic:		
Mass	kilogramme	kg
Length	metre	m
Time	second	s
Thermodynamic temperature	degree Kelvin	K
Electric current	ampere	A
Luminous intensity	candela	cd
Supplementary:		
Plane angle	radian	rad
Solid angle	steradian	sr

Table A1.2 Derived SI units

Quantity	Name of unit	Symbol	Basic equivalent units
Force	Newton	N	$\mathrm{kg\,m\,s^{-2}}$
Stored energy	Joule	J	$\mathrm{N\,m = kg\,m^2\,s^{-2}}$
Power	Watt	W	$\mathrm{J\,s^{-1} = kg\,m^2\,s^{-3}}$
Pressure	Pascal	Pa	$\mathrm{N\,m^{-2} = kg\,m^{-1}\,s^{-2}}$

Table A1.3 Examples of SI units for engineering quantities

	Symbol	SI units	Formulae
Shear stress	τ	$(\mathrm{N\,m^{-2}})$ $\mathrm{kg\,m\,s^{-2}\,m^{-2}}$	$c_f\,\rho\,u^2/2$ $(\mathrm{kg\,m^{-3}})(\mathrm{m^2\,s^{-2}})$ $\mathrm{kg\,m^{-1}\,s^{-2}}$
Dynamic viscosity	μ	$\tau/(\mathrm{d}u/\mathrm{d}y)$	$(\mathrm{N\,m^{-2}})/(\mathrm{m\,s^{-1}\,m^{-1}})$ $\mathrm{N\,s\,m^{-2}}$
Kinematic viscosity	v	μ/ρ	$(\mathrm{N\,s\,m^{-2}})(\mathrm{m^3\,kg^{-1}})$ $(\mathrm{kg\,m\,s^{-2}\,s\,m\,kg^{-1}})$ $\mathrm{N\,m^2\,s^{-1}}$
Thermal diffusivity	α	$k/\rho c_\mathrm{p}$	$(\mathrm{W\,m^{-1}\,K^{-1}})(\mathrm{m^3\,kg^{-1}})(\mathrm{kg\,K\,J^{-1}})$ $(\mathrm{J\,s^{-1}\,m^{-1}\,K^{-1}})(\mathrm{m^3\,kg^{-1}})(\mathrm{kg\,K\,J^{-1}})$ $\mathrm{m^2\,s^{-1}}$
Kinetic energy	$m\,v^2/2$		$(\mathrm{kg})(\mathrm{m^2\,s^{-2}})$ $(\mathrm{N\,m^{-1}\,s^{-2}})(\mathrm{m^2\,s^{-2}})$ $\mathrm{N\,m = J}$
Potential energy	$m\,g\,h$		$\mathrm{kg(m\,s^{-2})\,m}$ $(\mathrm{N\,m^{-1}\,s^2})(\mathrm{m\,s^{-2}})\mathrm{m}$ $\mathrm{N\,m = J}$
Thermal energy	$m\,c_\mathrm{p}\,T$		$\mathrm{kg\,(J\,kg^{-1}\,K^{-1})\,K}$ J
Pressure energy	$m\,P/\rho$		$(\mathrm{kg\,N\,m^{-2}})/(\mathrm{kg\,m^{-3}})$ $\mathrm{N\,m = J}$

Table A1.4 Energy equations

Quantity	Potential	+	Kinetic	+	Pressure	+	Thermal	=	Units
Energy	$m\,g\,h$	+	$m\,v^2/2$	+	$m\,P/\rho$	+	$m\,c_\mathrm{p}\,T$	=	Joules
Head/mass	h	+	$v^2/2g$	+	$P/\rho\,g$	+	$c_\mathrm{p}\,T/g$	=	m
Pressure head	$\rho\,gh$	+	$\rho\,v^2/2$	+	P	+	$\rho c_\mathrm{p}\,T$	=	$\mathrm{N\,m^{-2}}$
Temperature	gh/c_p	+	$v^2/2c_\mathrm{p}$	+	$P/\rho\,c_\mathrm{p}$	+	T	=	K

because it is a completely *coherent* system; the product or quotient of any two quantities leading to a unit of the resultant quantity with no need for multiplying factors.

As well as the benefits inherent in uniformity, the system has the additional advantage that analogies among different processes are not obscured by the use of different units. Thus, for example, amounts of energy associated with thermal, electrical, chemical, mechanical and other processes may be expressed in terms of a common unit: the Joule for stored energy and the Watt for power, or energy in transit.

The International System of Units is founded upon six basic units and two supplementary units shown in Table A1.1.

One grammè is the mass of 1 cc of water at $0\,°C$.

One metre was originally one ten-millionth part of the distance from the north pole through the equator through Paris, France. It was redefined in 1960 as the length equal to $1\,650\,763.73$ wavelengths in vacuo of the radiation corresponding to the transition between the levels $2p_{10}$ and $5d_5$ of the isotope 86/36 Kr.

Each other unit is built up from the appropriate basic units and so may be expressed in terms of the basic units, even though some special names and symbols are also recommended for some of the important derived units (Table A1.2).

The weight of a mass M kg is a force of $M\,g$ Newtons, where g is the local value of the acceleration due to gravity (≈ 9.81 m s^{-2}).

A pressure of 1 bar $\approx 10^5$ N m^{-2}.

Because some SI units are of inconvenient size, multiplying prefixes are available (Table A1.5). These prefixes are printed immediately adjacent to the unit symbols with which they are associated; they then become part of the symbol; i.e.

$$1\,\text{cm}^2 \equiv (10^{-2}\,\text{m})^2$$
$$1\,\text{MN m}^{-2} \equiv \text{M}(\text{N m}^{-2})$$
$$1\,\text{GJ m}^{-3} \equiv 10^9\,\text{J m}^{-3}$$

Table A1.6 details SI units for all other quantities used in the text.

Table A1.5 Recommended multiplying prefixes

Value	Name	
10^{15}	petra	P
10^{12}	tera	T
10^9	giga	G
10^6	mega	M
10^3	kilo	k
10^{-3}	milli	m
10^{-6}	micro	μ
10^{-9}	nano	n
10^{-12}	pico	p
10^{-15}	femto	f
10^{-18}	atto	a

Table A1.6 Nomenclature and definitions

A	Area		m^2
c	Specific heat		$J\,kg^{-1}\,K^{-1}$
c_f	Skin friction coefficient	$\tau/(\rho u^2/2)$	
c_p	Specific heat at constant pressure		$J\,kg^{-1}\,K^{-1}$
c_v	Specific heat at constant volume		$J\,kg^{-1}\,K^{-1}$
D	Diameter		m
F	Force		N
f	Fanning friction factor	$\Delta P/(L\rho u^2/2D)$ $=4\,c_f$ for a pipe	
G	Mass flow velocity	ρu	$kg\,m^{-2}\,s^{-1}$
h	Heat transfer coefficient		$W\,m^{-2}\,K^{-1}$
h_L	Overall average value of h over length L		$W\,m^{-2}\,K^{-1}$
h_x	Local value of h at x		$W\,m^{-2}\,K^{-1}$
k	Thermal conductivity		$W\,m^{-1}\,K^{-1}$
L	Length		m
m	Mass		kg
P	Pressure		$N\,m^{-2}$
Q, q	Heat content, heat flow rate	$m\,c\,T, \dot{m}c\,T$	J, W
T	Temperature		K
u	Velocity in the x-direction		$m\,s^{-1}$
v	Velocity in the y-direction		$m\,s^{-1}$
x, y	Cartesian coordinates		
α	Thermal diffusivity	$k/\rho c_p$	$m^2\,s^{-1}$
Δ	difference between (as prefix)		
δ	Thickness		
ε_m	Eddy viscosity		$m^2\,s^{-1}$
ε_h	Eddy diffusivity		$m^2\,s^{-1}$
γ	Specific humidity	kg_{water}/kg_{dryair}	
μ	Dynamic viscosity	$\tau = \mu\,du/dy$	$kg\,m^{-1}\,s^{-1}$
ρ	Density		$kg\,m^{-3}$
η	Kinematic viscosity	$=\mu/\rho$	$m^2\,s^{-1}$
ϕ	Relative humidity		%
τ	Shear stress		$N\,m^{-2}$

Other nomenclature is defined as it occurs in the text.

Table A1.7 Dimensionless groups

Bi	Biot modulus	Internal resistance/external resistance hL/k_{solid}
Ec	Eckert number	Gr/Re^2
Fo	Fourier modulus	Time dependence $\alpha t/L^3$
Gr	Grashof number	Buoyancy forces/viscous forces $\rho^2 g\beta(T-T_o)L^3/\mu^2$
j	Colburn 'j' factor	$St\,Pr^{2/3}$
Nu	Nusselt number	Conductive resistance/convective resistance hL/k_{fluid}

(*Contd.*)

Table A1.7 (*Continued*)

Pe	Peclet number	$Re\ Pr$
Pr	Prandtl number	Viscous transfer/heat transfer
		$v/\alpha = (\mu/\rho)(pc_p/k_{fluid}) = \mu c_p/k$
Pr_t	for turbulent flow	$\varepsilon_m/\varepsilon_h =$ approx. unity
Re	Reynolds number	Inertia forces/viscous forces
		$\rho u L/\mu = uL/v$
St	Stanton number	$Nu/Pe = Nu/Re\ Pr = h/\rho u c_p$

Table A1.8 Useful values in SI Units

Mean molecular weight of dry air	M_a	28.97	kg
Mean molecular weight of steam	M_v	18.02	kg
Density of dry air at 15.5 °C, $\phi = 60\%$, $P = 10^5\ N/m^2$	ρ_a	1.22	$kg\,m^{-3}$
Density of liquid water at 15.5 °C	ρ_{water}	1000	$kg\,m^{-3}$
Universal gas constant	$R_o(=MR_{gas})$	8314.3	$J\,kmol^{-1}\,K^{-1}$
Volume of 1 kmol of the permanent gases at 1.01325 bar and 0 °C	R_oT/P $8314 \times$ $273/101325$	22.4136	m^3
Characteristic gas constant for dry air	R_a	287	$J\,kg^{-1}\,K^{-1}$
Characteristic gas constant for steam	R_v	462	$J\,kg^{-1}\,K^{-1}$
Mean specific heat for air at room temperature at constant pressure	c_{pa}	1005	$J\,kg^{-1}\,K^{-1}$
Mean specific heat for air at room temperature at constant volume	c_{va}	718	$J\,kg^{-1}\,K^{-1}$
Mean specific heat for steam at room temperature at constant pressure	c_{pv}	4210	$J\,kg^{-1}\,K^{-1}$
Mean specific heat for steam at room temperature at constant volume	c_{vv}	1810	$J\,kg^{-1}\,K^{-1}$
Adiabatic index for air at room temperature and pressure	γ	1.4	
Latent heat of steam at 0 °C	h_{fg}	2500	$kJ\,kg^{-1}$

Table A1.9 Conversion factors (approximate)

Length	1 m	6.215×10^{-4} miles
		1.0936 yards
		3.281 ft
		39.372 in
		10^{10} Angstroms
Area	1 m^2	3.86×10^{-7} sq. miles
		2.47×10^{-4} acres
		1×10^{-4} hectares
		1.19 sq. yards
		10.76 sq. ft
		1565 sq. in
Volume	1 m^3	1.308 cu. yards
		35.32 cu. ft
		6.1×10^4 cu. in
		6.27 barrels
		220 gallons (imp.)
		264 gallons (US)
		1000 litres
		3.519×10^4 fl. oz
		1759 pints
Velocity	1 m s^{-1}	3.28 ft s^{-1}
		39.36 in s^{-1}
		6.215×10^{-4} miles s^{-1}
		197 ft min^{-1}
		2362 in min^{-1}
		3.73×10^{-2} miles per min
		11810 ft hr^{-1}
		1.42×10^5 in hr^{-1}
		2.24 miles per hour
Acceleration	1 m s^{-2}	3.28 ft s^{-2}
Mass	1 kg	9.84×10^{-4} ton (imp)
		11×10^{-4} ton (US)
		1×10^{-3} tonne
		1.96×10^{-2} cwt
		2.204 lb
Mass rate of flow	1 kg s^{-1}	2.2 lb s^{-1}
		9.8×10^{-4} ton (imp) s^{-1}
		133 lb min^{-1}
		5.88×10^{-2} ton min^{-1}
		7980 lb hr^{-1}
		3.53 ton hr^{-1}
Volume rate of flow	1 m^3 s^{-1}	35.3 cu.ft s^{-1} (cu.secs)
		220 gall s^{-1}
		1000 l s^{-1}
		2119 ft^3 min^{-1}
		13200 gall min^{-1}
		6×10^4 l min^{-1}
		1.27×10^5 ft^3 hr^{-1}
		7.92×10^{-5} gall hr^{-1}
		3.6×10^6 l hr^{-1}

(Contd.)

Table A1.9 (*Continued*)

Fuel consumption	$1 \, m^3 \, m^{-1}$	$220 \, gall \, m^{-1}$
		$1000 \, l \, m^{-1}$
		$2.2 \times 10^5 \, gall \, km^{-1}$
		$10^6 \, l \, km^{-1}$
		3.53×10^5 gallons per mile
Density	$1 \, kg \, m^{-3}$	$6.24 \times 10^{-2} \, lb \, ft^{-3}$
		1×10^{-2} lb per gallon
		$2.2 \times 10^{-3} \, lb \, l^{-1}$
Momentum	$1 \, kg \, m \, s^{-1}$	$7.23 \, lb \, ft \, s^{-1}$
Force	$1 \, N$	$1 \, kg \, m \, s^{-2}$
		$2.23 \times 10^{-1} \, lbf$
		$0.103 \, kgf$
		$10^5 \, dynes$
		$7.23 \, poundals$
		$10^{-4} \, tonf$
Pressure, stress	$1 \, N \, m^{-2}$	$1.45 \times 10^{-4} \, lbf \, in^{-2}$
	$1 \, Pa$	$6.47 \times 10^{-8} \, ton \, in^{-2}$
		$2.08 \times 10^{-2} \, lbf \, ft^{-2}$
		$9.32 \times 10^{-6} \, ton \, ft^{-2}$
		$0.01 \, mbar$
		$10^{-5} \, bar$
		$3.35 \times 10^{-4} \, ft \, water$
		2.95×10^{-4} in mercury
		7.5×10^{-3} mm mercury (torr)
		9.87×10^{-6} atmospheres
Energy, work, heat	$1 \, J$	$9.48 \times 10^{-9} \, therms$
	$1 \, Ws$	$2.78 \times 10^{-7} \, kWh$
	$1 \, Nm$	$3.73 \times 10^{-7} \, hp \, hr$
	$1 \, kg \, m^2 \, s^{-1}$	$9.48 \times 10^{-4} \, Btu$
		$0.738 \, ft \, lb$
		$0.239 \, cal$
		$10^7 \, erg$
		$6.242 \times 10^{18} \, eV$
Useful values	1 tce	$2.88 \times 10^4 \, MJ$
	1 tonne	
	brown coal	$8 \times 10^3 \, MJ$
	1 toe	$4.54 \times 10^4 \, MJ$
	$1 \, m^3$	$0.384 \times 10^2 \, MJ$
	1 barrel oil	$6.3 \times 10^3 \, MJ$
	1 tonne	
	uranium	$8.1 \times 10^{10} \, MJ$
Power, heat flow rate	$1 \, W$	$1.34 \times 10^{-3} \, hp$
	$1 \, J \, s^{-1}$	$0.738 \, ft \, lbf \, s^{-1}$
	$1 \, kg \, m^2 \, s^{-3}$	$44.25 \, ft \, lbf \, min^{-1}$
		$3.412 \, Btu \, hr^{-1}$
		$5.69 \times 10^{-2} \, Btu \, min^{-1}$
		$0.86 \, kcal \, hr^{-1}$
		$6.242 \times 10^{18} \, eV \, s^{-1}$

Table A1.9 (*Continued*)

Useful values	1 kWh	3.6 MJ
	1 kW day	24 kWh
		86.4 MJ
	1 kW week	168 kWh
		605 MJ
	1 kW year	8760 kWh
		31449 MJ
	1 MJ year	3.17×10^{-5} kW
Specific energy, specific latent heat	1 J per kg	4.3×10^{-4} Btu lb^{-1}
		2.388 kcal kg^{-1}
Calorific value volume basis	1 J m^{-3}	4.31×10^{-11} therm per gall
		2.68×10^{-5} Btu ft^{-3}
		4.31×10^{-6} Btu/gall
Specific heat capacity	1 J kg^{-1} K^{-1}	2.39×10^{-4} Btu lb^{-1} F^{-1}
Specific entropy	1 J m^{-3} K^{-1}	1.49×10^{-2} Btu ft^{-3} F^{-1}
Thermal conductance Heat transfer coefficient	1 W m^{-2} K^{-1}	0.176 Btu ft^{-2} hr^{-1} F^{-1}
		0.86 kcal m^{-2} hr^{-1} C^{-1}
Thermal conductivity	1 W m^{-1} K^{-1}	0.578 Btu hr^{-1} ft^{-1} F^{-1}
		6.93 Btu in ft^{-1} hr^{-1} F^{-1}
Combustion intensity	1 W m^{-3}	9.66×10^{-2} Btu ft^{-3} hr^{-1}
		0.86 kcal m^{-3} hr^{-1}
Temperature	°C	$5/9 \times (F - 32)$
Dynamic viscosity	1 kg m^{-1} s^{-1}	10 poise
		2419 lb ft^{-1} hr^{-1}
		0.02089 lbf s ft^{-2}

tce, tonne coal equivalent.
toe, tonne coal equivalent.

Table A1.10 Properties of some commonly-used metals at 20 °C

Metal	Density ρ kg m^{-3}	Thermal conductivity k W m^{-1} K^{-1}	Specific heat c_p J kg^{-1} K^{-1}	Volumetric heat capacity ρc_p 10^6 J m^{-3} K^{-1}	Thermal diffusivity α 10^{-6} m^2 s^{-1}
Aluminium	2790	164	883	2.46	66.6
Brass	8620	111	385	3.28	33.8
Cast Iron (4% C)	7270	52	419	3.05	17.6
Copper	8690	386	389	3.48	110.9
Magnetite Fe_2O_3	5177	1.9	752	3.85	0.5
Steel (1% C)	7800	43	473	3.69	11.7

Table A1.11 Average properties of some common non-metallic solids at 20 °C

Material	Density ρ kg m^{-3}	Thermal conductivity k W m^{-1} K^{-1}	Specific heat c_p J kg^{-1} K^{-1}	Volumetric heat capacity ρc_p 10^6 J m^{-3} K^{-1}	Thermal diffusivity α 10^{-6} m^2 s^{-1}
Ash	720	0.1	—	—	—
Brick (dry)	1785	0.45	837	1.49	0.3
9% moisture	1892	0.8	865	1.63	0.54
Corrugated cardboard	105	0.047	—	—	—
Cement	1700	0.8	—	—	—
Clay	1458	1.28	879	1.28	1.0
Concrete	2110	1.1	897	1.89	0.62
Stone	2304	0.93	837	1.93	0.48
10% moisture	2240	1.21	1172	2.63	0.46
Cotton	80	0.06	1300	0.1	0.59
cotton wool	80	0.04	—	—	—
Corkboard	160	0.043	2000	0.32	0.13
Diatomaceous earth	320	0.062	879	0.28	0.22
Coarse gravelly earth	2050	0.52	1840	3.77	0.14
Fibre insulating board	237	0.048	—	—	—
Fibreglass	10–150	0.04	—	—	—
Glass plate	2710	0.762	837	2.27	0.34
Glass wool	200	0.04	670	0.13	0.31
Granite	—	2.8	—	—	—
Ice (0 °C)	913	2.21	1930	1.76	1.26
Kapok	20	0.035	—	—	—
Marble	2600	2.77	808	2.1	1.32
Mineral wool	150	0.038	—	—	—
Mud	1840	0.43	—	—	—
Paper	—	0.128	—	—	—
Plaster	400	0.1	—	—	—
Expanded polystyrene	25	0.034	—	—	—
Rockwool	128	0.029	—	—	—
Roofing felt	960	0.19	—	—	—
Dry sand	1520	0.35	—	—	—
10% moisture	1600	0.38	—	—	—
Sandstone	2200	1.85	712	1.56	1.19
Sawdust	192	0.05	—	—	—
Slate	—	1.46	—	—	—
Snow (0 °C)	555	0.46	—	—	—
Dry soil	—	0.85	1841	—	—
Thatch	240	0.07	—	—	—
Urea formaldehyde foam	8	0.038	—	—	—
Wallboard	237	0.048	—	—	—
Wood–oak	700	0.19	239	0.17	1.12
Wood–Fir	419	0.19	272	0.11	1.28
Wool	110	0.036	—	—	—

Table A1.12 Properties of liquid water at 0.010 13 MN m^{-2}

Temperature T °C	Density ρ kg m^{-3}	Thermal conductivity k W m^{-1} K^{-1}	Specific heat c_p J kg^{-1} K^{-1}	Volumetric heat capacity ρc_p 10^6 J m^{-3} K^{-1}	Thermal diffusivity α 10^{-6} m^2 s^{-1}	Dynamic viscosity μ 10^{-3} kg m^{-1} s^{-1}	Kinematic viscosity ν 10^{-6} m^2 s^{-1}	Coefficient of volumetric expansion β 10^{-3} K^{-1}	Prandtl number Pr
0	1000	0.569	4217	4.217	0.134	1.755	1.755	−0.06	13.02
20	998	0.603	4182	4.173	0.145	1.002	1.004	0.207	6.95
40	992	0.632	4179	4.146	0.152	0.651	0.656	0.385	4.31
60	983	0.653	4185	4.113	0.159	0.462	0.476	0.523	2.96
80	972	0.670	4197	4.079	0.164	0.350	0.360	0.641	2.19
100	958	0.681	4206	4.029	0.169	0.278	0.290	0.750	1.72

Table A1.13 Properties of saturated water vapour at 0.1013 MN m^{-2}

Temperature T °C	Density ρ kg m^{-3}	Thermal conductivity k W m^{-1} K^{-1}	Specific heat c_p J kg^{-1} K^{-1}	Volumetric heat capacity ρc_p 10^5 J m^{-3} K^{-1}	Thermal diffusivity α 10^{-6} m^2 s^{-1}	Dynamic viscosity μ 10^{-3} kg m^{-1} s^{-1}	Kinematic viscosity ν 10^{-6} m^2 s^{-1}	Coefficient of volumetric expansion β K^{-1}	Prandtl number Pr	Saturation pressure P kN m^{-2}
0	0.004 85	0.017 3	1854	8.99	1900	0.008 8	1800	0.0036	0.942	0.61
20	0.017 3	0.019 1	1866	32.28	590	0.009 4	540	0.0034	0.918	2.34
40	0.015 3	0.0204	1885	96.70	210	0.0101	196	0.0032	0.930	7.38
60	0.1300	0.0217	1915	248.95	87	0.0107	82	0.0030	0.947	19.92
80	0.2930	0.0231	1962	574.87	40	0.0114	38	0.0028	0.966	47.36
100	0.598	0.0249	2028	1212.74	21	0.0121	20.2	0.0027	0.986	101.30

Table A1.14 Properties of dry air at 0.1013 MN m^{-2}

Temperature T °C	Density ρ kg m^{-3}	Thermal conductivity k W m^{-1} K^{-1}	Specific heat c_p J kg^{-1} K^{-1}	Volumetric heat capacity ρc_p 10^6 J m^{-3} K^{-1}	Thermal diffusivity α 10^{-6} m^2 s^{-1}	Dynamic viscosity μ 10^{-3} kg m^{-1} s^{-1}	Kinematic viscosity ν 10^{-6} m^2 s^{-1}	Coefficient of volumetric expansion β K^{-1}	Prandtl number Pr
−20	1.400	0.0223	1003.1	1404	15.8	0.0160	11.4	0.0039	0.720
−10	1.341	0.0213	1003.4	1345	15.8	0.0165	12.6	0.0038	0.716
0	1.286	0.0242	1003.8	1291	18.7	0.0170	14.9	0.0036	0.713
20	1.203	0.0250	1004.4	1208	20.6	0.0180	16.8	0.0034	0.710
40	1.130	0.0272	1005.6	1136	23.9	0.0190	18.4	0.0032	0.704
60	1.088	0.0290	1007.0	1096	26.4	0.0200	21.0	0.0030	0.697
80	1.000	0.0300	1008.0	1008	29.7	0.0210	23.3	0.0028	0.696
100	0.942	0.0318	1010.0	951	33.4	0.0220	23.35	0.0027	0.692

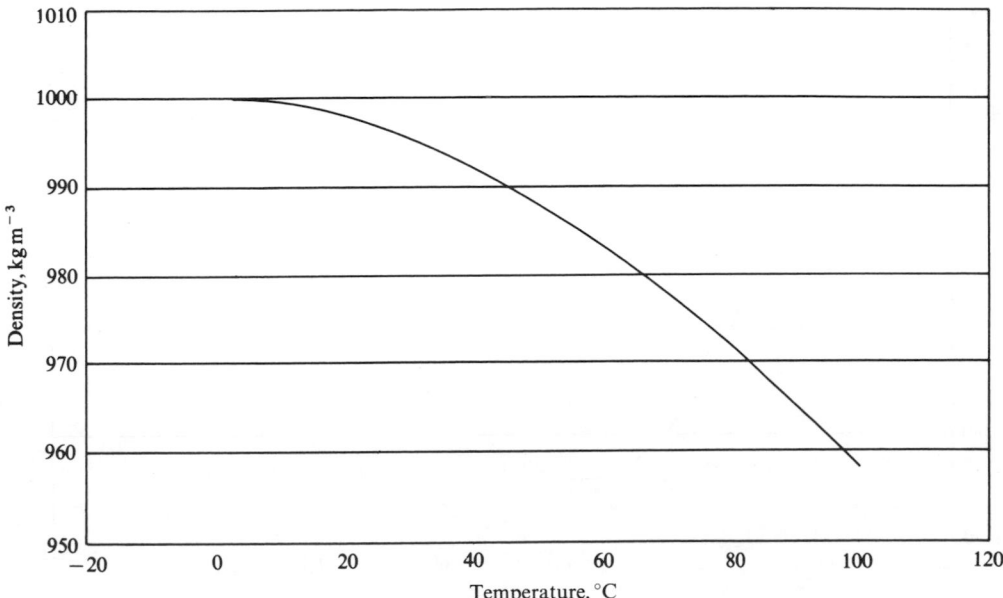

Figure A1.1 Density of liquid water at 0.1013 MN m^{-2}.

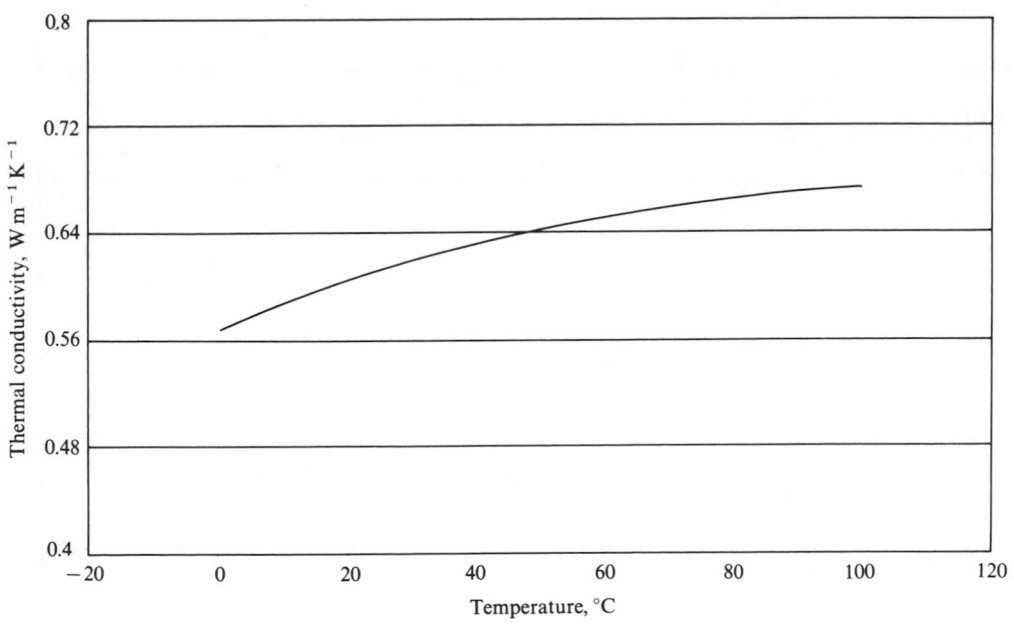

Figure A1.2 Thermal conductivity of liquid water at 0.1013 MN m^{-2}.

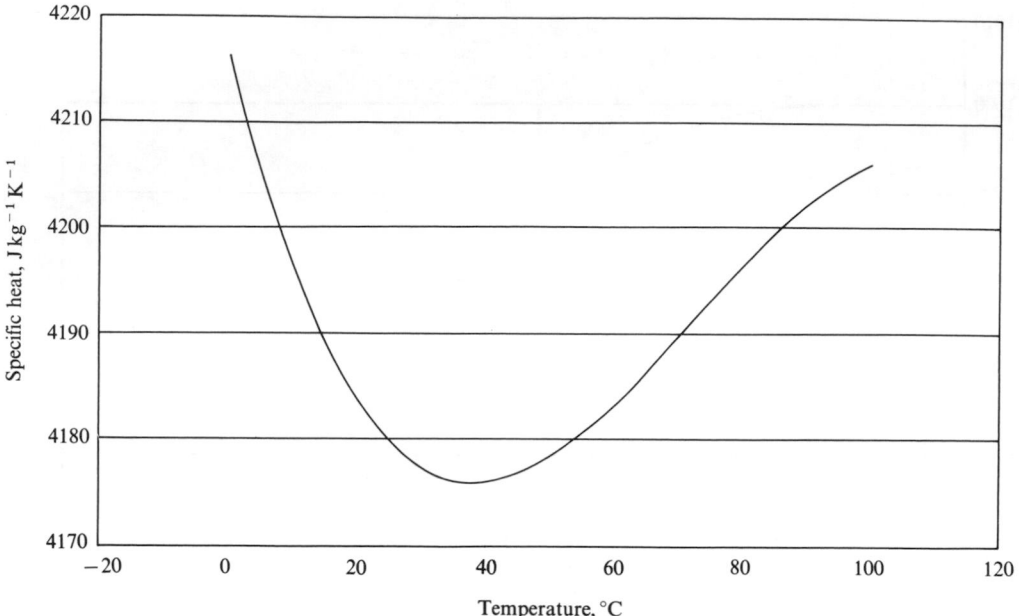

Figure A1.3 Specific heat of liquid water at 0.1013 MN m^{-2}.

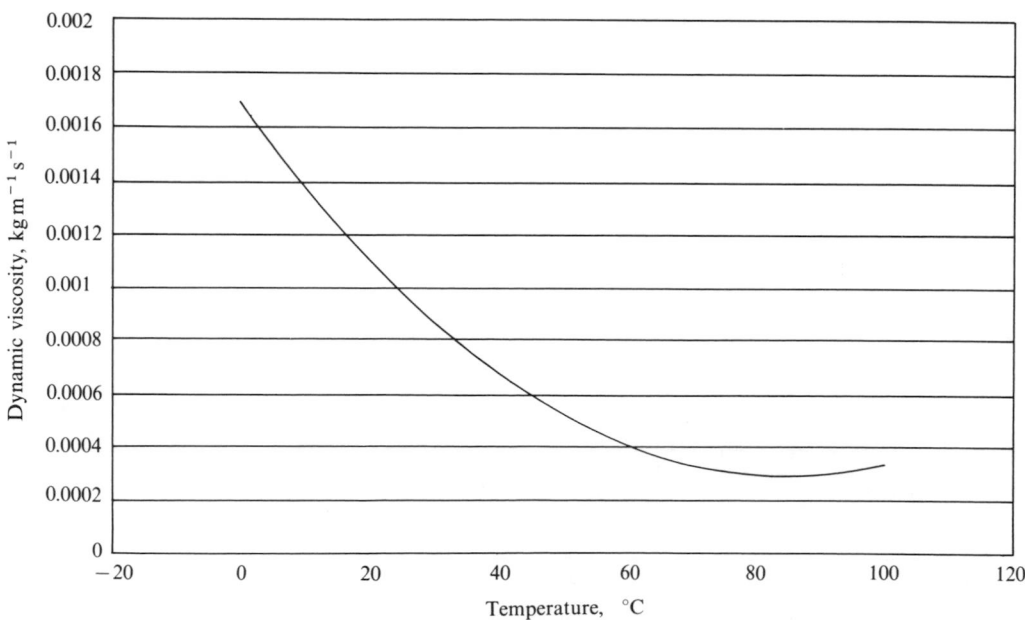

Figure A1.4 Dynamic viscosity of liquid water at 0.1013 MN m^{-2}.

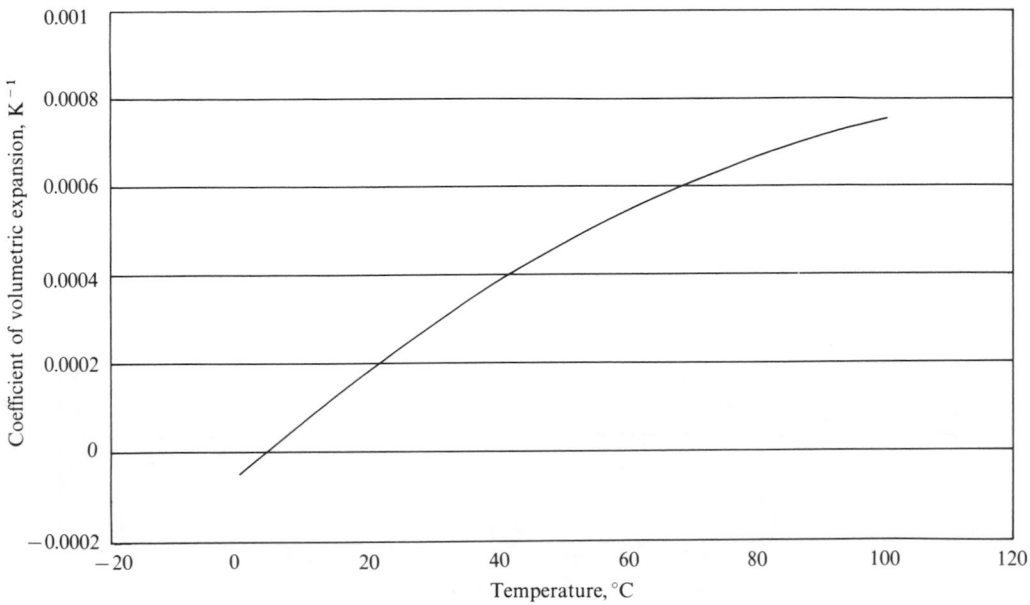

Figure A1.5 Coefficient of volumetric expansion of liquid water at $0.1013\ \text{MN m}^{-2}$.

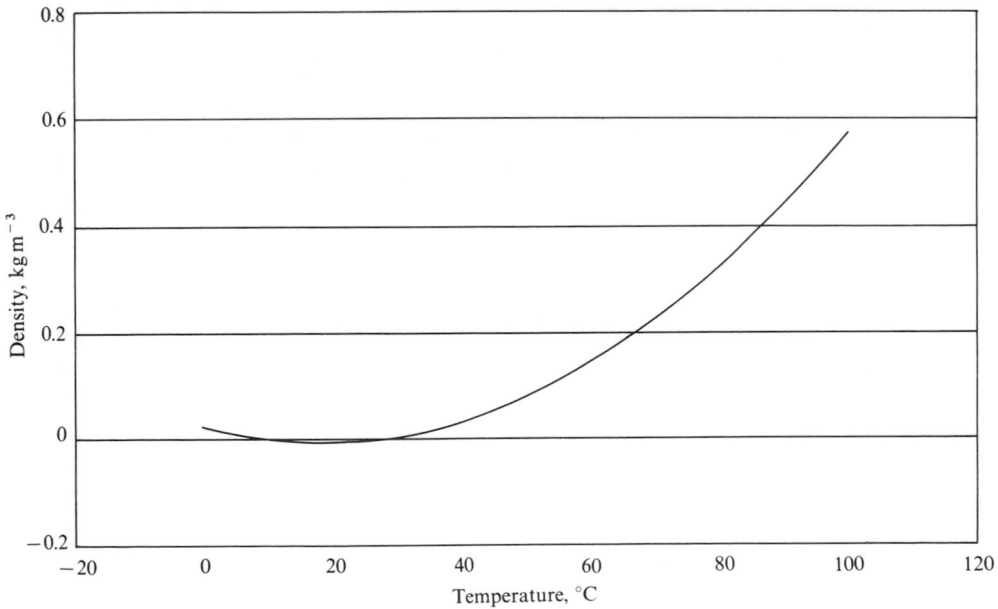

Figure A1.6 Density of water vapour at $0.1013\ \text{MN m}^{-2}$.

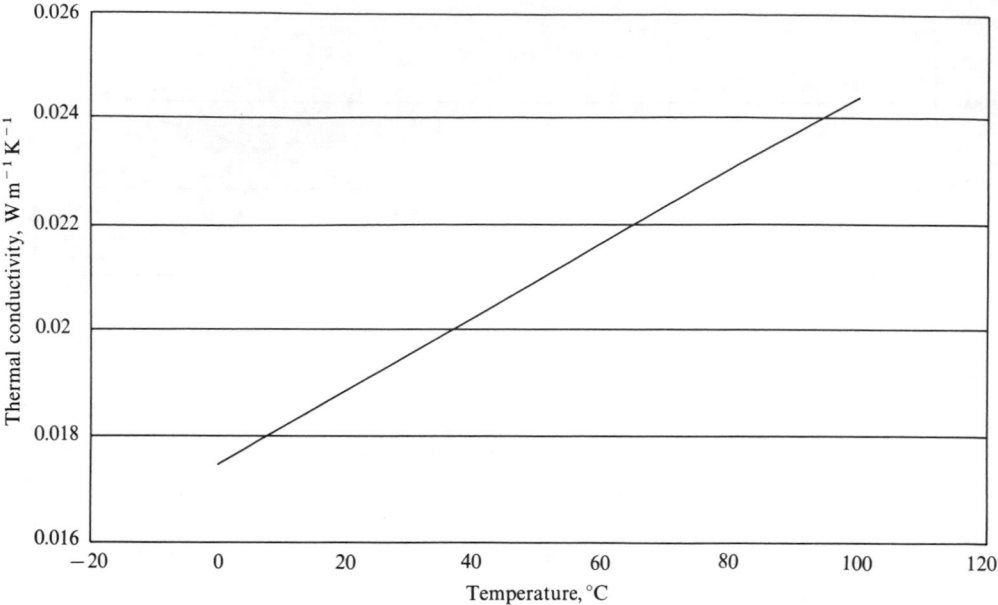

Figure A1.7 Thermal conductivity of water vapour at $0.1013 \, \text{MN} \, \text{m}^{-2}$.

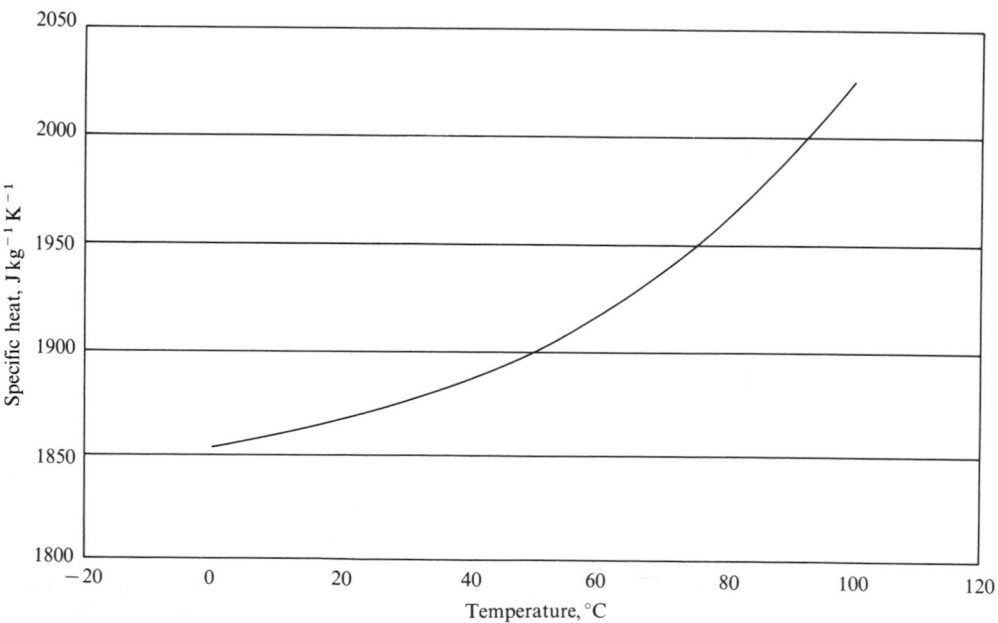

Figure A1.8 Specific heat of water vapour at $0.1013 \, \text{MN} \, \text{m}^{-2}$.

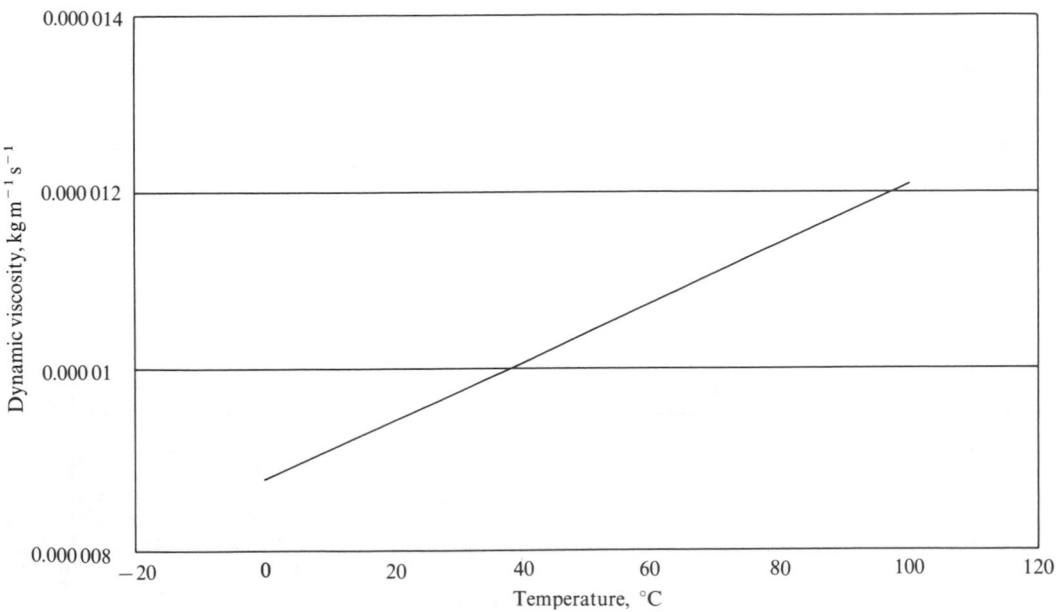

Figure A1.9 Dynamic viscosity of water vapour at 0.1013 MN m^{-2}.

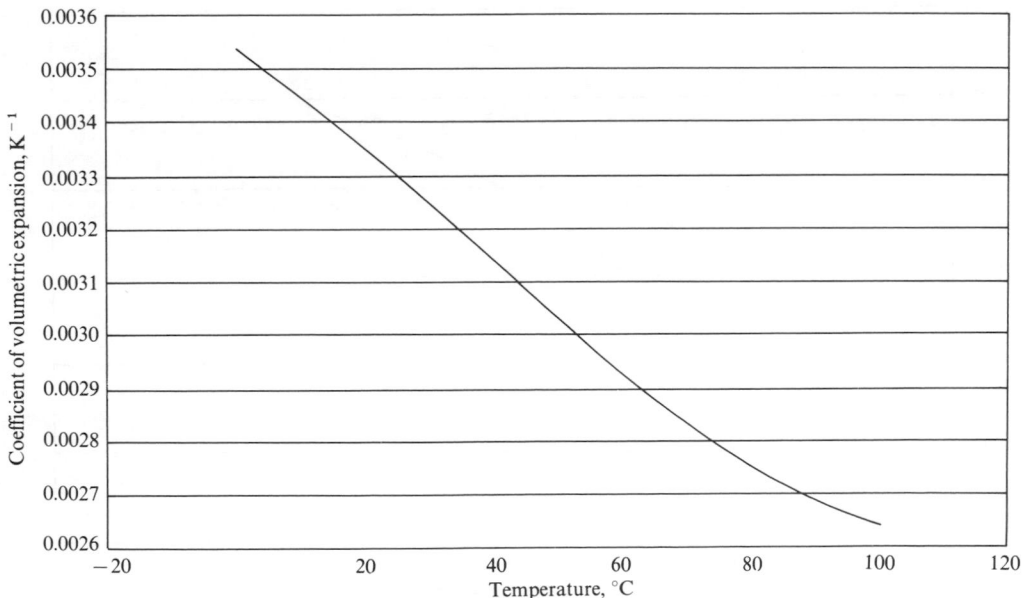

Figure A1.10 Coefficient of volumetric expansion of water vapour at 0.1013 MN m^{-2}.

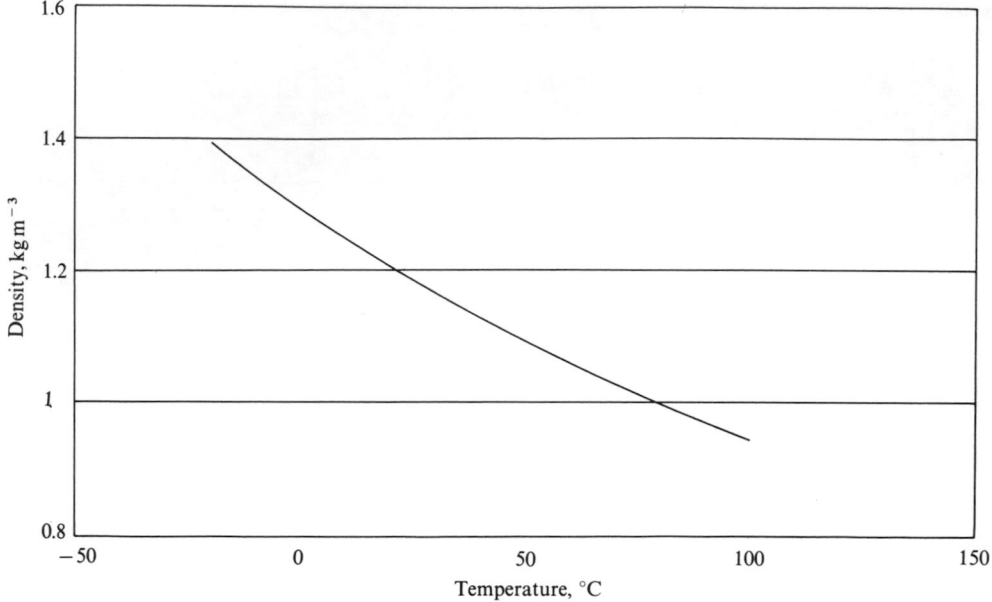

Figure A1.11 Density of dry air at 0.1013 MN m^{-2}.

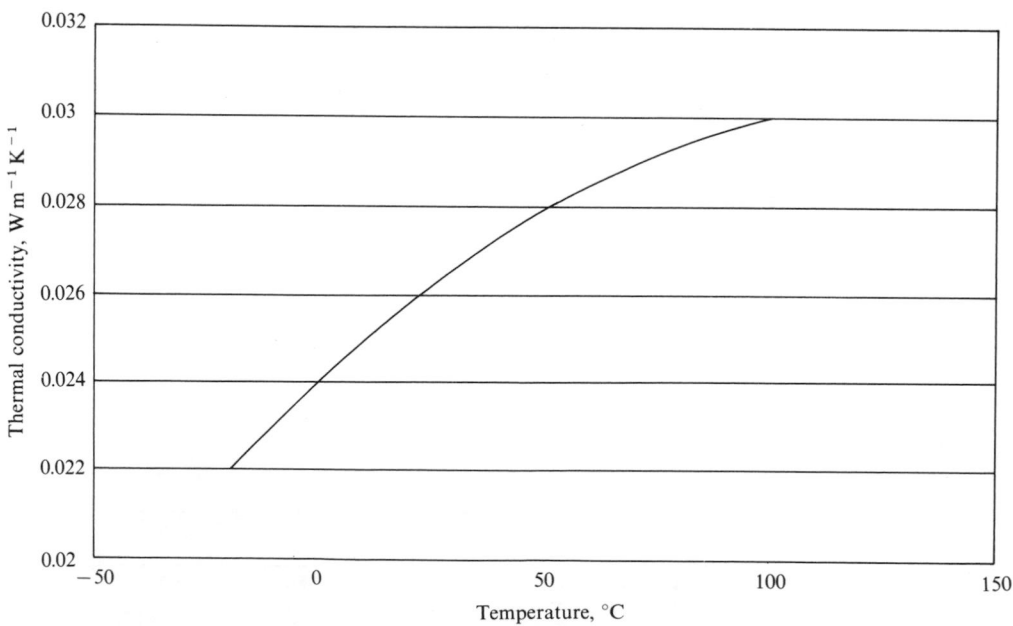

Figure A1.12 Thermal conductivity of dry air at 0.1013 MN m^{-2}.

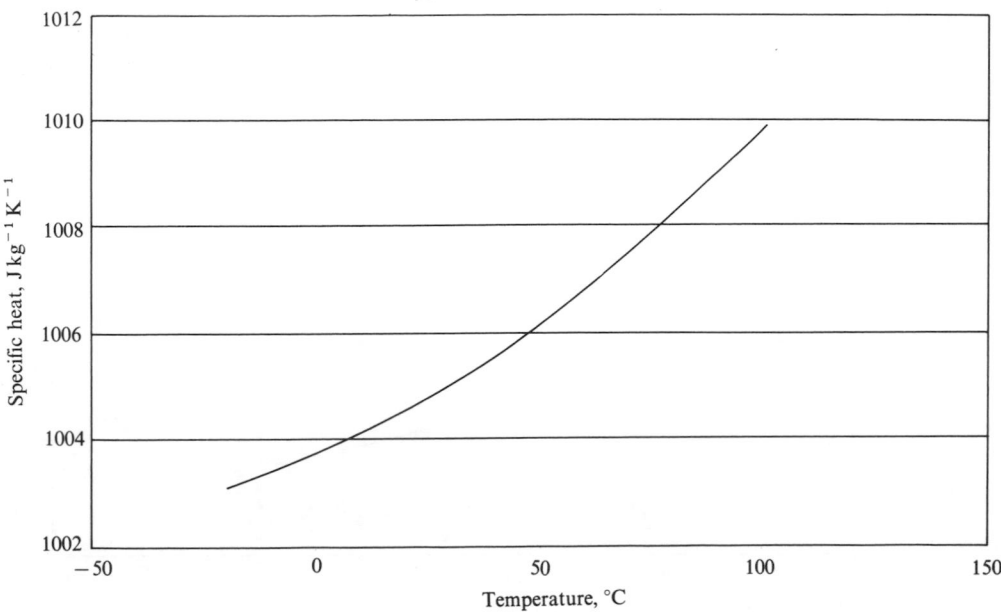

Figure A1.13 Specific heat of dry air at $0.1013 \, MN \, m^{-2}$.

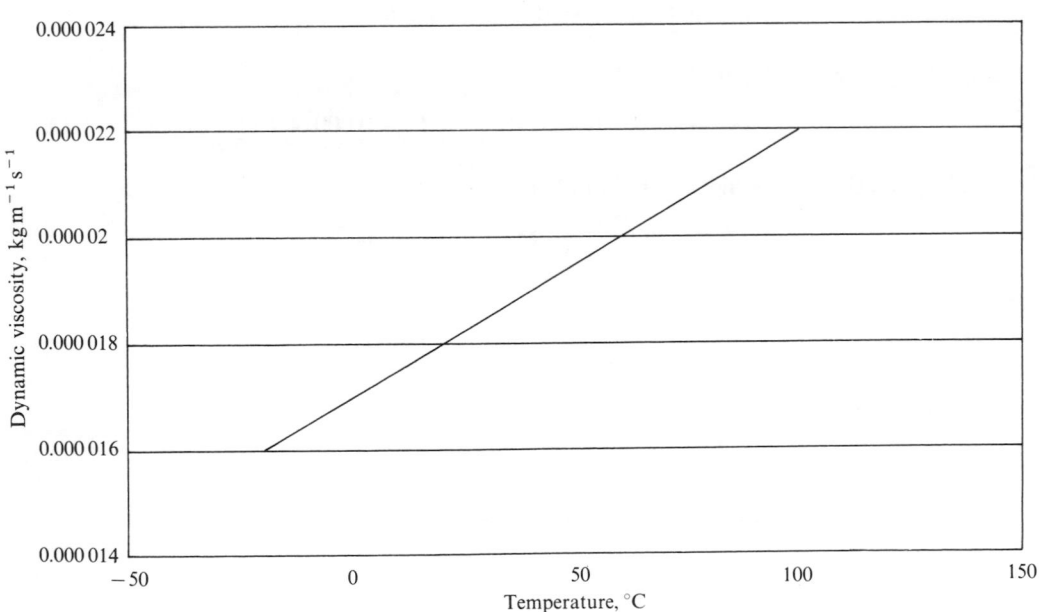

Figure A1.14 Dynamic viscosity of dry air at $0.1013 \, MN \, m^{-2}$.

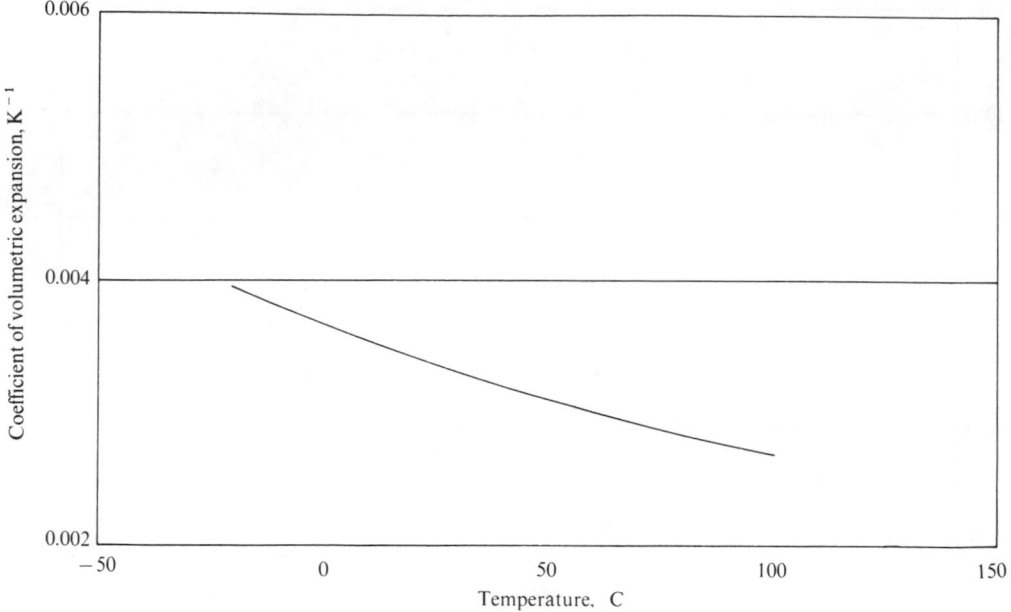

Figure A1.15 Coefficient of volumetric expansion of dry air at 0.1013 MN m^{-2}.

Relevant thermophysical properties of water, water vapour and dry air are given in Tables A1.12 to A1.14 and Figures A1.1 to A1.15.

The following equations are applicable for atmospheric pressure ($= 0.1013$ MN m^{-2}) in the temperature range 0 to 100 °C. Temperatures, T, are in °C.

Density of liquid water, kg m^{-3}

$$\rho = 1\,000.063 - 0.016\,4T - 0.004\,96T^2 + 0.000\,009\,3T^2 \tag{A1.1}$$

Thermal conductivity of liquid water, W m^{-1}K^{-1}

$$k = 0.569 + 0.001\,889T - 0.000\,008T^2 \tag{A1.2}$$

Specific heat of liquid water, J kg^{-1}K^{-1}

$$c_p = 4\,216.5 - 2.398\,4T + 0.042\,38\,T^2 - 0.000\,194T^3 \tag{A1.3}$$

Dynamic viscosity of liquid water, kg ms^{-1}

$$\mu = 1.69 \times 10^{-3} - 3.3 \times 10^{-5}T + 1.946 \times 10^{-7}T^2 \tag{A1.4}$$

Coefficient of volumetric expansion of liquid water, K^{-1}

$$\beta = -4.5 \times 10^{-5} + 1.25 \times 10^{-5}T - 0.628 \times 10^{-8}T^2 \tag{A1.5}$$

Density of water vapour, kg m^{-3}

$$\rho = 0.023\,551\,8 - 0.003\,303T + 0.000\,088\,312T^2 \tag{A1.6}$$

Thermal conductivity of water vapour, $W\,m^{-1}\,K^{-1}$

$$k = 0.017\,442\,9 + 0.000\,071\,5\,T - 1.79 \times 10^{-8}T^2 \tag{A1.7}$$

Specific heat of water vapour, $J\,kg^{-1}\,K^{-1}$

$$c_p = 1\,853.7 + 0.611\,9T + 0.000\,892\,9T^2 + 0.000\,104\,2T^3 \tag{A1.8}$$

Dynamic viscosity of water vapour, $kg\,m^{-1}\,s^{-1}$

$$\mu = (0.008\,796\,4 + 0.000\,030\,8T + 2.232 \times 10^{-8}T^2)/1000 \tag{A1.9}$$

Coefficient of volumetric expansion of water vapour, T^{-1}

$$\beta = 35.37 - 0.083T - 0.000\,645T^2 + 0.000\,005\,8T^3 \tag{A1.10}$$

The following equations are applicable for atmospheric pressure ($= 0.1013\,MN\,m^{-2}$) in the temperature range -20 to $100\,°C$. Temperatures, T, are in $°C$.

Density of dry air, $kg\,m^{-2}$

$$\rho = (0.1013 \times 10^6)/(287 \times (273 + T)) \tag{A1.11}$$

Thermal conductivity of dry air, $W\,m^{-1}\,K^{-1}$

$$k = 0.024\,038\,1 + 0.000\,096\,5T - 3.68 \times 10^{-7}T^2 \tag{A1.12}$$

Specific heat of dry air, $J\,kg^{-1}\,K^{-1}$

$$c_p = 1003.74 + 0.036T + 0.000\,223T^2 + 0.000\,000\,3T^3 \tag{A1.13}$$

Dynamic viscosity of dry air, $kg\,m^{-1}\,s^{-1}$

$$\mu = (17 + 0.05T) \times 10^{-6} \tag{A1.14}$$

Coefficient of volumetric expansion for dry air, K^{-1}

$$\beta = 1/(T + 273) \tag{A1.15}$$

Table A1.15 Net calorific values of representative fuels, $MJ\,kg^{-1}$

Bituminous Coal	28–35
Coke	33
Gasoline, petrol	44
Gas oil, diesel fuel	42
Hydrogen	120
Kerosene	43
Methane (natural gas)	50
Residual fuel oil	40–45
Town gas	32
Wood	17

Table A1.16 Combustion efficiency (%) tables for natural gas (ΔT = flue gas temperature minus combustion air inlet temperature, °C)

% Excess air	% O_2	% CO_2	77	82	88	93	99	104	110	116	121	127	132	138	143	149	232	238	243	248	254	260	266	271	277	282	288	293	299	305
0.0	0.0	11.8	86.3	86.1	85.9	85.7	85.5	85.3	85.1	84.9	84.7	84.5	84.2	84.0	83.8	83.6	80.4	80.1	79.9	79.7	79.5	79.3	79.0	78.8	78.6	78.4	78.1	77.9	77.7	77.5
2.2	0.5	11.5	86.3	86.1	85.9	85.6	85.4	85.2	85.0	84.8	84.6	84.4	84.1	83.9	83.7	83.5	80.2	80.0	79.7	79.5	79.3	79.1	78.8	78.6	78.4	78.2	77.9	77.7	77.5	77.2
4.5	1.0	11.2	86.2	86.0	85.8	85.6	85.3	85.1	84.9	84.7	84.5	84.2	84.0	83.8	83.6	83.4	80.0	79.8	79.5	79.3	79.1	78.9	78.6	78.4	78.2	77.9	77.7	77.5	77.2	77.0
6.9	1.5	11.0	86.1	85.9	85.7	85.5	85.2	85.0	84.8	84.6	84.4	84.1	83.9	83.7	83.5	83.2	79.8	79.6	79.3	79.1	78.9	78.6	78.4	78.2	77.9	77.7	77.5	77.2	77.0	76.8
9.5	2.0	10.7	86.1	85.8	85.6	85.4	85.2	84.9	84.7	84.5	84.2	84.0	83.8	83.6	83.3	83.1	79.6	79.4	79.1	78.9	78.7	78.4	78.2	77.9	77.7	77.5	77.2	77.0	76.7	76.5
12.1	2.5	10.4	86.0	85.7	85.5	85.3	85.1	84.8	84.6	84.4	84.1	83.9	83.7	83.4	83.2	83.0	79.4	79.1	78.9	78.7	78.4	78.2	77.9	77.7	77.4	77.2	76.9	76.7	76.5	76.2
15.0	3.0	10.1	85.9	85.7	85.4	85.2	85.0	84.7	84.5	84.2	84.0	83.8	83.5	83.3	83.0	82.8	79.2	78.9	78.7	78.4	78.2	77.9	77.7	77.4	77.2	76.9	76.7	76.4	76.2	75.9
18.0	3.5	9.8	85.8	85.6	85.3	85.1	84.8	84.6	84.4	84.1	83.9	83.6	83.4	83.1	82.9	82.6	78.9	78.7	78.4	78.2	77.9	77.6	77.4	77.1	76.9	76.6	76.4	76.1	75.9	75.6
21.1	4.0	9.6	85.7	85.5	85.2	85.0	84.7	84.5	84.2	84.0	83.7	83.5	83.2	83.0	82.7	82.5	78.7	78.4	78.1	77.9	77.6	77.4	77.1	76.8	76.6	76.3	76.1	75.8	75.5	75.3
24.5	4.5	9.3	85.6	85.4	85.1	84.8	84.6	84.3	84.1	83.8	83.6	83.3	83.1	82.8	82.6	82.3	78.4	78.1	77.9	77.6	77.3	77.1	76.8	76.5	76.3	76.0	75.7	75.4	75.2	74.9
28.1	5.0	9.0	85.5	85.2	85.0	84.7	84.5	84.2	84.0	83.7	83.4	83.2	82.9	82.6	82.4	82.1	78.1	77.8	77.6	77.3	77.0	76.7	76.5	76.2	75.9	75.6	75.4	75.1	74.8	74.5
31.9	5.5	8.7	85.4	85.1	84.9	84.6	84.3	84.1	83.8	83.5	83.3	83.0	82.7	82.4	82.2	81.9	77.8	77.5	77.2	77.0	76.7	76.4	76.1	75.8	75.6	75.3	75.0	74.7	74.4	74.1
35.9	6.0	8.4	85.3	85.0	84.7	84.4	84.2	83.9	83.6	83.3	83.1	82.8	82.5	82.2	82.0	81.7	77.5	77.2	76.9	76.6	76.3	76.0	75.7	75.5	75.2	74.9	74.6	74.3	74.0	73.7
40.3	6.5	8.2	85.1	84.9	84.6	84.3	84.0	83.7	83.4	83.2	82.9	82.6	82.3	82.0	81.7	81.5	77.1	76.8	76.5	76.2	75.9	75.6	75.3	75.1	74.8	74.5	74.2	73.9	73.6	73.3
44.9	7.0	7.9	85.0	84.7	84.4	84.1	83.8	83.5	83.2	83.0	82.7	82.4	82.1	81.8	81.5	81.2	76.7	76.4	76.1	75.8	75.5	75.2	74.9	74.6	74.3	74.0	73.7	73.4	73.1	72.8
49.9	7.5	7.6	84.8	84.5	84.2	83.9	83.6	83.3	83.0	82.8	82.5	82.2	81.9	81.6	81.3	80.9	76.3	76.0	75.7	75.4	75.1	74.8	74.5	74.1	73.8	73.5	73.2	72.9	72.6	72.2
55.3	8.0	7.3	84.7	84.4	84.1	83.8	83.5	83.1	82.8	82.5	82.2	81.9	81.6	81.3	81.0	80.7	75.9	75.6	75.3	74.9	74.6	74.3	74.0	73.6	73.3	73.0	72.7	72.3	72.0	71.7
61.1	8.5	7.0	84.5	84.2	83.9	83.6	83.2	82.9	82.6	82.3	82.0	81.6	81.3	81.0	80.7	80.4	75.4	75.1	74.8	74.4	74.1	73.8	73.4	73.1	72.8	72.4	72.1	71.8	71.4	71.1
67.3	9.0	6.7	84.3	84.0	83.7	83.3	83.0	82.7	82.3	82.0	81.7	81.4	81.0	80.7	80.4	80.0	74.9	74.6	74.2	73.9	73.6	73.2	72.9	72.5	72.2	71.8	71.5	71.1	70.8	70.4
74.2	9.5	6.5	84.1	83.8	83.4	83.1	82.8	82.4	82.1	81.7	81.4	81.0	80.7	80.3	80.0	79.7	74.4	74.0	73.7	73.3	73.0	72.6	72.2	71.9	71.5	71.2	70.8	70.4	70.1	69.7
81.6	10.0	6.2	83.9	83.5	83.2	82.8	82.5	82.1	81.8	81.4	81.1	80.7	80.3	80.0	79.6	79.3	73.8	73.4	73.0	72.7	72.3	71.9	71.6	71.2	70.8	70.4	70.1	69.7	69.3	68.9
89.8	10.5	5.9	83.6	83.3	82.9	82.5	82.2	81.8	81.5	81.1	80.7	80.3	80.0	79.6	79.2	78.8	73.1	72.7	72.4	72.0	71.6	71.2	70.8	70.4	70.0	69.6	69.2	68.9	68.5	68.1
98.7	11.0	5.6	83.4	83.0	82.6	82.2	81.8	81.5	81.1	80.7	80.3	79.9	79.5	79.1	78.7	78.3	72.4	72.0	71.6	71.2	70.8	70.4	70.0	69.6	69.2	68.8	68.4	67.9	67.5	67.1
108.7	11.5	5.3	83.1	82.7	82.3	81.9	81.5	81.1	80.7	80.3	79.9	79.4	79.0	78.6	78.2	77.7	71.6	71.2	70.8	70.4	69.9	69.5	69.1	68.6	68.2	67.8	67.4	66.9	66.5	66.1
119.7	12.0	5.1	82.7	82.3	81.9	81.5	81.1	80.6	80.2	79.7	79.3	78.9	78.5	78.1	77.7	77.2	70.7	70.3	69.8	69.4	68.9	68.5	68.1	67.6	67.2	66.7	66.3	65.8	65.4	64.9
132.0	12.5	4.8	82.4	81.9	81.5	81.0	80.6	80.2	79.7	79.3	78.8	78.4	77.9	77.5	77.0	76.6	69.7	69.3	68.8	68.3	67.9	67.4	66.9	66.5	66.0	65.5	65.1	64.6	64.1	63.6
145.8	13.0	4.5	82.0	81.5	81.0	80.6	80.1	79.6	79.1	78.7	78.2	77.7	77.3	76.8	76.3	75.8	68.6	68.1	67.6	67.1	66.6	66.2	65.7	65.2	64.7	64.2	63.7	63.2	62.7	62.2
161.5	13.5	4.2	81.5	81.0	80.5	80.0	79.5	79.0	78.5	78.0	77.5	77.0	76.5	76.0	75.5	75.0	67.3	66.8	66.3	65.8	65.3	64.7	64.2	63.7	63.2	62.6	62.1	61.6	61.1	60.5
179.5	14.0	3.9	81.0	80.4	79.9	79.4	78.8	78.3	77.8	77.2	76.7	76.2	75.7	75.1	74.6	74.0	65.9	65.3	64.8	64.2	63.7	63.1	62.4	62.0	61.5	60.9	60.3	59.8	59.2	58.7
200.2	14.5	3.7	80.3	79.8	79.2	78.6	78.1	77.5	76.9	76.4	75.8	75.2	74.7	74.1	73.5	72.9	64.2	63.6	63.1	62.5	61.9	61.3	60.7	60.1	59.5	58.9	58.3	57.7	57.1	56.5
224.3	15.0	3.4	79.6	79.0	78.4	77.8	77.2	76.6	76.0	75.3	74.7	74.1	73.5	72.9	72.3	71.7	62.3	61.7	61.0	60.4	59.7	59.1	58.5	57.8	57.2	56.5	55.9	55.3	54.6	54.0

Table A1.17 Heat exchanger U-values

Hot side to cold side	h_h $W\,m^{-1}\,K^{-1}$	v_h $m\,s^{-1}$	h_c $W\,m^{-1}\,K^{-1}$	v_c $m\,s^{-1}$	U $W\,m^{-1}\,K^{-1}$
Air to air	28.4	4.58	28.4	4.58	14.2
Air to air	56.8	12.2	56.8	12.2	28.4
Air to water	28.4	4.58	4090	1.52	28.1
Air to water	56.8	12.2	4090	1.52	56.8
Water to water	4090	1.52	4090	1.52	1760
Oil to oil	510	1.52	510	1.52	256
Condensing steam to boiling water	11 400	—	5678	—	2980
Condensing steam to water	11 400	—	1136	—	960
Condensing steam to oil	11 400	—	510	1.52	477
Oil to water	510	1.52	4090	1.52	440
Oil to air	510	1.52	28.4	4.58	25.8

Table A1.18 Common U-values for buildings, $W\,m^{-2}\,K^{-1}$

Component	U-value
Solid brickwork	
105 mm	3.3
220 mm	2.3
335 mm	1.7
260 mm cavity brick wall	0.8–1.5
5 mm corrugated sheeting	5.3[†]
150 mm solid concrete	3.4
Glazing	
6 mm single	5.6[†]
2 mm airspace double	2.9
Roofs	
uninsulated flat	≈2.0
uninsulated 35° slope pitched	≈1.5
Solid floors	≈0.3[‡]

[†] These large values are heavily dependent upon the values of the partial filmside heat transfer coefficients adopted. The solid U-value (i.e. neglecting boundary resistance) of 6 mm single glazing and 5 mm metal sheeting are ≈130 $W\,m^{-2}\,K^{-1}$ and ≈9000 $W\,m^{-2}\,K^{-1}$ respectively.

[‡] Referred to inside minus outside air temperature difference.

Table A1.19 Surface emissivities

Wavelength (μm)	9.30	5.40	3.60	1.80	0.60
Temperature (K)	311.83	537.04	805.56	1 611.11	Solar
Metals:					
Polished aluminium	0.04	0.05	0.08	0.19	0.30
Oxidized aluminium	0.11	0.12	0.18		
Polished brass	0.10	0.10			
Oxidized brass	0.61				
Polished iron	0.06	0.08	0.13	0.25	0.45
Oxidized iron	0.63	0.66	0.76		
Polished silver	0.01	0.02	0.03		0.11
Polished stainless steel	0.15	0.18	0.22		
Weathered stainless steel	0.85	0.85	0.85		
Building and insulating materials:					
Asbestos paper	0.93	0.93			
Asphalt	0.93		0.9		0.93
Red brick	0.93				0.70
Fire clay	0.90	0.70	0.75		
White paper	0.95		0.82	0.25	0.28
Plaster	0.91				
Paints:					
Aluminized lacquer	0.65	0.65			
Cream	0.95	0.88	0.70	0.42	0.35
Lampblack	0.96	0.97		0.97	0.97
Red	0.96				0.74
Yellow	0.95		0.50		0.30
White	0.95		0.91		0.18

Table A1.20 Emission and absorptance bands for common gases and vapours

Substance	Wavelength band λ, μm
Carbon dioxide	2.36–3.02
	4.01–4.80
	12.5–16.5
Water vapour	2.24–3.27
	4.80–8.50
	12.00–25.00

Orders of magnitude of heat transfer coefficients from surface-to-air

	h, W m^{-2} K^{-1}
Air in free convection	5–25
Air in forced convection	25–250
Oil in forced convection	50–150
Water in forced convection	250–10 000

Table A1.21 Absorption coefficients

Material or component		Colour or condition		
	White	Light	Dark	Dirty
Brick	0.2 to 0.5	0.4 to 0.5	0.6 to 0.9	0.5 to 0.9
Stone	0.3 to 0.5	0.3 to 0.5	0.5 to 0.6	0.5 to 0.9
Tiles	0.3 to 0.5	0.4	0.8	0.5 to 0.9
Asphalt			0.9	
Grey Slate			0.8 to 0.9	
Asbestos		0.6		
Aluminium		0.2		
Copper (tarnished)			0.6	
Water	1 m thick	2 m thick	3 m thick	
	0.56	0.61	0.64	

Characteristics of glazing

Table A1.22 Radiation transmitted and retransmitted

Type	Direct radiation for given angle of incidence (degrees)				Diffuse Radiation
	0	30	60	85	
Single glazing:					
Clear 4 mm	0.84	0.84	0.77	0.22	0.78
Clear 6 mm	0.80	0.79	0.72	0.20	0.73
Heat absorbing 6 mm	0.50	0.49	0.36	0.13	0.44
Heat reflecting	0.27	0.27	0.26	0.11	0.25
Double glazing:					
Clear 4 mm/clear 4 mm	0.75	0.73	0.65	0.12	0.67
Clear 6 mm/clear 6 mm	0.69	0.67	0.59	0.11	0.67
Heat absorbing/clear 4 mm	0.45	0.44	0.37	0.10	0.37
Heat absorbing/clear 6 mm	0.37	0.35	0.29	0.08	0.30
Heat reflecting/clear	0.33	0.28	0.25	0.07	0.28

Table A1.23 Typical solar gain factors, S

Type	S
Single clear glass	0.76
with internal white venetian blind	0.46
with internal white cotton curtain	0.41
Double clear glass	0.64
with internal white venetian blind	0.46
with internal white cotton curtain	0.40
Single heat absorbing glass	0.45
Double heat absorbing glass	0.31
Single heat reflecting glass	0.26
Double heat reflecting glass	0.25
Double clear glass with mid-pane white venetian blind	0.28
Single clear glass with external canvas roller blind	0.14
Double clear glass with external canvas roller blind	0.11
Single clear glass with external white louvred sun breaker	0.14
Double clear glass with external white louvred sun breaker	0.11

Table A1.24 Constants for use in solar equations

Month	Jan.	Mar.	May	Jul.	Sep.	Nov.
K_1	1.23	1.19	1.1	1.09	1.15	1.22
K_2	0.14	0.16	0.2	0.21	0.18	0.15
K_3	0.06	0.07	0.12	0.14	0.09	0.06

For calculation purposes, the solar equation constants, (Table A1.24) may be approximated as follows:

$$K_1 = 1.15 + 0.09 \sin((t - 295)2[PI]/365)$$

$$K_2 = 0.17 + 0.04 \sin((t - 100)2[PI]/365)$$

$$K_3 = 0.09 + 0.05 \sin((t - 100)2[PI]/365)$$

$$[PI] \equiv \pi$$

$$I_\delta = K_1 e^{-K_2/\sin \alpha'} \tag{A1.16}$$

$$I_{sc} = K_3 I_\delta \tag{A1.17}$$

Table A1.25 Sun angles (degrees) for the latitude 51.7 °N

Sun-time (= GMT)	Dec. 22	Jan. 21 and Nov. 21	Feb. 20 and Oct. 23	Mar. 22 and Sep. 22	Apr. 20 and Aug. 24	May 21 and Jul. 23	Jun. 21
			Solar altitude angle, α'				
0600 and 1800					9	15	18
0700 and 1700			1	10	18	25	27
0800 and 1600		2	10	19	28	34	37
1900 and 1500	6	10	17	27	37	44	46
1000 and 1400	12	15	24	34	44	52	55
1100 and 1300	15	19	28	39	50	58	61
1200	17	20	29	40	51	60	63
			Solar azimuth angle, z'				
0600				90	83	77	74
0700			108	101	94	88	85
0800		125	120	114	106	100	97
0900	139	138	133	127	120	114	110
1000	152	151	148	143	137	131	128
1100	166	165	163	161	157	153	151
1200	180	180	180	180	180	180	180
1300	194	195	197	199	203	207	209
1400	208	209	212	217	223	229	232
1500	221	222	227	233	240	246	250
1600		234	240	246	254	260	263
1700			252	259	266	272	275
1800				270	277	283	286

Table A1.26 Direct solar radiation intensities Wm^{-2}

Altitude angle	5	10	15	20	25	30	35	40	45	50	60	80
Normal to sun	210	388	524	620	688	740	782	814	840	860	893	920
On horizontal	18	67	136	212	290	370	450	523	594	660	773	907
On south-face	210	382	506	584	624	642	640	624	594	553	477	160

Table A1.27 Precipitation, mm water

Location	Maximum average monthly	Minimum average monthly	Average annual amount
Belfast	91	48	856
Dublin	76	48	754
Edinburgh	79	41	701
London	64	36	580

Table A1.28 Monthly average sol-air and outside air temperatures (°C) at 51.7 °N (UK)

Month	Mean Solar-air Temp., °C					Mean outside air temp.
	Flat roof	North wall	South wall	East wall	West wall	
Jan.	2.0	4.5	8.0	5.0	5.0	4.3
Feb.	3.6	5.5	9.6	7.0	7.0	5.0
Mar.	8.0	7.5	12.5	10.5	10.5	6.5
Apr.	14.0	11.0	16.5	16.5	16.5	9.0
May	20.5	16.5	20.5	21.5	21.5	13.5
Jun.	25.0	21.0	23.5	26.0	27.5	16.5
Jul.	27.0	22.5	26.0	27.6	27.5	19.0
Aug.	22.0	19.0	26.0	24.0	24.0	17.0
Sep.	15.5	15.0	21.5	19.0	19.0	13.0
Oct.	7.5	10.5	15.5	12.5	12.5	10.0
Nov.	4.0	7.0	11.5	7.5	7.5	6.5
Dec.	2.5	5.5	8.5	6.0	6.0	5.0
Annual mean	12.6	12.6	16.3	15.3	15.3	11.0

Table A1.29 UK average degree-days for the period 1968–1987

Region	Jan.	Feb.	Mar.	Apr.	May.	Jun.	Jul.	Aug.	Sep.	Oct.	Nov.	Dec.	Total
Thames	346	322	286	205	120	51	22	25	54	130	242	312	2115
SE	368	344	312	233	150	74	39	44	82	160	267	334	2407
Southern	345	327	301	229	148	72	39	43	79	150	251	312	2296
SW	293	285	271	207	137	63	28	28	55	116	206	258	1947
Severn	321	305	280	201	128	56	24	27	61	138	237	300	2078
Midland	376	359	322	243	162	83	44	48	90	178	275	343	2523
W Pennines	361	340	312	230	144	75	38	39	78	157	267	328	2369
NW	375	345	323	245	167	90	50	56	96	171	284	341	2543
Borders	376	349	330	271	206	117	66	68	104	182	282	339	2690
NE	381	358	322	247	168	87	46	49	88	175	281	346	2548
E Pennines	372	352	313	232	154	78	42	44	81	165	272	341	2446
E Anglia	378	349	317	239	149	73	40	39	71	154	269	341	2419
W Scotland	383	352	328	246	170	94	58	64	111	188	299	352	2645
E Scotland	388	357	332	263	197	109	62	67	109	192	301	354	2726
NE Scotland	401	368	346	277	206	120	74	78	127	203	311	362	2873
Wales	330	320	307	240	170	92	49	45	77	145	235	294	2304
N Ireland	365	334	320	242	171	92	53	59	99	173	282	329	2519

Degree-day users requiring actual figures or mean values for other periods may obtain these directly from UK Regional Energy Efficiency Officers whose addresses and telephone numbers may be found in any edition of *Energy Management*, Journal of the Energy Efficiency Office, UK Department of Energy, 1 Palace Street London SW1E 5HE, Telephone 071 215 0619 (London Region). The data are provided by the UK Meteorological Office, Advisory Services, Room 228, London Road, Bracknell, Berkshire, RG12 2ZS. Telephone 0344 856207.

The data are based upon the base temperature of 15.5 °C.

Table A1.30 UK average degree-days for the period 1959–1978

Region	Jan.	Feb.	Mar.	Apr.	May.	Jun.	Jul.	Aug.	Sep.	Oct.	Nov.	Dec.	Total
Thames	349	306	281	200	113	49	25	27	56	129	252	333	2120
SE	372	328	306	228	146	75	45	49	85	160	277	356	2427
Southern	343	311	293	217	142	70	42	43	79	143	255	327	2265
SW	295	275	267	201	130	60	33	31	57	112	212	276	1949
Severn	347	313	290	213	130	58	31	34	70	142	254	329	2211
Midland	376	339	317	236	153	78	48	52	92	170	287	359	2507
W Pennines	362	325	304	225	138	68	42	43	79	151	278	347	2362
NW	371	335	316	241	160	83	56	58	94	165	293	360	2532
Borders	379	345	328	262	193	107	76	73	107	180	298	361	2709
NE	378	339	314	237	155	78	50	51	88	167	292	361	2510
E Pennines	368	328	302	220	139	68	42	42	78	155	280	351	2373
E Anglia	383	340	316	235	146	73	45	44	75	152	281	361	2451
W Scotland	374	339	315	239	162	84	63	64	106	176	305	358	2585
E Scotland	384	348	324	257	190	101	72	72	107	184	310	370	2719
NE Scotland	399	362	340	274	203	112	86	86	125	196	321	382	2886
Wales	328	305	293	323	157	83	50	45	75	135	238	303	2244
N Ireland	363	327	311	240	166	86	62	64	101	167	290	345	2522
Average	363	327	306	232	154	78	51	51	86	157	277	345	2433
Difference	−1	12	7	6	7	5	−6	−3	0	6	−9	−17	7

Figures subject to integer rounding errors

Table A1.31 UK degree-days and average temperatures, °C

Region	Period 1968–1987	Mean temperature	Period 1959–1978	Mean temperature	1968–1987 Hotter by
Thames	2115	9.705	2120	9.691	0.013
SE	2407	8.905	2427	8.850	0.054
Southern	2296	9.209	2265	9.294	−0.084
SW	1947	10.16	1949	10.16	0.005
Severn	2078	9.806	2211	9.442	0.364
Midland	2523	8.587	2507	8.631	−0.043
W Pennines	2369	9.009	2362	9.028	−0.019
NW	2543	8.532	2532	8.563	−0.030
Borders	2690	8.130	2709	8.078	0.052
NE	2548	8.519	2510	8.623	−0.104
E Pennines	2446	8.798	2373	8.998	−0.2
E Anglia	2419	8.872	2451	8.784	0.087
W Scotland	2645	8.253	2585	8.417	−0.164
E Scotland	2726	8.031	2719	8.050	−0.019
NE Scotland	2873	7.628	2886	7.593	0.035
Wales	2304	9.187	2244	9.352	−0.164
N Ireland	2519	8.598	2522	8.590	0.008
UK Average	2438.1	8.820	2433.6	8.832	−0.012

During the period from 1968 to 1987 the UK was on average *colder* than it was during the period from 1959 to 1978.

Table A1.32 UK actual degee-days for 1989 and differences from 30-year average

Region	Jan.	Feb.	Mar.	Apr.	May.	Jun.	Jul.	Aug.	Sep.	Oct.	Nov.	Dec.	Total
Thames	285	254	216	228	63	51	11	21	30	78	247	271	1755
SE	311	271	238	252	90	74	21	36	45	106	280	295	2019
Southern	289	260	233	264	105	81	28	44	62	111	251	301	2029
SW	240	219	212	242	74	53	7	18	36	76	184	239	1600
Severn	249	226	209	230	62	48	5	14	30	87	224	280	1664
Midland	301	277	257	277	115	88	27	40	64	132	295	335	2208
W Pennines	285	269	262	274	110	81	23	41	65	128	273	351	2162
NW	260	266	276	266	134	104	35	39	76	117	289	411	2273
Borders	273	280	290	291	166	117	43	51	91	142	260	343	2347
NE	288	271	265	280	121	88	27	38	72	131	280	358	2219
E Pennines	291	270	252	278	122	88	29	40	52	127	289	339	2177
E Anglia	316	280	245	276	113	85	25	33	43	107	282	323	2128
W Scotland	257	291	308	280	152	105	45	59	115	152	309	421	2494
E Scotland	276	293	301	281	143	110	42	57	110	156	284	401	2454
NE Scotland	276	299	305	300	167	118	48	70	112	181	282	391	2549
Wales	272	261	261	272	124	89	28	32	63	118	230	316	2066
N Ireland	278	284	284	284	145	86	32	52	100	145	276	353	2319
Average	279	268	259	269	118	86	28	40	68	123	266	336	2144
Difference	83	71	54	−31	43	−3	17	8	18	40	2	−8	294
Diff. temp.	2.67	2.53	1.74	−1.03	−1.38	−0.1	0.54	0.25	0.6	1.29	0.06	−0.25	0.8

Table A1.33 Differences between 1989 degree-days and the 20-year average from 1968 to 1987

Region	Difference	deg C
Thames	360	0.98
SE	388	1.06
Southern	267	0.73
SW	347	0.95
Severn	414	1.13
Midland	315	0.86
W Pennines	207	0.56
NW	270	0.73
Borders	343	0.93
NE	329	0.9
E Pennines	269	0.73
E Anglia	291	0.79
W Scotland	151	0.41
E Scotland	277	0.75
NE Scotland	324	0.88
Wales	238	0.65
N Ireland	200	0.54
Average	289	0.79

Note that, in the United Kingdom, 1989 was hotter than the previous 20-year average in all regions.

Table A1.34 Pressure differences due to wind effects, Nm^{-2}

Building Height (m)	Open country ($9\,m\,s^{-1}$)	Suburban ($5.5\,m\,s^{-1}$)	City Centre ($3\,m\,s^{-1}$)
10	58	21	6
20	70	31	11
30	78	38	15
40	85	44	21
50	90	49	23
60	95	55	26
70	100	59	31
80	104	63	34

Typical occupation densities, per person

$10\,m^2$ in an office block
$20\,m^2$ in executive offices
$2\,m^2$ in restaurants
$0.5\,m^2$ in cinemas and theatres

Table A1.35 Metabolic rates for different activity levels, $W\,m^{-2}$

Activity	Rate of heat production
Sleeping	40
Seated quietly	60
Office work	60–80
Golf	80–150
Garage work	80–170
Vehicle driving	80–180
Domestic work	80–200
Teacher	90
Machine work	100–260
Carpentry	100–370
Light work	120
Shop assistant	120
Walking at 3 mph	150
Medium work	170
Foundry work	170–400
Tennis	200–270
Squash	290–420
Heavy work	300
Wrestling	400–500
Heaviest work possible	500

Table A1.36 Sundry heat gains

Activity	Hours/annum	Power rating, W	Average annual rating, W
Hot water tank	8760	50	50
Hot water use	8760	50 per capita	50 per capita
Cooking heat	1000	up to 1000 per capita	up to 120 per capita
People	1665	200	40
Lights	700	100	8
Television	1000	200	65
Tape recorder	200	40	1
Record player	200	40	1
Radio	1000	40	5
Washing machine	300	1000	35
Spin drier	300	125	4
Tumble drier	300	1000	35
Iron	300	400	14
Refrigerator	8760	40	40
Freezer	8760	200	200
Cooker hood	1000	100	11
Kettle	100	3000	35
Toaster	50	1000	6
Dish washer	500	750	42
Vaccum cleaner	300	250	10
Electric blanket	300	75	0.5
Hair drier	150	700	12

Table A1.37 Heat dissipations from lamps and luminaires

Illumination on the working plane, lux	W m^{-2} floor area	
	Filament with reflector	Lamps with diffuser
150	19–28	28–36
200	28–36	36–50
300	38–55	50–69
500	66–88	
	80 W white in diffusing fitting	Fluorescent lamps in louvred ceiling panel
150	8	8–11
200	11	11
300	11–16	14–19
500	22–28	22–33
1000	36–55	44–66

Table A1.38 Fresh air supply requirements

Application	litres s^{-1} person	litres s^{-1} m^2	Occupancy unknown Air change per hr
Private dwellings	8–12	—	—
Board rooms	18–25	—	—
Bars	12–18	—	—
Stores	6–8	—	—
Factories	16–28	0.8	—
Garages	—	8.0	5
Operating theatres	—	16.0	—
Hospital wards	8–12	—	—
General offices	6–8	1.3–2.0	3–8
Private offices	8–12	1.3–2.0	3–8
Restaurants	12–18	—	5–10
Theatres, cinemas	6–8	—	5–10
Schools	14	—	—
Engine rooms	—	—	4
Baths	—	—	5–8
Lavatories	—	—	5–10
Kitchens	—	—	10–40

Table A1.39 Some physical dimensions of the earth

Equatorial diameter	12 756.776 km
Mass	5.98×10^{24} kg
Relative density	5.52
Period of axial rotation	23 hours 56 minutes and 4 seconds
Period of revolution around the sun	365.256 days
Mean distance from the sun	149 598 000 km
Inclination of Equator	$23°27'$ to the ecliptic
Volume $(4\pi r^3/3)$	1.09×10^{21} m^3

Table A1.40 Areas versus populations

Areas		1990 Population (millions)
Surface area $(4\pi r^2)$	5.11×10^{14} m^2	
Water (70.79%)	3.62×10^{14} m^2	
Land (29.21%)	1.49×10^{14} m^2	5292.2
Africa	0.303×10^{14} m^2	642.1
Antarctica	0.133×10^{14} m^2	negl.
Asia	0.439×10^{14} m^2	3112.7
Australia	0.077×10^{14} m^2	26.5
Europe and USSR	0.105×10^{14} m^2	787.0
North America	0.243×10^{14} m^2	275.9
South America	0.176×10^{14} m^2	448.1

THE EARTH

- Third planet from the sun in the Solar System
- Central core surrounded by a mantle overlaid by a thinnish outer crust
- Seventy per cent surface covered by water
- Whole planet surrounded by a gaseous atmosphere
- SUPPORTS LIFE

The carboniferous Period occurred 345–280 million years ago and lasted 200 million years. Today's annual fossil fuel combustion corresponds to 123 333 of those carboniferous years. As a result, the atmospheric concentration of carbondioxide is increasing, reducing rates of heat loss via long-wave radiation to the vault of deep space. This may lead to global warming.

Reference

Editorial Announcement, The International System of Units, *International Journal of Heat and Mass Transfer*, **9**, 837–844, 1966.

ENERGY MANAGEMENT CHECKLISTS

FUELS

All leaks of fuels, materials and energy should be prevented.

Coal

- Avoid manual handling.
- Use recent deliveries first.

Oil

- Minimize the use of trace lines.
- Control the temperatures of tanks and pipelines.
- Optimize atomizing temperatures.

Gas

- Check continuously for leaks.
- Optimize storage arrangements.

ELECTRICITY

- Examine tariff structure.
- Meter electricity use to all sectors.
- Select optimal tariffs.

- Attempt to balance load factors.
- Identify equipment contributing to peak demand.
- Check and correct power factors.
- Peak lop.
- Stagger start-up times.
- Reschedule peak activities.
- Convert to thermal energy (heat or cold) at off-peak periods and introduce thermal storage.
- Consider the use of standby generators to peak lop.
- Try to use nightrate electricity, i.e. for charging batteries of electric vehicles.
- Introduce compressed air storage:

  (1) to peak lop
  (2) to use off-peak electricity.

- Select electrical motors so that they run at near full load.
- Maximize power factors—introduce capacitances.
- Switch off plant and lighting when not required.
- Pay attention to lighting.
- Consider the use of standby generators for peak lopping.
- Consider the introduction of a *total energy* system.
- Invest in an *energy management control system.*

Electricity should only be used as a last resort

This statement arises from the fact that, for every kWh of electricity consumed, 3 to 4 kWh of fossil fuel is consumed at the power station. For certain industrial heating applications, however, it may be more efficient to heat by induction or microwave heating in which the heat can be precisely directed to the component to be heated.

ENERGY RELEASE—FURNACES

These include batch ovens, rotary kilns, tunnel kilns, etc.

- Check and review maintenance and operating procedures.
- Check conditions of plant and equipment.
- Check that plant if operating efficiently.
- Check the control arrangements.
- Check adequacy and operation of monitoring instruments and controls.
- Check air/fuel ratio.
- Carry out combustion performance tests at regular intervals.
- Evaluate performance by comparing the fuel input to perform a specific task under standard conditions.
- Construct an energy balance over the furnace.
- Perform efficiency checks—indirect (flue gas losses).
- Check furnace insulation.
- Check state of furnace lining.
- Optimize insulation levels.

- Look for air leaks into furnace—confounding air/fuel ratios.
- Eliminate infiltration—a slightly positively internal pressure will assist this.
- Use doors for as short a time as possible—consider the use of chain barriers, air curtains and the like.
- Check burners and combustion conditions.
- Avoid fluctuations in fuel or air supplies.
- Select proper firing and control equipment.
- Minimize excess air—do not produce CO unless a reducing (non-oxidizing) atmosphere is required.
- Optimize flame temperatures.
- Avoid flame impingement (sting) on refractory surfaces.
- Aim for a uniform flame distribution filling the combustion chamber.
- Optimize conditions to give maximum heat transfer rates by convection or radiation (or by both mechanisms according to the purpose of the furnace and the method used to achieve this purpose—drying verses sensible heating).

To maximize radiative heat transfer:

- Gas and surface temperatures should be high.
- The flame should be highly luminous.
- The distance between flame and stock should be short.
- The flame coverage of the hearth should be good.

To maximize convective heat transfer:

- Gas temperatures should be high.
- Gas velocities should be high.
- Stagnant pockets should be avoided by good circulation and mixing.
- All heat transfer surfaces should be maintained clean.

- Warm waste gases rather than cold inlet air should be used for dilution (i.e. for flue gas temperature control).
- Maintain full load if possible.
- Investigate load patterns and operating cycles.
- Attempt to balance load factors and hence avoid periodic modulation or intermittent operation (switching) when much of the energy supplied heats up the furnace to the working temperature, the furnace room and the surroundings.
- Use lightweight carriers—especially in highly intermittent plant or tunnel kilns.
- Look for heat recovery opportunities.
- Check possibilities for heat recovery.

ENERGY CONVERSION—BOILERS, AUTOCLAVES AND LIQUID HEATERS

- Check flow and return temperatures.
- Check conditions of plant, equipment and flues.

- Check steam pressures and temperatures.
- Check for leaks.
- Check levels of insulation.
- Check insulation of hot wells.
- Investigate loading schedules.
- Check load patterns and operating cycles.
- Investigate sequencing of modular boilers.
- Check the control arrangements.
- Investigate adequacies and operations of controls:

    —for start-up
    —for modulation
    —for sequencing of modular boilers.

- Check maintenance procedures.
- Carry out boiler efficiency checks, i.e. direct (e.g. steam generated/fuel supplied).
- Carry out combustion performance tests at regular intervals.
- Construct an energy balance over the device.
- Before increasing device capacity, seek every opportunity to reduce demand, smooth the load and increase thermal efficiency.
- When demands fluctuate, use energy accumulation to smooth firing rates.
- Always operate in accordance with design specifications.
- Check cleanliness of heat transfer surfaces.
- Ensure that correct feedwater treatment is carried out—this reduces inefficiencies due to scaling and blowdown losses.
- Check blowdown arrangements.
- Recover heat during blowdown if possible.
- Recover as much condensate as possible.
- Ascertain effects of energy-conserving measures on boiler loads and performances.
- Look for heat recovery opportunities.

BOILERHOUSE AUXILIARIES

- Check pump glands for leakages.
- Check motors, bearings and belts.
- Check boiler feedwater treatment.
- Look for heat recovery opportunities:

    —from blow-down
    —to pre-heat air or feedwater
    —from economizers.

- Check make-up water quantities.
- Check insulation of heated pipework and storage vessels.
- Check all lines for leaks.
- Check ventilation arrangements.

INDIRECT HEATED VESSELS

- Check insulation and covers.
- Check heating supply conditions.
- Check load patterns and operating cycles.
- Check heat transfer surfaces.
- Check controls.
- Check steam trap operations.
- Look for heat recovery prospects.
- Consider thermal accumulation for ballasting mis-matches between supply and demand.

DRIERS

- Check water quantities (too much–too little).
- Check insulation of drying plant.
- Check drier gas circulation patterns.
- Check load patterns and operating cycles.
- Check controls and monitoring procedures.
- Check waste heat recovery operations and possibilities.

HEAT DISTRIBUTION SYSTEMS

- Check conditions of plant and equipment.
- Check for leaks of hot water or steam.
- Leaks should be prevented.
- Check amount and condition of thermal insulation on equipment and pipework.
- Estimate distribution losses.
- Apply optimal levels of insulation and ensure that the insulant does not become dirty, compressed, water-logged or degraded.
- Ensure that direct losses from uninsulated pipelines, heaters or hot surfaces through building boundary walls do not occur.
- Insulate all hot (cold) storage tanks to optimal economic levels.

HEATING SYSTEMS

- Check heating control arrangements.
- Check building/system response to controls.
- Check that temperatures and ventilation rates are not excessive.
- Check supply and return temperatures.
- Check conditions of plant and equipment.
- Check insulation levels and conditions throughout.
- Check maintenance procedures.

- Specify the exact purpose for which heating is required (sensible heating versus drying).
- Analyse load profiles.

Space heating:

- Stick to minimum fresh air requirements—consider the use of obscuration meters—question the minimum air change rate adopted.
- Ensure that the required air change rate is attained in working areas but do not air change volumes where fresh air changes are not required (e.g. roof spaces, storage area, unoccupied rooms or regions).
- When minimum air change rates are being achieved, consider the introduction of waste heat recovery.
- Isolate and vent-off processes which contaminate the working atmosphere.
- Analyse area per capita.
- Ensure adequate air distribution.
- Check for unoccupied heated parts of the building.
- Check zonal heating requirements.
- Minimize infiltration and uncontrolled air flows.
- Measure zonal airchanges rates.
- Split space-heated areas into zones having different requirements—consider partitioning.
- Check that ceiling heights are not excessive—consider the use of false ceilings.
- Maximize the extraction of contaminant and minimize the supply and extraction of fresh air.
- Ensure that the ventilation system is controlled.
- Ensure that opening windows or doors is not used to control temperatures.
- Eliminate random infiltration.
- Eliminate arbitrary manual adjustment of air flows and temperatures.
- Balance inlet and extract fan sets.
- Maintain a slight positive pressure to eliminate draughts.
- Introduce self-closing exterior doors, plastic or air-curtains, vestibules and air-locks.

- Do not overcapitalize plant—reduce the demand before increasing the amount of heat supplied.
- Consider thermal accumulation to avoid maximum demand and low-or-intermittent-fire inefficiencies.
- Keep heat transfer surfaces clean.
- Look for heat recovery opportunities.

ENERGY STORAGE SYSTEMS

- Check heating arrangements for storage tanks.
- Check temperatures of storage vessels.
- Check temperature control arrangements.
- Check amount and condition of insulation.

- Check maintenance procedures.
- The use of energy accumulation should be considered:

  —to balance load factors
  —to peak lop and use off-peak electricity
  —to increase overall energy efficiencies of boilers and distribution systems
  —to harness ambient energy.

PLANT AND EQUIPMENT

- Check conditions of plant and equipment.
- Check functions and efficiencies of electrical equipment.
- Measure process temperatures and pressures and check if appropriate.
- Check efficiencies of plant and equipment.
- Check maintenance procedures.
- Check control arrangements.
- All hardware should be matched to the purpose for which it is required.
- All systems should be operated at rates corresponding to maximum efficiency (normally fully-loaded in continuous operation).
- Intermittent operations and fluctuations should be avoided.
- Efficiency checks should be carried out frequently using standardized procedures.
- Plant should be selected on sensible extreme conditions.
- Look for heat recovery opportunities.

LIGHTING

- Check zonal lighting requirements.
- Check conditions and cleanliness of luminaires and windows.
- Check the maintenance procedures.
- Check lighting controls.
- Check that parts of the building are not being lit unnecessarily.
- Challenge the need for large areas of glazing.
- Obtain the economic balance of artificial versus daylighting.
- Check colours of room surfaces.
- Eliminate glazing.
- Keep windows and rooflights clean.
- Eliminate luminaires.
- Keep luminaires clean.
- Replace lamps when their efficiency drops.
- Avoid dark background colours.
- Use automatic controls.
- Zone lighted areas.
- Do not light unoccupied areas—use infra-red detecting switches.
- Use separate circuits for cleaners and for times outside working hours.

- Use separate circuits at the daylighted peripheries.
- Never use filament lamps.
- Use low-energy fluorescent or discharge lamps.
- Maintain lighting systems in good order.
- Look for heat recovery opportunities.

THERMAL INSULATION

- Check conditions of buildings, plant and equipment.
- Check weather-stripping of external doors and windows.
- Check seals on doors leading to stairwells and vertical shafts.
- Check self-closures.
- Check loading bays.
- Check structural integrity.
- Seal unused stacks and vents.
- Check effective use of building space.
- Check thermal insulation of roofs, walls, windows and floors.
- Is insulation and draughtproofing adequate and in good repair?—roof, walls, floors, inter-zones, doors, windows?
- Insulate all high temperature surfaces according to economic optima, bearing in mind maximum refractory temperatures allowable.
- The 'law of diminishing returns' applies.
- Apply insulation to the inside of intermittently heated enclosures.
- Apply insulation to the outside of continuously heated systems.
- Conduct a comprehensive transmission heat loss survey.
- Specific all insulating opportunities and evaluate cost-effectivenesses.
- Produce a schedule describing the optimal sequence for the application of insulating options and compare this with schedules for other energy saving options.
- Evaluate the diseconomies ensuing from reduced firing rates arising from lessening the demand.
- Consider the additional diseconomic effects of the likelihood of mechanical damage, moisture or vapour ingress, leading to condensation within the insulant or structure, infestation, fire hazard or deterioration which may occur.
- Check vapour barriers and look for interstitial condensation.
- Wet insulation acts as a heat pipe thermal fin!
- Insulation also masks problems which may occur in the future.
- Incorporate vapour barriers and ensure adequate ventilation to vent off moisture and drainage arrangements should water infiltrate the insulant.

VENTILATION

- Check air handling plant.
- Check thermal insulation of plant.
- Look for blockages.

- Check filters for cleanliness.
- Check settings and operations of dampers.
- Check cleanliness and operations of heating/cooling coils.
- Check ventilation rates.
- Check ventilation arrangements.
- Check how air gets in.
- Check how air gets out.
- Measure ventilation rates in different zones.
- Check air distributions.
- Check zonal ventilating requirements.
- Check local extract requirements.
- Check extract flow rates.
- Check operations of fans.
- Check if parts of the building are being ventilated unnecessarily.
- Check for vertical stratification.
- Check maintenance procedures for fans, ducts and filters (sick building syndrome).
- Check control arrangements
- Check for infiltration of outside air:

  —at loading and delivery bays
  —at external doors
  —at openable windows
  —at shafts and flues—consider the introduction of controlled dampers
  —at other openings
  —at broken glazing.

- Consider the introduction of vestibules, air curtains and the like.
- Consider the introduction of automatic door closures.
- Minimize infiltration and uncontrolled air flows.
- Divide the building into zones having different temperature and air change requirements—consider the use of partitions and local variable-speed fans.
- Check that ceiling heights are not excessive—causing stratification and presenting a large unoccupied volume to heat or cool—consider the use of false ceilings.
- Minimize vertical stratification by ensuring adequate mixing throughout the volume.
- Consider the use of destratifiers.
- Ensure adequate air distribution.
- Stick to minimum fresh air requirements—consider the use of obscuration meters—question the minimum air change rate adopted.
- Ensure that the required air change rate is attained in working areas but do not air change volumes where fresh air changes are not required (e.g. roof spaces, storage areas, unoccupied rooms or regions).
- When minimum air change rates are being achieved, consider the introduction of waste heat recovery.
- Isolate and vent-off processes which contaminate the working atmosphere.
- Maximize the extraction of contaminant and minimize the supply and extraction of fresh air.
- Ensure that the ventilation system is controlled.
- Check start-up shut-down and sequencing.

- Check controls.
- Eliminate random infiltration.
- Eliminate arbitrary manual adjustment of air flows and temperatures.
- Balance inlet and extract fan sets.
- Maintain a slight positive pressure to eliminate draughts.
- Draughts lead to demands for higher temperatures, which lead to calls for higher ventilation rates to combat excess heating, the opening of windows and doors, activating an insidious spiral towards excess air changes and room temperatures—stop this happening.
- Introduce self-closing exterior doors, plastic or air-curtains, vestibules and air-locks.
- Check that ceiling heights are not excessive—causing stratification and presenting a large unoccupied volume to heat or cool—consider the use of false ceilings.
- Look for heat recovery opportunities.

AIR CONDITIONING SYSTEMS

- Check cooling control arrangement.
- Check building/system response to controls.
- Check that temperatures and ventilation rates are not excessive.
- Check supply and return temperatures.
- Check conditions of plant and equipment.
- Check that heating and cooling systems cannot conflict.
- Specify the exact purpose for which air conditioning is required.
- Analyse load profiles.
- Do not overcapitalize plant—reduce the demand before increasing the amount of cold supplied.
- Consider ice accumulation to avoid maximum demand changes and on-peak tariffs.
- Keep heat transfer surfaces clean.
- Check zonal cooling requirements.
- Measure zonal air changes rates.
- Split areas into zones having different air conditioning requirements—consider insulated partitioning.
- Check insulation levels and conditions throughout.
- Check for unoccupied air conditioned parts of the building.
- Check maintenance procedures for fans, ducts and filters (sick building syndrome).
- Minimize infiltration and uncontrolled air flows.
- Measure zonal air changes rates.
- Check that ceiling heights are not excessive—consider the use of false ceilings.
- Ensure adequate air distribution.
- Stick to minimum fresh air requirements—consider the use of obscuration meters—question the minimum air change rate adopted.
- Ensure that the required air change rate is attained in working areas but do not air change volumes where fresh air changes are not required (e.g. roof spaces, storage areas, unoccupied rooms or regions).
- When minimum air change rates are being achieved, consider the introduction of waste heat recovery.

- Isolate and vent-off processes which contaminate the working atmosphere.
- Maximize the extraction of contaminant and minimize the supply and extraction of fresh air.
- Ensure that the ventilation system is controlled.
- Ensure that opening windows or doors is not used to control temperatures.
- Eliminate random infiltration.
- Eliminate arbitrary manual adjustment of air flows and temperatures.
- Balance inlet and extract fan sets.
- Maintain a slight positive pressure to eliminate draughts.
- Introduce self-closing exterior doors, plastic or air-curtains, vestibules and air-locks.
- Solar gains, lighting dissipations and high temperature thermal loads, emanating from electronics and electrical systems, should be extracted by cooling windows, louvres, shutters, luminaires or equipment, using air or water at outside environmental temperatures. This avoids the wasteful practice in air conditioning systems of allowing such energy to infiltrate into and so disturb the thermal equilibrium of a room, for which it is necessary to use high grade chilled water or refrigerant to remove the excess heat via a large heat transfer surface in order to regain comfort conditions.
- Look for heat recovery opportunities.

DOMESTIC HOT WATER SYSTEMS

- Investigate uses for hot water.
- Chronicle water usage patterns.
- Check for leads of hot water.
- Check the maintenance procedures.
- Check the control arrangements.
- Temperatures should be controlled and optimized.
- Leaks should be prevented.
- Check condition and insulation of hot water storage tanks.
- Pipes and storage tanks should be adequately lagged.
- Look for heat recovery opportunities.

COMPRESSED AIR SERVICES

- Check conditions of plant and equipment.
- Check compressor efficiency.
- Check the position of the inlet air duct.
- Check the maintenance procedures.
- Check the control arrangements.
- Check the amount of compressed air supplied.
- Check delivery temperature and pressure.
- Check for leaks.
- Leaks should be prevented.
- Check the uses of compressed air.

- Check pressures at points of use.
- Challenge every use of compressed air—this is the most expensive energy commodity.

The first law of thermodynamics states:

$$\text{heat} = \text{work} + \text{change in internal energy} \tag{A2.1}$$

In compression, work in = pressure energy + change in internal energy, i.e.

$$mc_p\Delta T = pV + mc_v\Delta T$$
$$mc_p\Delta T = mR\Delta T + mc_v\Delta T$$
$$c_p = R + c_v \tag{A2.2}$$

For air

$$1005 = 287 + 718 \,\text{J}\,\text{kg}^{-1}\,\text{K}^{-1} \tag{A2.3}$$

Thus 1005 units of work are required to produce 287 units of pressure energy, even with 100 per cent efficient compression. Furthermore, the work (electricity) has been produced in the first place in the conversion of heat to work at 30 per cent efficiency at best. Thus it requires at least 3350 units of heat to produce 287 units of pressure energy or 11.7 units of heat to produce 1 unit of pressure energy.

- Ensure that minimum pressure is utilized for the required operation: compressed air is invariably generated at the highest pressure needed in multifarious activities and then throttled down the pressure of each activity.

For example, paint spraying is best accomplished at 40 psi although air is supplied to the adjustable spray guns at 100 psi. Table A2.1 shows the *minimum* savings resulting from reduced delivery pressures.

Table A2.1 Savings by reducing compressed air delivery pressures

Delivery pressure, psi($10^5\,\text{Nm}^{-2}$)	Adiabatic delivery temperature, K	Total work done, MJ kg^{-1}	Savings by reduction, MJ kg^{-1}	% Saving	£ saving per £1000 annual bill
100(6.90)	489	0.020	0	0	0
90(6.21)	477	0.019	0.001	5	50
80(5.52)	461	0.018	0.002	10	100
70(4.83)	443	0.016	0.004	20	200
60(4.14)	407	0.013	0.007	35	350
50(3.45)	403	0.012	0.008	40	400
40(2.76)	377	0.009	0.011	55	550
30(2.07)	348	0.007	0.013	65	650
20(1.38)	310	0.003	0.017	85	850
14.7(1.013)	283	0	0.02	100	1000

Opportunities should be identified for using the waste heat from the compressed air cooling system, such as for space heating.

- Reduce the generating pressure to a minimum.
- Consider interstage bleed-off.
- Consider the use of localized booster compressors when higher pressures are unavoidable, especially where usage is intermittent.
- Switch-off compressors when not in use.
- Consider the introduction of compressed air accumulation:

    —to peak lop to reduce maximum demand payments
    —to use off-peak electricity when possible
    —to balance the load.

Consider the use of pressure stabilizing heated bellows and bags containing phase change (liquid-to-vapour) materials in accumulators.

There are two ways to obtain compressed air—heating at constant volume to increase pressure, and by mechanical compression.

- Site air inlets in cool, dry positions—up to 7 per cent of electricity costs can be saved by supplying cold denser air from outside the building. Often reject air from coolers is recycled through the compressor getting hotter and hotter, lighter and lighter and hence more expensive to compress, as well as worsening cooling and intercooling effectiveness.
- Recover heat from cooling and intercooling systems.
- Supply outside air for cooling and intercooling systems.
- Avoid condensation in pipelines—this must be forced along with the air and wastes pressure energy.
- Reheat compressed air where possible to increase discharge pressures.
- *Compressed air should only be used as a last resort*. It is not uncommon in manufacturing systems for the cost of compressed air to be of the order of 50 per cent of the entire electricity bill, which itself constitutes the major proportion of the overall fuel bill.
- Never use compressed air for swarf-blowing and cleaning purposes.
- Meter compressed air usage.
- Look for heat recovery opportunities.

REFRIGERATION PLANT AND CHILLED WATER DISTRIBUTION SYSTEMS

- Check maintenance and operating procedures.
- Evaluate load patterns and operating cycles.
- Check conditions of plant and equipment.
- Check for evidence of inefficient compressor operation.
- Calculate coefficient of performance and energy efficiency.
- Check for leaks.
- Leaks should be sealed.
- Check for leaks of refrigerant or chilled water.
- Check the maintenance procedures.

- Check operation of condenser fans.
- Check cleanliness of air-cooled condenser coils.
- Check cooling tower spray water system and water treatment.
- Check cooling tower performance.
- Check cooling tower outlet-to-inlet by-pass air circulation.
- Check operations of pumps and valves.
- Check the control arrangements.
- Check operating pressures and temperatures.
- Temperatures and pressures should be controlled and optimized.
- Provide monitoring instruments.
- Ensure that adequate controls are provided.
- Check effective operation of controls.
- Check temperature settings.
- Check controls for cooling tower and condensers.
- Check condition and insulation of cold water storage tanks.
- Check condition of insulation and vapour seals on cold lines.
- Pipes and storage tanks should be adequately lagged.
- Look for heat recovery opportunities.
- Check whether useful heat recovery from the condenser might be accomplished.

STEAM PLANT

- Check conditions of plant and equipment.
- Check efficiency.
- Check for steam leaks.
- Check for condensate recovery.
- Check condition of steam traps.
- Check maintenance procedures.
- Check control arrangements.
- Look for heat recovery opportunities.

WASTE HEAT AND MATERIALS RECLAMATION

- High-grade energy (which may be 'hot' or 'cold' with respect to the environmental datum) should not be allowed to be dissipated directly to the environment.
- The energy rejected from a high-grade process should be collected and redirected via heat exchangers (or simply fans or pumps) to be employed at another place, collected and stored to be employed at another time, or concentrated for another higher-grade purpose using a heat pump or other thermal transformer, as long as these operations are economically justifiable.
- Attempts should be made to introduce feedback from energy loss centres to higher-grade stations in the energy flow sequence (e.g. by recycling materials, heat pumping or incinerating waste).
- Attempts to reduce or reuse waste should be made before any attempts at recycling or recovery.

- Waste energy and materials should be reused wherever economically possible ensuring that practical grade, time and space-matched uses have been found for the reclaimed amounts.
- The value of the savings must clearly exceed the cost of recovery.
- Before attempting to recover 'waste' heat, ensure that a matched need exists for it.
- Attempt to balance the load factor between recovered heat and utilization.
- Evaluate the economics of using flue gas to pre-heat combustion air.
- Consider the use of thermal accumulation at the interface to compensate for phase mismatches.
- Consider the grade of energy recovered—match this to the grade of the energy required—the use of heat pumps may be considered.
- Evaluate diseconomic effects (i.e. the reduction of plume buoyancy and hence flue gas dispersal effectiveness resulting from recuperating heat from (and hence lowering the temperature of) exhaust gases, and the onset of condensation (especially from sulphurous fuels producing corrosive sulphuric acids) inside the flues).
- Optimize the amount of heat recovered.
- Greater energy efficiency always requires an expenditure of materials and vice versa (e.g. the greater the area of a heat exchanger, the more effective the transfer of heat).
- The 'law of diminishing returns' applies.
- Consider the direct use of exhaust gases (i.e. for drying or for secondary combustion when the first combustion is reductive).
- Consider the advantages and diseconomies of latent heat recovery from flues.
- Consider recuperative or regenerative heat exchange (with possibly a latent heat exchange facility), heat piping or heat pumping to reclaim waste heat from extract ventilating air in order to pre-heat fresh air.
- Be careful to maintain minimum fresh air requirements if regenerators are adopted.

CONTROLS

That which is not measured cannot be controlled

- Energy or materials cannot readily be conserved unless accurate and comprehensive measurements in consistent units are first obtained for all activities within the system boundary.
- Ensure that all sensors are situated in sensible positions.
- Use the widest possible bandwidths for heating/cooling systems.
- Use the minimum setting for heating systems.
- Use the maximum setting for cooling systems.
- Control separately areas with different heat (cold) demands (e.g. sundry gains due to solar irradiance, people, lights, equipment, etc.)
- Use optimum start controllers.
- Use optimum stop controllers.
- Ensure that systems are operated on optimized time schedules.
- Switch off equipment when not in use.
- Install a computerized *energy management control system* to monitor all electricity use and maximum damand.

Remember to track back energy saved to that fuel (money) saved at the boilerhouse.